Powder Metallurgy

and

Advanced Materials

Selected papers from the 5th International Conference on Powder Metallurgy & Advanced Materials,

RoPM&AM 2017,

17-20 September, Cluj-Napoca, Romania

Edited by

Traian Florin MARINCA, Bogdan Viorel NEAMȚU, Florin POPA

Technical University of Cluj-Napoca

Peer review statement
All papers published in this volume of "Materials Research Proceedings" have been peer reviewed. The process of peer review was initiated and overseen by the above proceedings editors. All reviews were conducted by expert referees in accordance to Materials Research Forum LLC high standards.

Published under License by **Materials Research Forum LLC**
Millersville, PA 17551, USA

Published as part of the proceedings series

Materials Research Proceedings
Volume 8 (2018)

ISSN 2474-3941 (Print)
ISSN 2474-395X (Online)

ISBN 978-1-945291-98-2 (Print)
ISBN 978-1-945291-99-9 (eBook)

Distributed worldwide by

Materials Research Forum LLC
105 Springdale Lane
Millersville, PA 17551
USA
http://www.mrforum.com

Manufactured in the United State of America
10 9 8 7 6 5 4 3 2 1

R○PM
Advanced Materials

**5th INTERNATIONAL CONFERENCE ON
POWDER METALLURGY & ADVANCED MATERIALS**

17-20 September, Cluj-Napoca ROMANIA

TECHNICAL UNIVERSITY OF CLUJ-NAPOCA ROMANIA

TU WIEN — TECHNISCHE UNIVERSITÄT WIEN — Vienna University of Technology

TOPICS

- Powder and PM Products
- Advanced Materials Processing
- New Materials and Applications
- Functional Materials
- Nanomaterials and Nanotechnologies
- Health, Safety and Environmental Aspects of Particulates

RoPM
Advanced Materials

5th**INTERNATIONAL CONFERENCE ON POWDER METALLURGY & ADVANCED MATERIALS**

17-20 September, Cluj-Napoca, ROMANIA

http://www.sim.utcluj.ro/ropm-am2017/

Conference partners

SARTOROM

UNIVERS

VERDER scientific

NAPOSINT

The aim of the RoPM-AM2017 conference is to offer a forum for scientists in the field of Materials Science and Engineering to discuss and promote advances in knowledge, research and practice in the field of the Powder Metallurgy and Advanced Materials (Processing and Technologies).

The International Conference on Powder Metallurgy and Advanced Materials, RoPM-AM2017, represents the joint of the traditional RoPM (International Conference on Powder Metallurgy) and MATEHN (International Conference on Materials and Manufacturing Technologies) conferences organized by Technical University of Cluj-Napoca.

The Conference includes plenary sessions, key notes, oral and poster presentations, informative actions and social events.

International Advisory Committee

D. Banabic – Cluj-Napoca, Romania
M. Campos - Madrid, Spain
M. Călin – Dresden, Germany
H. Chiriac – Iaşi, Romania
C. Munteanu – Iaşi, Romania
H. Danninger – Vienna, Austria
C. Gierl-Mayer – Vienna, Austria
O. Gîngu – Craiova, Romania
O. Isnard – Grenoble, France
W. Kappel – Bucharest, Romania
S. Moisa – Beer-Sheva, Israel
A. Molinari - Trento, Italy
D. Munteanu – Braşov, Romania
R.L. Orban – Cluj-Napoca, Romania
J.M. Torralba – Madrid, Spain
M. Turker – Ankara, Turkey
V.A. Şerban – Timişoara, Romania
I. Vida-Simiti – Cluj-Napoca, Romania

Organizing Committee

Herbert Danninger

Honorary Chairman

Ionel Chicinaş

Chairman

Mariana Pop	Cătălin Ovidiu Popa
Vice chairman	Vice chairman

Organizing committee

Gabriel Batin	Dan Noveanu
Marius Bodea	Violeta Paşcalău
Dana Bota	Codruţa Pavel
Victor Cebotari	Florin Popa
Horea Florin Chicinaş	Călin Virgiliu Pricǎ
Ana Cotai	Monica Sas-Boca
Traian Florin Marinca	Nicolina Ocuhel
Violeta Murle	Lidia Adriana Sorcoi
Radu Mureşan	Gyorgy Thalmaier
Bogdan Viorel Neamţu	

Conference secretariat

Marius Bodea	Bogdan Viorel Neamţu
Florin Popa	Lidia Adriana Sorcoi
Traian Florin Marinca	

Foreword

The 5th International Conference on Powder Metallurgy & Advanced Materials, 5th RoPM&AM 2017, was held in Cluj-Napoca at Univers T Hotel between 17-20 September 2017. Conference was organized by Technical University of Cluj-Napoca in collaboration with Vienna University of Technology.

The International Conference on Powder Metallurgy & Advanced Materials, RoPM-AM2017, represents the joint of the traditional RoPM (International Conference On Powder Metallurgy) and MATEHN (International Conference on Materials and Manufacturing Technologies) conferences, organized before by Technical University of Cluj-Napoca. The aim of the International Conference on Powder Metallurgy & Advanced Materials (RoPM&AM2017) is to offer a forum for scientists in the field of Materials Science and Engineering to discuss and promote advances in knowledge, research and practice related to the Powder Metallurgy and Advanced Materials (Processing and Technologies).

The Conference attracted more than 110 participants from 11 countries (Austria, Belgium, England, Italy, French, India, Israel, Moldova, Romania, Serbia, Turkey) from prestigious universities, research centers and industry. The Conference Committee has decided to publish the presented papers, after revision process, in one of the following journals: Powder Metallurgy, Materials Research Forum LLC (Open Access, ISI indexed), Acta Tehnica Napocensis (ISI indexed).

The present volume Powder Metallurgy and Advanced Materials provides up-to-date, comprehensive and worldwide state-of-the-art knowledge in the Powder Metallurgy and Advanced Materials fields, including: Powder and PM Products, Advanced Materials Processing, New Materials and Applications, Functional Materials, Nanomaterials and Nanotechnologies, Health, Safety and Environmental Aspects of Particulates.

The conference secretariat received a number of 148 abstracts from 15 countries. The organising committee carried out the selection of invited and contributed papers, and their arrangements into sessions. In conference have been presented a number of 103 papers (invited papers: 9, oral presentations: 22, poster presentation: 72). For publication in this volume were considered 60 papers (Invited papers: 6, New theoretical aspects and products in PM: 8, Advanced Materials with special properties: 22, Characterisation of powder and powder metallurgy products: 15, New developments in powder production and processing technologies: 9). All the papers were evaluated by two peer reviewers and 5 papers were accepted in the original form sent by the authors, 25 papers were accepted with minor modifications, 30 papers were accepted with major modifications and 14 papers were rejected.

The editors were responsible for selection of reviewers for the papers, whom we thank for their promptitude and accurate work. The communication between authors and referees was managed by the editors.

The organising committee expresses its entire gratitude to all the authors who presented their works at the Conference RoPM&AM 2017 and contributed in this way to the success of this event. Special thanks are due to the authors from abroad for attending the conference and to the reviewers for their support in improving the quality of the papers and finally for the assurance the quality of this volume. The organising committee also addresses warmest thanks to all the members of the International Advisory Committee for their support, to the generous sponsors for their financial assistance and too many others for their contribution in organising RoPM&AM 2017.

We hope that the contents of this volume will prove useful for researchers in powder metallurgy and advanced materials field and practitioners in developing and applying new materials and processes.

Cluj-Napoca, September 2018   Professor Ionel Chicinaş

                                                              Conference Chairman

Table of Contents

Powder Metallurgy and Advanced Materials – RoPM&AM 2017   Materials Research Forum LLC
Materials Research Proceedings 8 (2018) 1-10   doi: http://dx.doi.org/10.21741/9781945291999-1

A study of the ferromagnetic microwires retention in cellulose matrix in the security papers

Mirela Maria CODESCU[1,a], Wilhelm KAPPEL[1,b], Eugen MANTA[1,c,*], Eros PATROI[1,d], Delia PATROI[1,e], Remus ERDEI[1,f], Valentin MIDONI[2,g], Ion ZĂPODEANU[3], Marilena BURLACU[3,h]

[1] Research & Development National Institute for Electrical Engineering ICPE-CA
313 Splaiul Unirii, Bucharest - 3, Romania

[2] SC MEDAPTEH PLUS CERT SA, 27 Selimber St, Magurele, Romania

[3] SC CEPROHART SA Braila, 3 Al. I. Cuza Blvd, Braila, Romania

[a]mirela.codescu@icpe-ca.ro;[b]wilhelm.kappel@icpe-ca.ro; [c]eugen.manta@icpe-ca.ro;
[d]eros.patroi@icpe-ca.ro; [e]delia.patroi@icpe-ca.ro; [f]remus.erdei@icpe-ca.ro; [g]medapteh@mail.ru;
[h]marilena_burlacu@yahoo.com

Keywords: Security papers, Ferromagnetic microwires, Retention, Cellulose matrix

Abstract. Prepared by the Taylor – Ulitovsky technique, the glass-coated microwires are formed from a metallic core, with the diameter 3 to 50 μm, surrounded by an insulating layer from glass, with the thickness of 1 to 20 μm. Embedded in the cellulose matrix, the ferromagnetic glass-coated microwires allow their use as security element for the authentication of valuable papers in the electronic validation process. The authentication of the security paper is realised with a special detector, by "YES" or "NO" answer. This paper can be used as anti-shoplifting or validating elements to identify the counterfeit products. The paper presents the experimental results related to the retention of ferromagnetic microwires in the cellulose matrix, a complex process characterised by specific features, primarily due to the shape and diameter/length ratio of the microwires. The ferromagnetic retention yield was $\eta = 65 - 90\%$, for the prepared papers with basis weight more than 50 g/m^2.

Introduction

Faced with increasing of goods counterfeiting, a wide range of methods are currently used to protect consumer goods, bank, state and commercial documents. Thanks to impossibility to produce security elements without proper equipment and under special conditions imposed by the very high degree of accuracy, the advanced technologies offer the solution, ensuring a high degree of protection against falsification. Investment and research efforts are being made to diversify the field of high security elements. The moment of launching the technology for glass-coated microwires (GMW) fabrication [1,2] has become revolutionary on the high-tech technology market, opening up the gates of a large variety of technological benefits for the existing applications and also setting the foundation for new applications [3-9]. The advantages of ferromagnetic GMWs securing [10] were: possibility of identifying at distance; stable magnetic properties even at high temperatures and corrosive media; wide range of functional temperatures; stability at shielding – the codes shielded by metallic panels can be read; stability at the mechanical action; small sizes and low consumption and, for the microfibbers from the last generation, with special properties, allowing the possibility to the information magnetic encoding): very large amount of the generated codes; the information can be read both from a stationary source and from a source in motion; the encoding is impossible to destroy, both in the

Powder Metallurgy and Advanced Materials – RoPM&AM 2017   Materials Research Forum LLC
Materials Research Proceedings 8 (2018) 1-10   doi: http://dx.doi.org/10.21741/9781945291999-1

continuously and in variable magnetic field, (reliable encoding); possibility to read the information from any code randomly oriented in space.

The structure of the paper consists of vegetable fibers (wood or non-wood), in which auxiliary materials, such as fillers, gluing agents, pigments, additives etc. are incorporated. Depending on the application field of paper, some structural features are imposed to the network: number of fiber-fiber contacts and of sizes of interfibrillar spaces, density and roughness of the surfaces. These properties are depending on the fibbers nature, on their processing degree, on the amount and properties of the auxiliary, and also on the processing technique used for forming and finishing of network. The term of filler defines any non-fibrous material added to the paper pulp to improve the optical properties of the paper, but also other features such as porosity, smoothness, printing ability etc. By incorporating of pigments into the paper pulp, the papers optical inhomogeneity increases, the amount of reflected and refracted light in the paper sheet increases, and the whiteness and opacity is improved. At the same time, the pigment particles retained in the sheet structure increase the interfibrillar spaces and reduce the possibility to set-up interfibrillar bonds having negative effects on the paper resistance indices. The fillers retention into the paper sheet is realised mainly by filtration for the particles with large sizes and by colloidal phenomena for fine particles. The introduction of filling is primarily determined by technical considerations, since certain characteristics of the paper, particularly optical indices and printing ability, are limited if only fibrous materials are used. Currently fillers can also serve as partial substitutes for fibrous materials in some cases, thereby helping to reduce the production costs.

Developed by the Taylor-Ulitovsky process, the GMWs consist of a cylindrical metal core that is covered with a glass-insulating layer, the diameter of the metal core is 3 - 50 µm, and the thickness of the glass insulation is 1 - 20 µm. The length of such microwires, under laboratory conditions, reaches approx. 1 km. The ferromagnetic glass-coated microwires, cut at ca. 7 mm lengths, is included in the paper composition also as filling material, but in the paper pulp and in the paper sheet structure, the microwire segments have a certain behaviour that differentiates them from the classical materials of filling. Unlike these materials, the ferromagnetic microwires introduced into the paper in very small amounts do not significantly influence the rheological characteristics of the paste and the paper resistance characteristics. The appearance of wires, the diameter, length and the microwires density are also specific characteristics that differentiate the materials currently used to fill the paper. The importance of retention efficiency in the case of ferromagnetic microwires is primarily due to the need to achieve a certain microwires density in the paper sheet, in order to ensure its security without affecting the paper quality and functionality.

2. Experimentals

The Taylor-Ulitovsky technique for GMWs preparation consists in placing in a high-frequency inductor of a glass tube with a metallic rod inside (Fig. 1). Under the influence of the generated electromagnetic field, the metal melts, forming a drop. In contact with the molten metal, a part of the glass tube softens and a coating is formed from the glass covering the drop. For a particular working regime [11], this glass soaked by pulling also trains the metal, leading to microwires formation, which is collected on the spool. Different metal core structures can be obtained: polycrystalline crystals of different sizes (microcrystalline, nanocrystalline) or amorphous. For experimental research were used $Fe_{77}B_{13}Si_{10}$ GMWs, which are structurally, by X-ray diffraction and magnetically, by vibrating sample magnetometry, characterized.

Achieving certain density of GMW in the paper sheet structure, as in the fillers case, depends on the action of factors with a particular influence on intelligent material retention in the papermaking process. Knowing and controlling these influences will ultimately allow finally reaching the density that is sufficient for paper securing. In this respect, have been experimented several programs in which the basic recipe for realisation of the GMWs secured paper has been supplemented with several variables specific to each influenced factor studied.

Fig. 1. Aspect during the $Fe_{77}B_{13}Si_{10}$ ferromagnetic glass-coated microwires drawing.

Table 1. The studied parameters and the experimented recipes used to for paper preparation

Parameter	S/H ratio, [wt. %]	S/H Schopper – Riegler degree, [°SR]	Microwires amount, [g]	Retentor amount, [%]	Paper weight, [g/m²]
Nature of the fibrous materials	100 S or 100 H	30	0.005; 0.007; 0.009		
Schopper – Riegler degree °SR of the cellulosic material	60/40	30/20; 40/30; 50/40; 60/50	0.009		
GMWs amount	60/40	30/20; 40/30; 50/40; 60/50	0,005; 0,007; 0,009; 0,011		
Softwood / hardwood cellulosic pulp ratio from the fibrous composition of paper	20/80; 30/70; 40/60; 50/50	45/30	0.007		
Amount of retention emulsion, dosed in the paper manufacturing receipts	60/40	30/20		0; 0.2; 0.4; 0.6; 0.8	
Paper weight	60/40	30/20			50; 70; 90; 110

The GMWs with 7 mm lengths are embedded as filler into the cellulose matrix (the pulp), the main receipt of the mixture, in wt.%, being: bleached cellulose sulphate from softwood (S) pulp (different amounts: 30 - 100%; Schopper – Riegler degree: 30 - 60 °SR); bleached cellulose sulphate from hardwood (H) pulp (different amounts: 30 - 100%; Schopper – Riegler degree 20 - 50 °SR); paper filling material: 15% calcium carbonate; gluing emulsion: 1,5% alkyl–dimercetene (AKD); retention additive: 0,5% polyamide–amine and different amounts of GMW (for 10 sheets with paper weight q = 75 g/m^2) – 0.005g; 0.007g; 0.009g and 0.011g. The particularities of the experimented recipes for paper sheets preparation are chosen to highlight the influence of different process parameters (Table 1). The Schopper-Riegler test provides a measure of the rate at which a dilute suspension of pulp may be dewatered. It has been shown that the drainability is correlated to the surface conditions and swelling of the fibbers, and constitutes a useful index of the amount of mechanical treatment to which the pulp has been subjected. The retention efficiency (η) was expressed as the ratio of the amount of GMW initially used to prepare the cellulosic paste and the remaining GMW amount in the laboratory prepared sheet (in each experiment, the retention yield was determined for 10 sheets of paper).

3. Results and discussions

3.1 Structural characterization for $Fe_{77}B_{13}Si_{10}$ ferromagnetic glass-coated microwires

After preparation, the GMWs were structurally characterized by X-ray diffraction investigations. The glass-coating was eliminated using as etchant a solution of 50% hydrogen fluoride (HF) in water. As expected, the ferromagnetic microwires are quasi-amorphous. The samples show a crystalline structure of Fe_3Si in the cubic system, the structure that is very similar to that of α-Fe, but with a lower cell parameter, in this case the network parameter being $a = 2.837$ Å compared to $a = 2.866$ Å in the case α-Fe. It is considered that the effect is due to the ultra-rapid solidification phenomenon of the alloy, but a small contribution to this reduction can also be brought about by the induction of a local stress factor by the preparation conditions, due to the presence of glass covering the metallic core. Other phase present in the $Fe_{77}B_{13}Si_{10}$ GMW is Fe_8B, which crystallizes in a tetragonal system. The mean crystallite size, determined for the Fe_3Si phase using the Debye-Scherrer formula is D = 19 nm. The calculated crystallinity of the analysed sample is around 74,4%.

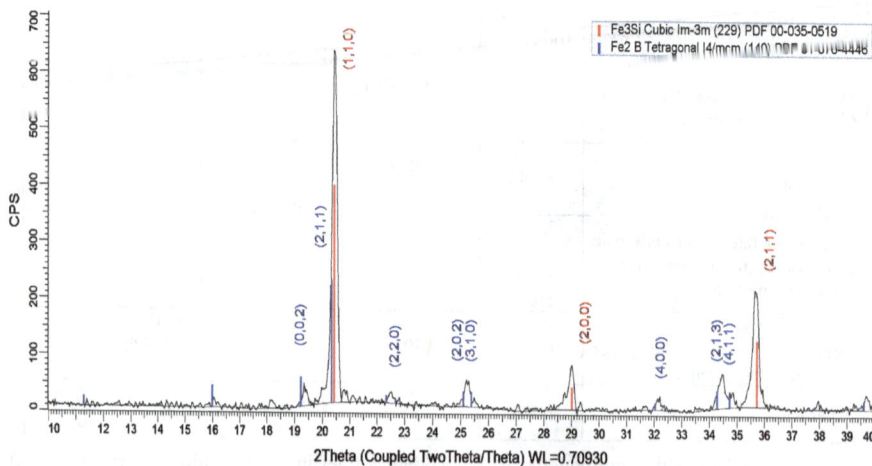

Fig. 2. X-ray diffraction diffractogram of the $Fe_{77}B_{13}Si_{10}$ ferromagnetic GMWs.

3.2 Magnetic characterization for $Fe_{77}B_{13}Si_{10}$ ferromagnetic microwires

The shape of hysteresis curve (Fig. 3), plotted for the $Fe_{77}B_{13}Si_{10}$ GMWs confirms that the microwire metallic core is ferromagnetic. The magnetic values, obtained by evaluating the magnetic core volume, were compared with data obtained on bulk material. From the point of view of the magnetic properties, the structure of the ferromagnetic material (evidenced by the X-ray diffractogram in Fig. 2) favours the magnetic properties suitable for application as soft magnetic material, i.e. high magnetization (1,2 T) and low coercivity $H_c = 17.73$ (A/m).

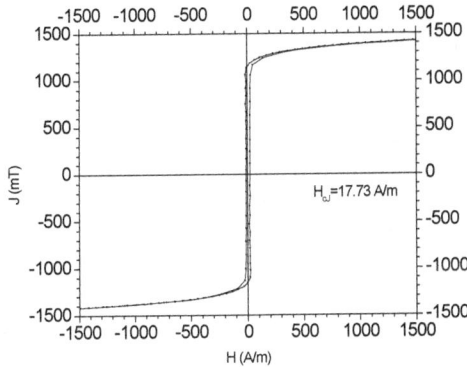

Fig. 3. Hysteresis curve of the $Fe_{77}B_{13}Si_{10}$ ferromagnetic GMWs.

3.3 Influence of fibrous material nature

The results obtained during this experiment show that both for softwood pulp and hardwood pulp, the number of microwires retained in the paper sheets increases with the ferromagnetic GMWs amount (Table 2).

Table 2. Number of ferromagnetic microwires retained in paper sheets at various additions of ferromagnetic microwires (paper weight $q = 75$ g/m^2).

Specification	Microwires addition for 10 sheets, [g]		
	0,005	0,007	0,009
The number of microwires retained in paper sheets obtained from 100% softwood cellulose	258	366	472
The number of microwires retained in paper sheets obtained from 100% hardwood pulp	239	330	424

When the retention efficiency is determined, it can be seen that the increase of the initial GMWs addition, occur differences of 4 – 9% between the two types of cellulose, both having the same Schopper-Riegler degree: 30 °SR, the lower values being obtained in the case of the hardwood cellulose. Depending on the GMWs amount added to the pulp, in the case of the softwood cellulose, the retention efficiency increases slightly, whereas for the hardwood cellulose, this efficiency decreases (Fig. 4). For the same type of cellulose, the decrease or the increase of the efficiency based on the ferromagnetic GMWs addition is ca. 2% (in the studied range, for addition of 0.005 – 0.009 g GMWs / 10 sheets).

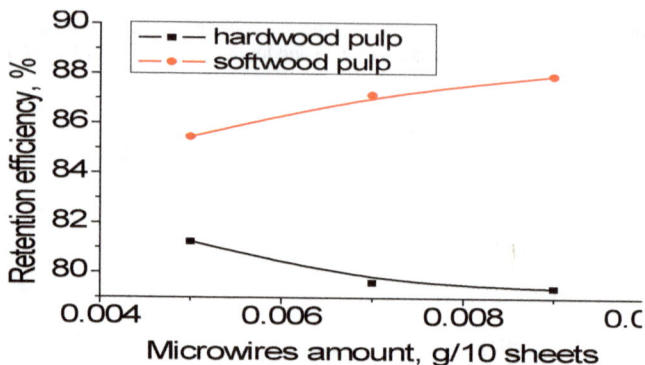

Fig. 4. Dependence of retention efficiency on the nature of the fibbers material,
for various amounts of ferromagnetic microwires.

The ferromagnetic GMWs, having an average length of ca. 7 mm and a much higher density than the cellulose fibbers, are better retained by the greater fibber (in diameter and length) of softwood pulp than those of hardwood pulp. Also, the softwood pulp fibbers, with 30 °SR, already exhibit a certain degree of fibre-lisation, and the structure of paper, at stage of network formation, retains better the ferromagnetic GMWs. In the case of the hardwood cellulosic fibbers, due to their smaller sizes, the GMWS retention efficiency is lower, these ones "penetrating" easily at filtering the fibrous network of the paper sheets.

3.4 Influence of sizes of the ferromagnetic microwires and cellulosic fibbers
In order to estimate the compatibility of the ferromagnetic GMWs with the cellulose matrix in which they will be embedded, were prepared samples showing that the microwires sizes are compatible with the dimensions of the cellulose fibbers of the matrix. The comparison was always done with a blank sample, i.e. a sample without filler, prepared from cellulose only. The micrographs of these samples are shown in Fig 5a) and b).

3.5 Influence of the Schopper-Riegler degree of the cellulosic material
By increasing of the milling degree, the fibbers specific surface increased and their ability to be felt is improved. Consequently, the amount of filler material retained in the paper sheet by adsorption and filtration should increase proportionally to the increasing of the milling degree. Additionally, it should be added that the dehydration rate of the paste on the forming screen decreases with the advance of milling process, which also has a positive effect on the retention rate. Figure 7 shows the variation of GMWs retention efficiency (at an addition of 0.009 g/10 sheets), depending on the increase of the milling degree of the cellulose used in the production of the laboratory paper sheets (for softwood cellulose, with 30; 40; 50 and 60 °SR and hardwood cellulose, with 20; 30; 40 and 50 °SR).

Powder Metallurgy and Advanced Materials – RoPM&AM 2017   Materials Research Forum LLC
Materials Research Proceedings 8 (2018) 1-10   doi: http://dx.doi.org/10.21741/9781945291999-1

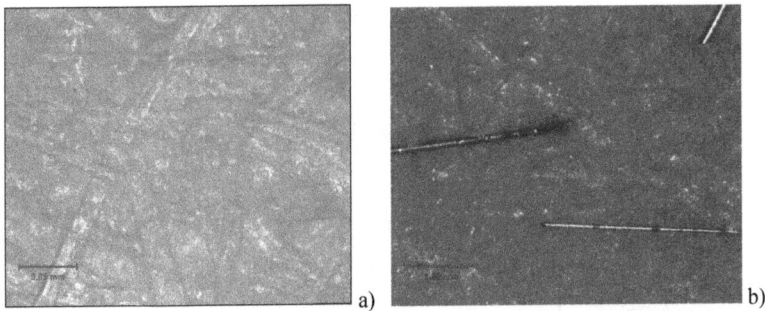

Fig. 5. Optical micrographs of the cellulose matrix without a) and with randomly inserted $Fe_{77}B_{13}Si_{10}$ microwires filler b).

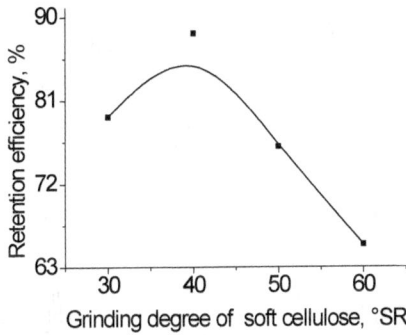

Fig. 7. Microwires retention efficiency *versus* milling rate of the paper pulp.

The ferromagnetic GMWs behaviour is the same to that of the classic filler up to a certain milling degree, after which it begins to decrease rather abruptly. The highest value was obtained on paper sheets in which softwood pulp had a milling degree of 40 °SR, and the hardwood pulp 30 °SR. Considering the specific characteristics of the ferromagnetic GMWs (shape, dimensions and specific weight), it may be possible, that starting from a certain milling degree, the under developing paper fibrous network to be more easily traversed thereby resulting a decreased retention efficiency. The same evolution of retention efficiency was found for other values of the GMWs amount.

3.6 Influence of the amount of ferromagnetic microwires

The data reported in literature shows that the retention efficiency of the classical fillers used in the paper manufacturing process is influenced both by the properties of the filler material, in particular the shape and size of the particles and by the amount of filler material used in the preparation of paper pastes. It has a growing trend, with the increase of the amount of filler up to a certain value, then the retention efficiency decreases. The retention efficiency dependence on the GMWs amount was studied so far only separately, for the two types of cellulose, namely S pulp and H pulp. Using a certain combination of these two types of celluloses, the results presented in Fig. 8 show that indeed the retention efficiency is depending on the ferromagnetic GMWs amount. The retention efficiency increases with the increase of the GMWs amount (in the range of 0.005 – 0.011 g/10 sheets), even when the milling degree of fibrous material was modified (30 - 60 °SR

Powder Metallurgy and Advanced Materials – RoPM&AM 2017   Materials Research Forum LLC
Materials Research Proceedings 8 (2018) 1-10   doi: http://dx.doi.org/10.21741/9781945291999-1

for the S cellulose and 20 - 50 °SR for the H cellulose). The growth trend is not substantial, ranging from 1 to 4%.

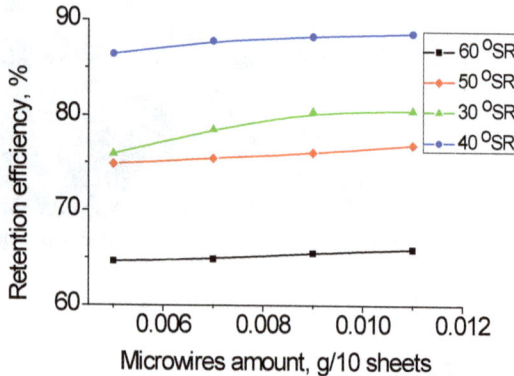

Fig. 8. Dependence of retention efficiency on the amount of microwires, for various milling degrees of the cellulose pulp.

3.7 Influence of the hardwood and softwood celluloses ratio in the paper composition

It is known that each assortment of paper must be realised with a certain fibrous composition that will provide the desired quality characteristics, be the cheapest and to require the simplest manufacturing technology. In this regard, the paper fibrous composition will usually consist on two types of celluloses, usually hardwood and softwood pulps. Table 3 shows the GMWs retention efficiency, resulted in the case of paper sheets whose fibrous compositions were obtained from H and S celluloses, combined in various ratios. In the case of the paper sheets having a higher content of softwood pulp (from 20% to 50%), should increase the GMWs retention efficiency. On the other hand, the softwood pulp has a high milling degree, much higher than that of hardwood pulp, which makes that the fibrous material mixture to present an increased milling degree from ca. 30 °SR to 40 °SR.

Table 3. The GMWs retention efficiency for fibrous composition consisting of different ratios of H and S celluloses

Specification	Cellulosic ratio H/S, [%]			
	80/20	70/30	60/40	50/50
Retention efficiency, [%]	78,38	76,28	74,00	72,48

In these conditions, the values determined for the retention efficiency decrease slightly as the softwood pulp amount increases in the paper pulp composition.

3.8 Influence of the retentor amount dosed in the manufacturing prescriptions

From the data presented in Fig. 9, results that the retention efficiency of the GMWs depends on the retentor amount used to prepare the paper paste, and in this case the increase of efficiency is proportional with the retentor content. If the investigation field of experiment had been extended to larger retentor contents, it was probable that the retention performance would not have significantly increased. In practice, a retentor overdose is not recommended, to avoid the formation of very large flocks that disturb the process of paper formation. The increase in retention efficiency with the retentor amount indicates that the behaviour of microwires from the paper pulp aligns to the main behaviour of the fillers and of the cellulosic fibbers. It means that microwires retention is achieved by both mechanisms, namely mechanical retention and colloidal adsorption. The results show also that in the overall retention the greatest share is the mechanical retention. Without retentor addition, the retention efficiency already has an appreciable value (60.86%), although the cellulosic milling degrees used in the experimental recipes are quite low (30 °SR for the softwood pulp, respectively 20 °SR for hardwood pulp).

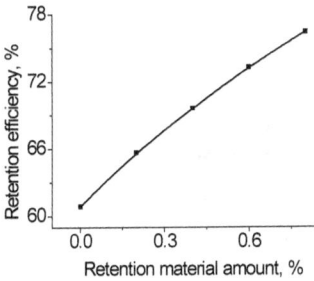

Fig. 9. Variation of retention efficiency [η] on the retentor content.

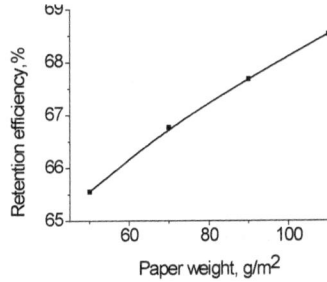

Fig. 10. Influence of paper weight on retention efficiency [η] of ferromagnetic GMWs.

3.9 Influence of the paper weight
Another factor that significantly influences the retention efficiency is the paper weight. With the increase of the weight and the thickness of the filter layer, it is also expected to increase the cross-sectional strength of the filler material, in this case of the microwires. At higher pass resistance, the amount of microwires remaining in the paper sheet will increase, which means an increase of the retention efficiency. The retention efficiency determined under these conditions is presented in Fig. 10. By increasing of the paper weight from 50 to 110 g/m^2, the retention efficiency increased with ca. 3%, from 65.55% to 68.53%. The values determined for the retention efficiency include both mechanical retention and colloidal retention, since on the paper sheets realisation was used also retentor.

Summary
The embedding of GMWs in the cellulosic matrix of the paper aims to develop a new type of paper, with a novel securing element – the ferromagnetic GMW - as a field sensor, for applications in the field of electronic detection of valuable documents validation. The retention of GMWs in the paper sheet, unlike the usual filler materials, has a certain specificity given mainly by the shape and size of the microwires (6 to 8 mm long and diameter smaller or comparable with the cellulose fibbers diameter), of their density (over 3 kg/m^3 versus 1,300 kg/m^3 per paper and 2,500-2,700 kg/m^3 for calcium carbonate) and the possibility of superficial loading. In order to be used as security paper, depending on its destination, a certain density of FM microwires must be achieved in the paper. To the achievement of this density acts the retention efficiency of the ferromagnetic GMWs, influenced by technological factors as well as factors specific to the paper manufacturing equipment. The paper analyzes the main technological factors involved in ferromagnetic GMWs retention, such as: fibbers nature, grinding degree of the cellulosic pulp, GMWs amount, the hardwood and the softwood cellulose ratio in the fibrous paper composition, the amount of retention emulsion dosed in the paper manufacturing receipts, the paper weight. Through the performed experimental work was realised the experimental model for the retention of

ferromagnetic GMWs in the cellulosic paper structure, defined by the following manufacturing receipt: 30-70% softwood bleached cellulose, with Schopper-Riegler degree: 35-65 °SR; 30-70% hardwood bleached cellulose, with Schopper-Riegler degree: 30-50 °SR; calcium carbonate: 10 - 30%, as filling material; 0.3-0.5% polyamide-amine, as retention emulsion; 1-1.5% alkyl – dimercetines, as gluing agent; 0.02 - 0.05% GMWs. Under these conditions, the retention yield of the GMWs for a paper with a weight greater than 50 g/m^2 will be: η = 65 - 90%. Taking into account the variation limits of the indicated parameters, this model contains practically several experimental versions of the GMWs secured paper.

References

[1] G. F. Taylor, A method of drawing metallic filaments and discussion of their properties and uses, Phys. Rev. 23 (1924) 665-60. https://doi.org/10.1103/PhysRev.23.655

[2] A. Zhukov, J. Gonzalez, J. M. Blanco, M. Vazquez, V. Larin, Microwires coated by glass: a new family of soft and hard magnetic materials, J. Mat. Res 15 (2000) 2107-2113. https://doi.org/10.1557/JMR.2000.0303

[3] A. Talaat, J. Alonso, V. Zhukova, E. Garaio, J. A. García, H. Srikanth, M. H. Phan, A. Zhukov, Ferromagnetic glass-coated microwires with good heating properties for magnetic hyperthermia, Scie. Rep. 6 39300 (2016) 1-6. https://doi.org/10.1038/srep39300

[4] A. Zhukov, Glass-coated magnetic microwires for technical applications, *J. Magn. Magn. Mater. 242 – 245 (2002) 216-223. https://doi.org/10.1016/S0304-8853(01)01258-6*

[5] Jing-Shun Liu, Xiao-Dong Wang, Fa-Xiang Qin, Fu-Yang Cao, Da-Wei Xing, Hua-Xin Peng, Xiang Xue, Jian-Fei Sun, GMI Output stability of glass-coated co-based microwires for sensor application, Piers online, 7 7 (2011) 661-665.

[6] R. Hudak, I. Polacek, P. Klein, R Sahol, D. Varga, J. Zivcak, M. Vazquez, Nanocrystalline Magnetic Glass Coated Microwires Using the effect of superparamagnetism are usable as temperature sensors in biomedical applications, IEEE Trans. on Magn. 53 4 (2017) 5300305.

[7] J. Nabias, A. Asfour, J-P Yonnet, Temperature dependence of giant magnetoimpedance in amorphous microwires for sensor application, IEEE Trans. on Magn. 53 4 (2017) 4001005.

[8] M. Karimov, M. Gruzdev, Method and device for protecting products against counterfeiting, US 20070291988 A1, 2007.

[9] E. Manta, M. M. Codescu, M. Petrescu, Ferromagnetic glass-coated microwires and their applications, U.P.B. Sci. Bull. Series B 74 4 (2012) 177-184.

[10] M. M. Codescu, E. Manta, E. A. Pătroi, W. Kappel, I. Zăpodeanu, M. Burlacu, P. Nechita, V. Midoni, Securing elements with ferromagnetic microwires, Optoel. & Adv. Mat. – Rapid Comm. 4 10 (2010) 1361-1365.

[11] S. A. Baranov, O. V. Yaltychenko, E. Yu. Kanarovskii, M. M. Codescu, Preparation of the cast glass-coated amorphous magnetic microwires, Proc. SPIE 10010, Advanced Topics in Nanoelectronics, Microelectronics and Nanotechnologies VIII, 1001016 (2016) 117-123.

Powder Metallurgy and Advanced Materials – RoPM&AM 2017
Materials Research Proceedings 8 (2018) 11-17

Materials Research Forum LLC
doi: http://dx.doi.org/10.21741/9781945291999-2

The properties of bimetallic multi-layer (C45 and S235JR) and the multi-layer steel made by forging

Ioana Monica SAS-BOCA [1, a *], Dan Ioan FRUNZA [1,b], Dana Adriana ILUȚIU-VARVARA [1,c], and Istvan TOMA [1,d]

[1]Technical University of Cluj-Napoca, 28 Memorandumului Street, 400114, Cluj-Napoca, Romania

[a]monica.sas.boca@ipm.utcluj.ro, [b]dan.frunza@ipm.utcluj.ro, [c]dana.adriana.varvara@insta.utcluj.ro, [d]toma_istvan@yahoo.com

Keywords: Multi-layer, Bimetallic material, Forging.

Abstract. In this paper are presented the results of the research on properties and behavior of hot-forged bimetallic multi-layer material (C45-S235JR) compared to the properties and behavior of hot-forged multi-layer materials C45 and S235JR. The material was layered by successive manual hot forging to form 36-layers of the billets. Thus, it has been attempted to obtain superior materials in terms of properties, to withstand the demands they are subjected to. It has also been tried by stratification, obtaining results particularly relevant for resilience testing, where the different layer breakage occurs at higher strengths and has a high malleability. The microstructure of multi-layered materials was investigated in this paper and the mechanical properties were studied by tensile testing and Charpy impact testing. The Brinell micro-hardness has also been studied.

Introduction

Hot plastic deformation processes are the most common and used method for generating metallurgical metals [1, 2]. These processes are based on the characteristics of the metals obtained by high-temperature processing. By heating metals, we get less mechanical strength and increased malleability [3, 4]. Thus, raw materials can be processed with low material losses and low energy consumption in a form close to the finished piece.

The need to obtain the most effective and safe materials leads to a reorientation of the research to the ancient techniques [5] and technologies applicable to modern areas, such as cycling and, in particular, the acrobatics on the bicycle.

The main objective of the paper was to determine the characteristics [6 - 8] of the layered metal [9 - 11]. Another approach was to obtain bimetallic layered steel by forging [12]. This type of material combines the advantages of both materials and reduces the major inconvenience of each one taken separately. Often, the outer part of the piece is made of other metallic material to provide outstanding properties, as well as to reduce the cost price. This is possible due to the understanding of the function of the piece because the piece needs a certain mechanical strength [4, 13, 14], which does not mean that the whole piece will be made of that material but only that part inner or the outer part of the piece. The rest of the material can be a cheaper material also, in accordance with the requirements of the finished product.

In this work was aimed at making multi-layered steel bars and sandwich bars (C45- S235JR-C45).

The mechanical and microstructural properties were determined as follows: the tensile testings, Brinell hardness measurement, Charpy impact testing and microstructures were performed.

Powder Metallurgy and Advanced Materials – RoPM&AM 2017   Materials Research Forum LLC
Materials Research Proceedings 8 (2018) 11-17   doi: http://dx.doi.org/10.21741/9781945291999-2

Material and method

The materials used are the C45 band according to SR EN 10083 and the S235JR steel band according to EN 10204/2004. The chemical composition according to the current standards is presented in Table 1.

Table 1. Chemical composition

	C	Mn	Si	P	S	Cu	Ni	N	Cr
C45	0.45	0.7	≤0.30	≤0.035	≤0.4				
S235JR	0.13	0.5	0.15	0.018	0.007	0.05	0.03	0.012	0.05

The sandwich multi-layer resulted from manual hot forging. The billets were made of 36 layers.

The materials that we have been used in the study (not only C45 but even S235JR) were came from commercial source, laminate bar of 30x30x100. These were hot forged on the supporting bar which has a handling purpose. The heating at about 1100 °C has been done within a coal forge. After the heating process the sample has been immersed into borax and are-put into the forge. When the sample reached the 1100 °C again then it has been manually forged in order to achieve the welding between layers. The process has been about 3 – 4 times repeated until it has been accomplished a product which has not any fissures; after that the material has been stretched and bent in order to obtain the 12 layers.

The bent processes and the forges for the C45 multi-layer samples and for the S235JR multi-layer samples have been another 2 times repeated in order to obtain some bars semi-finished products which have a square section of 36 layers. In order to achieve the sandwich shape product there have been together forged about 24 of C45 layers which have in the middle of them about 12 layers of S235JR which have been accomplished in the same way.

The material has been stretched through free forging process and cut into bars of 100mm length each; from these bars there have been done some specimens for traction, resilience and hardness tests. The multi-layers samples which belong to the same semi-finished product has been labeled with 1,2,3 numbers as: multi-layer C45_1, multi-layer C45_2, multi-layer C45_3 for the specimens obtained from the C45 multi-layer; multi-layer S235JR 1, multi-layer S235JR 2, multi-layer S235JR 3 for the specimens obtained from the S235JR multi-layer and sandwich 1, sandwich 2, sandwich 3 for the specimens obtained from the 24 layers of C45 which have in the middle 12 layers of S236JR.

By bending, twelve-layer billets of the same material were made. In the final stage, three types of billets were made by layer overlays, such as 36-layer C45 steel specimen, 36-layer S235JR steel specimen and sandwich specimen 12 x C45 - 12 x S235JR - 12 times x C45.

The borax was used to clean the oxides on the three sandwich packages, ensuring better bonding of the layers.

The strength and tensile ductility, toughness, and brittle-to-ductile transition have been the main thrust for multi-layer investigations. The tensile testings were performed on a 200kN Heckert-EDZ-20S testing machine. Brinell hardness measurement was performed with a Amsler OTTO Wolpert-WERKE GMBA Hardness Tester Typ. Dia Testor 2 Rc-S type with ⌀5 mm ball, the ductility were determinated using a 300N. Instrumented Charpy impact tests were performed

Powder Metallurgy and Advanced Materials – RoPM&AM 2017   Materials Research Forum LLC
Materials Research Proceedings **8** (2018) 11-17   doi: http://dx.doi.org/10.21741/9781945291999-2

according to the standard ASTM A370 on impact testing machine, the microstructures were performed using a Jenoptik Prog Res C10 photodigital microscope.

Results and test methods

The Charpy test specimens (Fig. 1a) of dimensions 10x10 mm with a length of 55 mm and a U-notch with a radius of 1 mm were made from billets. It can be noticed that the outer layers of carbon with higher carbon content (C45) have cracked and the inner layers are only strongly deformed (plastic deformation) (Fig. 1b).

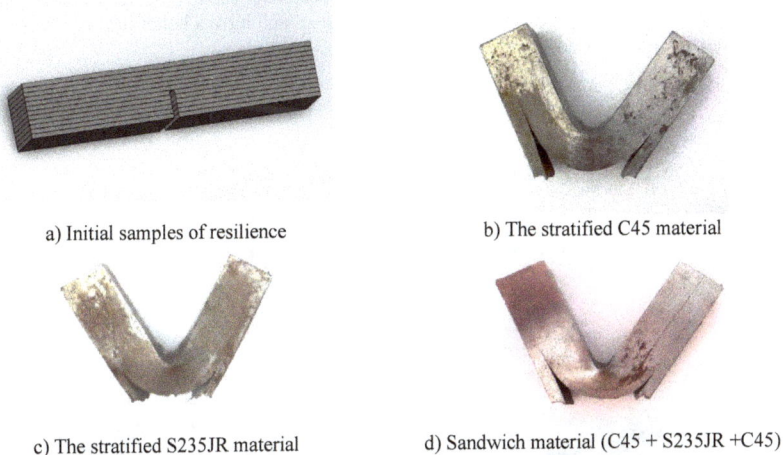

a) Initial samples of resilience

b) The stratified C45 material

c) The stratified S235JR material

d) Sandwich material (C45 + S235JR +C45)

Fig. 1. Charpy impact test samples.

The obtained results (Fig. 2) of the impact tests are superior for the sandwich material (Fig. 1d) versus S235JR multi-layer material (Fig. 1c) or the C45 multi-layer material (Fig. 1a). Thus, assumptions were made that the tougher exterior of the sample (C45) and the softer interior (S235JR) represent a malleability characteristic that requires more energy compared to the single type of layered material (S235JR or C45).

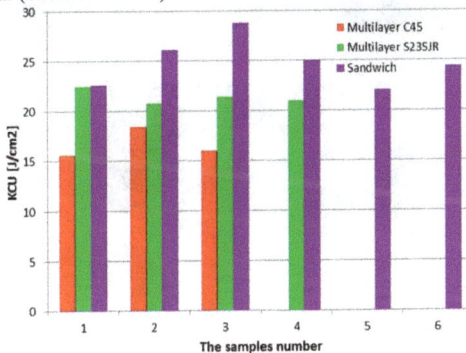

Fig. 2. Charpy impact test results.

Powder Metallurgy and Advanced Materials – RoPM&AM 2017   Materials Research Forum LLC
Materials Research Proceedings 8 (2018) 11-17   doi: http://dx.doi.org/10.21741/9781945291999-2

When assessing the tenacity of a material, the macroscopic appearance of the breakage section must also be taken into account. This aspect generally has two distinct parts: an outer part with a crystalline, fibrous and matte appearance, corresponding to a fragile crack, and the other central part, grunt and shiny, corresponding to a plastic deformation (breaking tenacity).

It can be seen from Fig. 2, for sandwich models (magenta) the resilience was obtained by 15,7% higher than for S235JR steel specimens (green) and also by 49,3% higher than for specimens of C45 with 36 layers (red).

From Figure 3 can be observed that sandwich specimens have an intermediate hardness between C45 multi-layer steel and S235JR multi-layer steel and have a better uniformity in results than the other samples, from 183 HB to 197 HB.

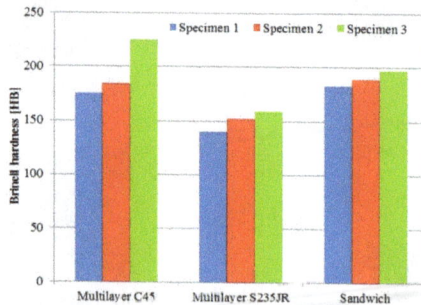

Fig. 3. Brinell hardness test.

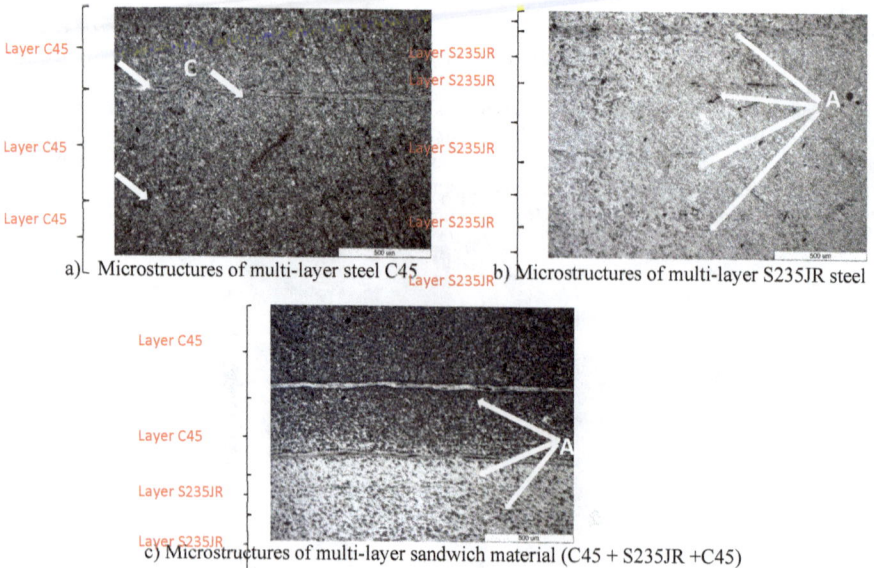

a) Microstructures of multi-layer steel C45   b) Microstructures of multi-layer S235JR steel

c) Microstructures of multi-layer sandwich material (C45 + S235JR +C45)

Fig. 4. The microstructures of the samples 100x.

Powder Metallurgy and Advanced Materials – RoPM&AM 2017   Materials Research Forum LLC
Materials Research Proceedings **8** (2018) 11-17   doi: http://dx.doi.org/10.21741/9781945291999-2

The forged specimens had a smooth and homogeneous structure, but due to many forging defects, such as decarburization, inclusions and complete welding failure, the results were not as homogeneous as we expected.

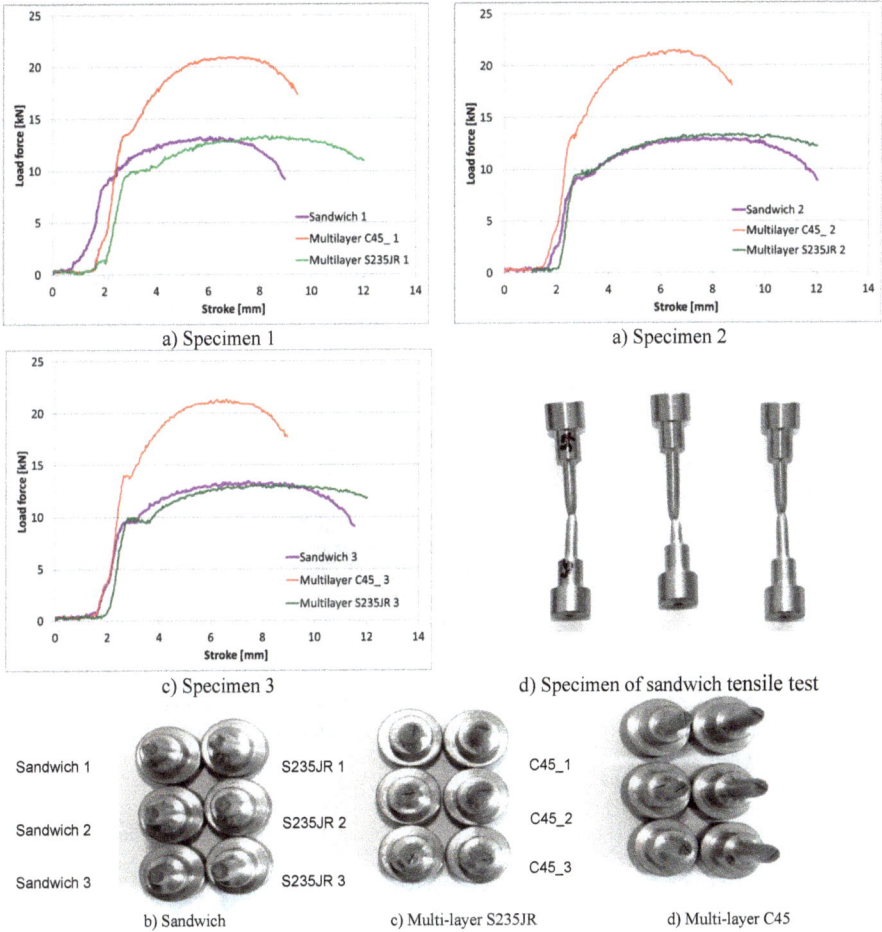

a) Specimen 1

a) Specimen 2

c) Specimen 3

d) Specimen of sandwich tensile test

b) Sandwich   c) Multi-layer S235JR   d) Multi-layer C45

Fig. 5. Tensile test.

The impact test specimen needed an identification of position on the steel layers to perform resilience tests correctly and uniformly on all samples. Was prepared and studied the specimens to be observed under a metallographic microscope, the attack was performed with nital.

Powder Metallurgy and Advanced Materials – RoPM&AM 2017   Materials Research Forum LLC
Materials Research Proceedings 8 (2018) 11-17   doi: http://dx.doi.org/10.21741/9781945291999-2

The specimens were study by optical microscopy (Fig. 4), the layers were observed and analyzed. The boundaries between the layers it was searched, so was managed to identify the overlapping planes.

The ferrite-pearlite structure is dominating. The microstructures show the interface (A) between layers shows discontinuities and, in their proximity,, a slight decarburization (C) can be observed in the C45 steel layer below 50 micrometers. At the same time, in the case of this sample, it can be seen that in the C45 steel area at the bending interface (B) the presence of an excess of borax about 10 micrometers thick is observed.

For sandwich samples: (Fig. 5) it can be seen that the force (14 kN) determined in the tensile testing is close to that of the S235JR steel samples. At tensile testing, all the sandwich samples had a ductile behavior. The sandwich specimens behaved like a medium ductility with ductile fracture type, which showed the separation of the forged layers in some cases.

The C45 specimens (multi-layer C45_1,2, 3) had superior S235JR (multi-layer S235JR 1,2,3) characteristics with a shorter neck and force required to achieve higher fracture. The rupture is of the ductile fracture type and does not reveal the detachment of the forged layers. It can be seen in Figure 5 that sandwich 1,2,3 specimens show shorter cracks than the S235JR steel specimens, but the maximum forces are similar.

Summary

From the Charpy impact test results, it can be concluded that the sandwich material has superior characteristics, is more ductile than C45 multi-layer material. The C45 material steel layers (from sandwich) are more fragile and cracks, also and for the S235JR steel layers (from sandwich) have high ductility, so they deform and absorb more energy but do not fracture (it is there a plastic deformation without cracks deflection). The fact that the C45 steel layers are broken is a technological advantage. From the point of view of requesting a bicycle frame made of the bimetallic multi-layer material (sandwich), when at the exterior we have strongest layers from more fragile material (multi-layer C45) and the inner ductile (S235JR multi-layer) exhibits a higer malleability (about 15.3% to specimens S235JR multi-layer, and about 49,3% to specimens C45 multi-layer). So that may be a warning to the user and it can avoid serious injuries caused by its totally break-down.

The hardness analysis has revealed that the exterior layers of the sandwich, C45 multi-layer ones are harder and give better resistance. The sandwich specimens have an intermediate hardness between C45 multi-layer steel and S235JR multi-layer steel and have a better uniformity in results than the other samples, from 183 HB to 197 HB.

The optical microstructures have helped to properly study the bounding of the layers and samples.

It can be clearly noticed (Fig. 1) that the brittle-to-ductile transition of the C45 multi-layer from outside to the S235JR multi-layer inside.

It can also be concluded that the other two layered materials show similarities in the test stress. A future research could be comparing the results obtained from testing C45 and S235JR with the layered materials of the same materials.

References

[1] H. Berns, The history of hardening, Harterei Gerster AG, 2013.

[2] R. L. Apps, D. R. Milner, Introduction to welding and brazing, Pergamon Press, 1994.

[3] K. C. John, Metal casting and joining, PHI Learning, 2015.

Powder Metallurgy and Advanced Materials – RoPM&AM 2017   Materials Research Forum LLC
Materials Research Proceedings 8 (2018) 11-17   doi: http://dx.doi.org/10.21741/9781945291999-2

[4] ***ASM Handbook Volume 8: Mechanical testing and evaluation, ASM International, 2000.

[5] T. Fumon, Samurai fighting arts: the spirit and the practice. Kodansha International, 2003.

[6] G.E. Dieter, Introduction to ductility, in Ductility, American Society for Metals, 1968.

[7] ***ASM Handbook Volume 9: Metallography and microstructures, ASM International, 2004.

[8] *** ASM Handbook Volume 14: Forming and forging, ASM International, 1998.

[9] J. Reisera, L. Garrisonb, H. Greunerc, J. Hoffmanna, T. Weingärtnera, U. Jäntscha, M. Klimenkova, P. Frankea, S. Bonka, C. Bonnekoha, S. Sickingera, S. Baumgärtnera, Daniel Bolicha, M. Hoffmanna, R. Zieglera, J. Konrada, J. Hohed, A. Hoffmanne, T. Mrotzeke, M. Seisse, M.l Rietha, A. Möslanga: Ductilisation of tungsten (W): Tungsten laminated composites, Int. J Refractory Metals Hard Mater. 69 (2017) 66–109.

[10] Y. Zhang, T. Ouyang, D. Liu, Y. Wang, J. Du, C. Zhang, S. Feng, J. Suo: Effect of thickness ratio on toughening mechanisms of Ta/W multi-layers, J All. Compd. 666 (2016) 30-37. https://doi.org/10.1016/j.jallcom.2016.01.084

[11] Y. Zhang, G. Xu, Y. Wang, C. Zhang, S. Feng, J. Suo: Mechanical properties study of W/TiN/Ta system multi-layers, J All. Compd. 725 (2017) 283-290. https://doi.org/10.1016/j.jallcom.2017.07.115

[12] N. A. Mara, I. J. Beyerlein: Interface-dominant multi-layers fabricated by severe plastic deformation: stability under extreme conditions, Curr. Opin. Solid State Mater. Sci. 19 (2015) 265–276. https://doi.org/10.1016/j.cossms.2015.04.002

[13] X.P. Zhang, S. Castagne, T.H. Yang, C.F. Gu, J.T. Wang: Entrance analysis of 7075 Al/Mg–Gd–Y–Zr/7075 Al laminated composite prepared by hot rolling and its mechanical properties, Mater. Des. 32 (2011) 1152–1158. https://doi.org/10.1016/j.matdes.2010.10.030

[14] G. Annea, M.R. Ramesha, H. S. Nayakaa, S. B. Arya: Investigation of microstructure and mechanical properties of Mg—Zn/Al multi-layered composite developed by accumulative roll bonding, Perspectives in Science 8 (2016) 104-106. https://doi.org/10.1016/j.pisc.2016.04.008

Powder Metallurgy and Advanced Materials – RoPM&AM 2017
Materials Research Proceedings 8 (2018) 18-27

Materials Research Forum LLC
doi: http://dx.doi.org/10.21741/9781945291999-3

Reactive mechanical milling of Fe-Ni-Fe$_2$O$_3$ mixtures

Traian Florin MARINCA [1,a*], Horea Florin CHICINAȘ[1,b],
Bogdan Viorel NEAMȚU [1,c], Florin POPA [1,d], Niculina Argentina SECHEL [1,e] and
Ionel CHICINAȘ [1,f]

[1]Materials Science and Engineering Department, Technical University of Cluj-Napoca, 103-105
Muncii Avenue, 400641 Cluj-Napoca, Romania

[a] traian.marinca@stm.utcluj.ro, [b]horea.chicinas@stm.utcluj.ro, [c]bogdan.neamtu@stm.utcluj.ro,
[d]florin.popa@stm.utcluj.ro, [e]niculina.sechel@stm.utcluj.ro, [c]ionel.chicinas@stm.utcluj.ro

* traian.marinca@stm.utcluj.ro

Keywords: Reactive mechanical milling, Magnetic composite, Ferrite, Magnetite.

Abstract. Fe-Ni-Fe$_2$O$_3$ mixture in various ratios has been milled in a high energy mill for synthesis of Ni$_3$Fe/Fe$_3$O$_4$ type magnetic nanocomposite up to 10 h. The samples have been investigated in the light of X-ray diffraction, scanning electron microscopy, energy dispersive X-ray spectrometry and laser particle size analysis. The formation of the composite begins after 2 h of milling. After 10 h of milling the nanocomposite with the high amount of metallic phase consists in a mixture of Ni$_3$Fe and Fe$_3$O$_4$ alongside of a small amount of residual Fe$_2$O$_3$. Both phases are formed progressively upon increasing the milling time. Upon increasing the amount of oxide in the starting mixture at the end of the milling time the phases present in the nanocomposite material are changing. In the sample with the higher amount of oxide at the end of the milling time are present: Fe$_3$O$_4$, fcc Ni-based structure and unreacted Fe$_2$O$_3$ and Fe phases. The mean crystallite size estimated for the main phases of the nanocomposite, Ni$_3$Fe and Fe$_3$O$_4$, are in the nanometric range at the end of milling time independent on starting ratio among starting materials. For both, Ni$_3$Fe and Fe$_3$O$_4$, the mean crystallite size is ranging from 9 to 12 nm for all compositions. The nanocomposite particles shape is irregular. The powder consists in two types of particles: very fine particles with the size of less than 100 nm and larger particles with the size of a few micrometers. The micrometric particles are formed by fine agglomerated particles and by smaller particles that are welded together. The median diameter d$_{50}$ present a significant increase in the first stage of milling and is maintained at values close to 5 μm, having small variations on each milling time.

Introduction

The development of the magnetic materials is in growth due to the large demand from various industry branches such as telecommunication, computer technologies or electronics. Therefore the development of performant magnetic materials is one of the most important and interesting research subjects [1-4]. The spinel ferrites are one the most important class of magnetic materials with a large field of applications. The ferrites are used, for example, in high frequencies applications due to their high electrical resistivity. One of the main drawbacks of the spinel ferrites when are used in high frequencies applications is the relatively low magnetic induction and magnetic permeability [5-9]. The Fe-Ni alloys have very good magnetic induction and permeability and are used nowadays in multiple applications. The use in high frequencies applications of these alloys is limited do to their low electrical resistivity as compared to the ferrites one which favour the increase of the eddy currents [10 - 12]. The synthesis of a composite

Powder Metallurgy and Advanced Materials – RoPM&AM 2017   Materials Research Forum LLC
Materials Research Proceedings **8** (2018) 18-27   doi: http://dx.doi.org/10.21741/9781945291999-3

material consisting in two phases, one of Ni-Fe alloy and other in spinel ferrite can combine the above maintained advantages of each phase.

The paper presents the synthesis of composite magnetic material of Ni_3Fe/Fe_3O_4 type. The main purpose for the synthesis of this type of magnetic composite is obtain an material that can combine the high magnetic permeability of the Permalloy with the high electrical resistivity of the magnetite (electrical resistivity of magnetite is with three order of magnitude higher than the one of iron) using accessible precursors. The paper is focus on the synthesis conditions in order to obtain such type of composites.

Experimental details

As raw materials have been used Ni carbonyl, Fe NC100.24 and Fe_2O_3 powders. The ratio among the initial powder was calculated in order to correspond to fourth atom/molecule ratio between metal (both Ni and Fe) and oxide (Fe_2O_3) as follow: 0.9/0.1, 0.8/0.2, 0.7/0.3 and 0.6/0.4. These ratios correspond to the reaction of the starting materials in order to obtain Ni_3Fe and Fe_3O_4 in the following mole ratios: 100/8.5 (noted S1), 100/20 (noted S2), 100/36 (noted S3) and 100/60 (noted S4). In the above mentioned ratios the mixture has been milled in a planetary ball mill Pulverisette 4 manufactured by Fritsch up to 10 h using air as milling atmosphere. An amount of 100 g of powder has been loaded in a 500 cm^3 hardened steel vial alongside of 80 balls of 14 mm in diameter. The ball to powder mass ratio (BPR) was 9:1. The vial speed was set to 800 rpm and the disc speed was set to 400 rpm. The vial and disc were rotated in the opposite directions. The milled powders structural evolution was checked by X-ray diffraction (XRD). An INEL EQUINOX 3000 diffractometer that use $CoK\alpha$ radiation ($\lambda = 1.7903$ Å) was used. The investigated angular interval was 20-110°. The mean crystallite size was estimated using Scherrer formula [14]. The morphology of the powder was studied by scanning electron microscopy using a JEOL JSM-5600LV scanning electron microscope (SEM). The local chemical homogeneity was verified by energy dispersive X-ray spectrometry (EDX) using an Oxford Instrument EDX detector (INCA 200 software) with which it is equipped above mentioned scanning electron microscope. By a Laser Particle Size Analyzer (Fritsch Analysette 22—Nanotec) the samples particle size distribution and d_{50} parameter (the diameter for which 50 vol.% of particles are smaller than the d_{50} value) were studied. Before analysis an ultrasound treatment was applied for deagglomeration and dispersion of the particles.

Results and Discussion

In Fig. 1 are presented the X-ray diffraction patterns of the $Fe_2O_3+Fe+Ni$ mixture milled for 0, 0.5, 1, 2, 3, 4, 6 and 10 h for S1. The positions of the peaks of each phase that was observed in the samples are indicated on the graph alongside of the Miller indices of the diffraction planes. It can be noticed in the patterns of the unmilled samples (0 min MM) the peaks characteristic of the Fe bcc structure from space group Im-3m (JCPDS file no. 06-0696), Ni fcc structure from space group Fm-3m (JCPDS file no. 04-0850) and Fe_2O_3 rhombohedral structure from space group R-3c (JCPDS file no. 33-0664). After only 0.5 h of milling it can observed that the diffraction peaks becomes broadened, especially the peaks of iron oxide-hematite. The stresses induced in material and the decrease of the crystallite size are the causes of the peaks broadening. Due to the fragile character of the hematite its peaks are broader as compared with the diffraction peaks of Ni and Fe. After 2 h of milling one can observe is that the change in the intensity of the peaks of iron oxide – Fe_2O_3. The diffraction plane (104) and (110), which are the most intense ones, are changing the intensity. Before milling the (104) reflection is the most intense and after 2 h of milling the (110) reflection becomes the most intense. The (110) diffraction line of Fe_2O_3 is overlapping on the (311) diffraction line of magnetite – Fe_3O_4 (JCPDS file no. 19-0629). Upon

Powder Metallurgy and Advanced Materials – RoPM&AM 2017   Materials Research Forum LLC
Materials Research Proceedings **8** (2018) 18-27   doi: http://dx.doi.org/10.21741/9781945291999-3

increasing the milling time the most intense peak of hematite – (104) becomes less and less intense and the other diffraction peaks are vanishing. After 10 h of milling it is hardly identifiably hematite (104) diffraction line (the most intense peak of this phase). This indicates the progressive formation of magnetite upon increasing the milling time [9, 14].

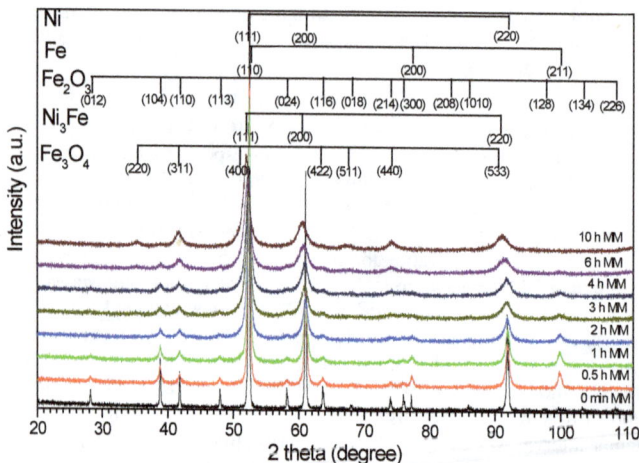

Fig.1. X-ray diffraction patterns of the Fe_2O_3+Fe+Ni mixture milled for 0, 0.5, 1, 2, 3, 4, 6 and 10 h for S1.

The iron diffraction lines becomes broadened, decreasing in intensity upon increasing the milling time and are vanishing after 10 h of milling. The Ni diffraction lines become broadened and are shifted to lower angles upon increasing the milling time. This indicates the presence of the fcc structure in material for each milling time. The displacement to lower angles of the fcc structure corroborated with the disappearance of the Fe diffraction lines is assigned to the formation of a Ni-Fe alloy. The alloy is identified as being of Ni_3Fe type (JCPDS file no. 65-3244) taken in consideration the ratio among the atoms in the starting mixture and the formation of Fe_3O_4 during milling. The Ni_3Fe is progressively form as the Fe_3O_4 and the formation of this alloy is evidenced in the Fig. 2. At the end of milling time for this composition the material consists in Ni_3Fe and Fe_3O_4 alongside to very small amount of a residual Fe_2O_3 phase. The X-ray diffraction patterns of the Fe_2O_3+Fe+Ni mixture milled for 0, 0.5, 1, 2, 3, 4, 6 and 10 h for S2 have been presented elsewhere [14]. In the case of this ratio between the starting powders it was observed the similar behaviour as in the case of S1. Although are some differences: at the end of milling time the presence of Fe_2O_3 is much more pregnant as compared to S1 and due to the larger amount of oxide the magnetite diffraction peaks are more visible in the diffraction pattern as compared to the one of Ni_3Fe.

In Fig. 3 are presented the X-ray diffraction patterns of the Fe_2O_3+Fe+Ni mixture milled for 0, 0.5, 1, 2, 3, 4, 6 and 10 h for S3. In the case of this mixture the phases that are identified at the end of milling time are in part similar with the ones observed in the case of the other two presented above. Due to the presence in the starting mixture of a larger amount of oxides thus resulting in a

Powder Metallurgy and Advanced Materials – RoPM&AM 2017   Materials Research Forum LLC
Materials Research Proceedings 8 (2018) 18-27   doi: http://dx.doi.org/10.21741/9781945291999-3

larger amount of oxide also at the end of milling time and the Bragg reflections of these are better defined as compared to the other. The phases present in material at the end of milling time are: Fe_3O_4, Fe_2O_3 (larger amount as compared to S1 and S2), fcc Ni-based structure and residual Fe. It can be observed that in this case fcc structure does not have the composition closer to the one of the Ni_3Fe intermetallic compound as in the case of the other two compositions. The proves are the very low displacement of the Ni diffraction peaks to lower angles which can be assigned to the internal stress induced by milling and the presence of the residual Fe.

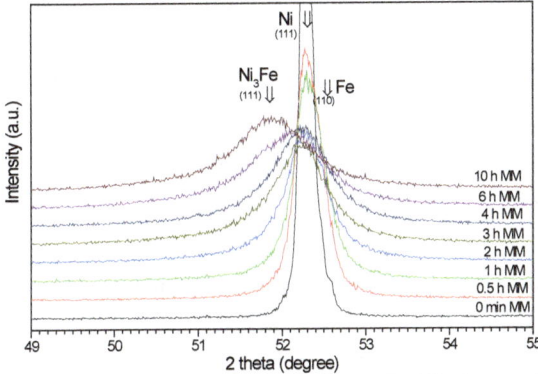

Fig. 2. Detail of the X-ray diffraction patterns of the $Fe_2O_3+Fe+Ni$ mixture milled for 0, 0.5, 1, 2, 3, 4, 6 and 10 h for x=S1 in the zone of the most intense diffraction peaks of the newly formed Ni_3Fe compound.

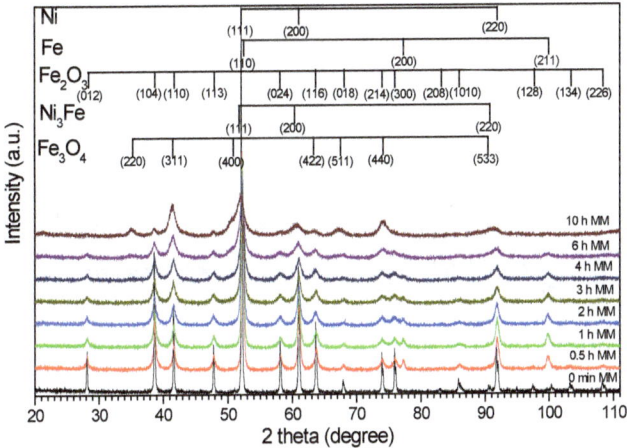

Fig. 3. X-ray diffraction patterns of the $Fe_2O_3+Fe+Ni$ mixture milled for 0, 0.5, 1, 2, 3, 4, 6 and 10 h for S3.

Powder Metallurgy and Advanced Materials – RoPM&AM 2017   Materials Research Forum LLC
Materials Research Proceedings 8 (2018) 18-27   doi: http://dx.doi.org/10.21741/9781945291999-3

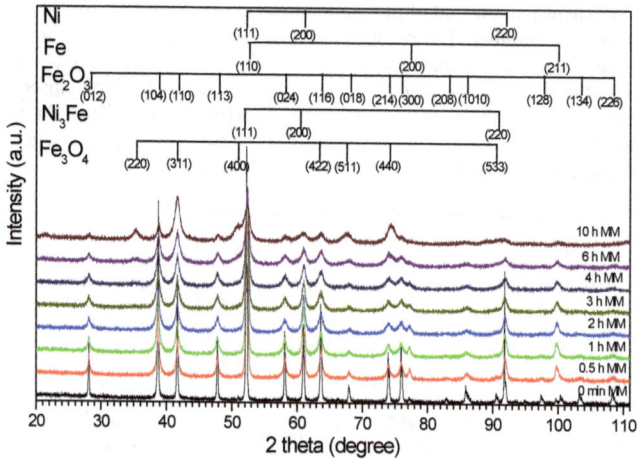

Fig. 4. X-ray diffraction patterns of the $Fe_2O_3+Fe+Ni$ mixture milled for 0, 0.5, 1, 2, 3, 4, 6 and 10 h for S4.

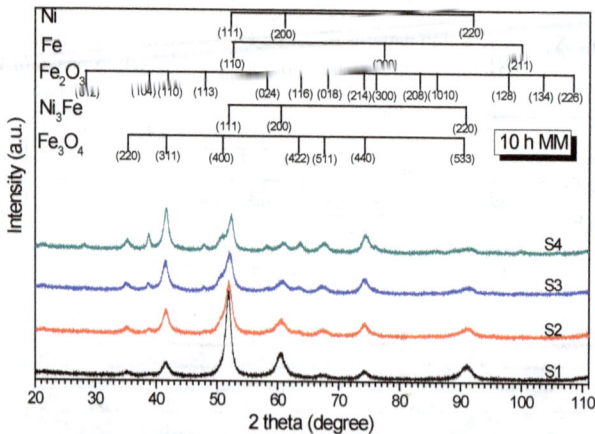

Fig. 5. X-ray diffraction patterns of the $Fe_2O_3+Fe+Ni$ mixtures milled for 10 h for S1, S2, S3 and S4.

Fig. 4 presents the X-ray diffraction patterns of the $Fe_2O_3+Fe+Ni$ mixture milled for 0, 0.5, 1, 2, 3, 4, 6 and 10 h for S4. The phases identified in the diffraction patterns are similar with the ones identified in the case of the samples S3. For a better view of the difference between the phases present in the samples S1, S2, S3 and S4 after 10 h of milling the X-ray diffraction patterns are presented in the Fig. 5.

Powder Metallurgy and Advanced Materials – RoPM&AM 2017   Materials Research Forum LLC
Materials Research Proceedings 8 (2018) 18-27   doi: http://dx.doi.org/10.21741/9781945291999-3

Table 1. Mean crystallite size of the Ni_3Fe and Fe_3O_4 after 10 h of milling for samples S1, S2, S3 and S4.

Phase	Crystallite size (nm)			
	S1	S2	S3	S4
Ni_3Fe	11	9	10	12
Fe_3O_4	9	10	10	12

In the case of each composition the phases are in nanocrystalline state and the powder is a nanocomposite one of Ni_3Fe/Fe_3O_4 type. In the table 1 are presented the mean crystallite size of the Ni_3Fe and Fe_3O_4 (main phases of the composite) after 10 h of milling for samples S1, S2, S3 and S4. One can observe for both phases is the mean crystallite size ranging from 9 to 12 nm for all four compositions. The width of the diffraction peaks is increasing upon increasing the milling time this indicating a decrease of the crystallite size upon increasing the milling time for both phases (metallic and oxide) and increasing of the internal stresses. The crystallite size decreasing upon increasing the milling time is in good relation with the earlier reported results both on ferrites and Ni-based alloys [11, 14-16].

Fig. 6. SEM images of a sample milled for 1h for the composition S3 at magnification of 200x, 1500x and 10,000x.

Fig. 6 shows the SEM images of a sample milled for 1h for the composition S3 at three magnifications: 200x, 1500x and 10,000x. The particles shape is irregular. At a first look it can be remarked two types of particles: very fine particles with the size of less than 100 nm and larger particles with the size of a few micrometers. At a closer look it can be remarked that the micrometric particles are formed by fine agglomerated particles and by small particles that are welded together upon processing the powder in the mill.

Fig. 7. SEM images of a sample milled for 3h for the composition S3 at magnification of 200x, 1500x and 10,000x.

Powder Metallurgy and Advanced Materials – RoPM&AM 2017   Materials Research Forum LLC
Materials Research Proceedings 8 (2018) 18-27   doi: http://dx.doi.org/10.21741/9781945291999-3

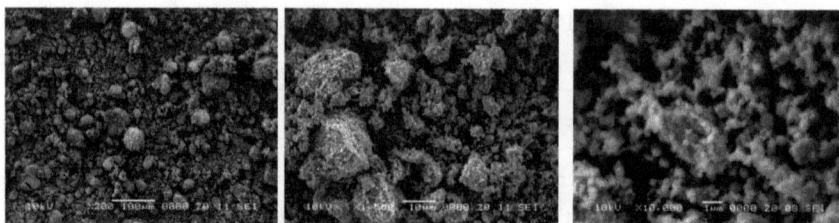

Fig. 8. SEM images of a sample milled for 6h for the composition S3 at magnification of 200x, 1500x and 10,000x.

SEM images of a sample milled for 3h for the composition S3 at magnification of 200x, 1500x and 10,000x are presented in Fig. 7. The powder morphology does not present significant change upon increasing milling time from 1h to 3h. Also after 6h of milling the powder present similar characteristics as in the case of the powder milled for smaller milling time. In Fig. 8 are displayed SEM images of the sample milled for 6h for the composition S3 at magnification of 200x, 1500x and 10,000x.

The EDX local chemical elemental mapping for a sample milled for 6 h from the composition S3 is revealed in the Fig. 9. The Ni, Fe and O elements are uniformly distributed in the analysed sample. This suggests a very fine distribution of the metallic and oxide phases. This type of investigation comes to confirm the formation of the nanocomposite by mechanical milling by formation of an intimate mixture between the metallic phase and oxide phase at nanoscale.

Fig. 9. EDX local chemical elemental mapping for a sample milled for 6 h from the composition S3.

Powder Metallurgy and Advanced Materials – RoPM&AM 2017   Materials Research Forum LLC
Materials Research Proceedings 8 (2018) 18-27   doi: http://dx.doi.org/10.21741/9781945291999-3

The evolution of the d_{50} parameter as a function of the milling time for the S2 samples as it results from the particle size analysis effectuated with the laser particle size analyser is displayed in Fig. 10. It can be observed that the starting sample presents a median diameter less than 1 micrometer. This is due to the very fine particles of Ni and Fe_2O_3 present in the starting mixture. The iron particles are larger as compared to the ones of Ni and Fe_2O_3 but due to the low amount in this composition it does not influence in deep the median size of the particles of the starting mixture. On can note after 0.5h of milling a large increase of the median diameter occurs up to about 8 μm. This is assigned to the cold welding of the particles in the early stage of mechanical milling [16]. Increasing the milling time at 1h leads to a decrease of the median diameter up to around 5 μm. This decrease is assigned to the formation of the intimate mixture between the oxide and metallic phase which creates composite particles and which becomes more fragile due to the presence of oxide. Increasing the milling time up to 10 h of milling keeps the median diameter close to the value of 5 μm, having small variations on each milling time.

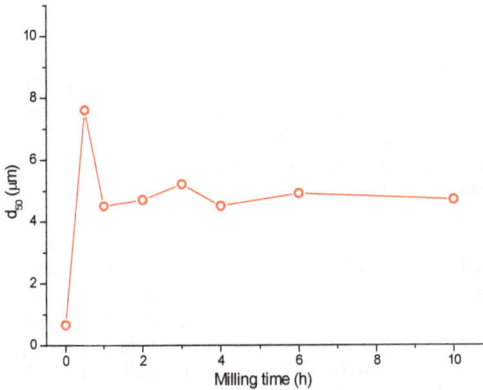

Fig. 10. Evolution of the d50 parameter as a function of the milling time for the S2 samples.

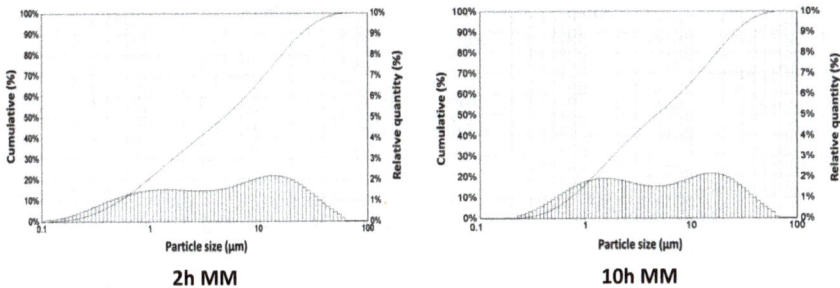

2h MM   10h MM

Fig. 11. Particle size distribution of the samples milled for 2 and 10 h in the case of the composition S2 (the cumulative and relative quantity represents vol. %).

The particle size distribution (vol. %) of the samples milled for 2 and 10 h in the case of the composition S2 is shown in Fig. 11. The particles size distribution of these two samples is similar. It can be observed that in both cases the particle size distribution is bimodal. The particle size distribution is in the same manner for all the milling time from 1 to 10 h of milling. The presence of two types of particles is assigned to the two types of process that are present in the milled samples: cold welding and fragmentation. The smaller particles come from the fragmentation of the bigger one and the bigger one is provided by cold welding of the smaller ones. Is assumed that these two types of processes are in a quasi-equilibrium for these milling times and for this composition of the material. The particle size distribution doe not reveal the presence of very fine particles (less than 100 nm), particles that have been evidenced by SEM analyses. It is assumed that the very fine particles are attached by the larger ones and cannot be separate during ultrasound treatment applied in particle size distribution analysis procedure.

Summary

Nanocomposites of the Ni_3Fe/Fe_3O_4 type have been successfully obtained using accessible precursors, Ni, Fe and Fe_2O_3, in various ratios, by high energy reactive mechanical milling. The ratio among starting powders was chosen in such way in order to vary the ratio between Ni_3Fe intermetallic compound and Fe_3O_4 in the final products. The formation of the nanocomposite has been studied by X-ray diffraction technique. By scanning electron microscopy, energy dispersive X-ray spectrometry and laser particle size analysis have been investigated the powder morphology, local chemical homogeneity and particles size distribution. The nanocomposite is forming continuously during the entire period of milling. The nanocomposite formation begins after 2 h of milling in the case of all the compositions. In the case of composition with the highest amount of metallic phases after 10 h of milling the nanocomposite consists in Ni_3Fe and Fe_3O_4. A small amount of unreacted Fe_2O_3 is evidenced. The increase of the amount of oxide in the starting mixtures leads to the formation of a larger amount of magnetite in the final product and also leads to the changes of the phases in composite after the end of milling time. Fe_3O_4, fcc Ni-based structure and unreacted Fe_2O_3 and Fe phases are founded in the nanocomposite with the higher amount of oxide in the starting mixture. The mean crystallite size of the composite main constituent phases, Ni_3Fe and Fe_3O_4, is in the range of 9-12 nm independent on the ratio between the metallic and oxide phases. The particles shape of the nanocomposite has no regular shape. The nanocomposite powder has two types of particles: very fine particles (size less than 100 nm) and large particles (a few micrometers in size). The micrometric particles consist in fine agglomerated and welded particles. The median diameter d_{50} has large variation in the first stage of milling and a small variation on each milling time is evidenced after that.

Acknowledgement

This work was supported by a grant from the Romanian National Authority for Scientific Research, CNCS and UEFISCDI, project number PN-III-P2-2.1-PED-2016-1816.

References

[1] J. Xiang, X. Shen, F. Song, M. Liu, G. Zhou, Y. Chu, Fabrication and characterization of Fe-Ni alloy/nickel ferrite composite nanofibers by electrospinning and partial reduction, Mater. Res. Bull. 46 (2011) 258-261. https://doi.org/10.1016/j.materresbull.2010.11.004

[2] S. Wu, A. Sun, W. Xu, Q. Zhang, F. Zhai, P. Logan, A. Volinsky, Iron-based soft magnetic composites with Mn-Zn ferrite nanoparticles coating obtained by sol-gel method, J. Magn. Magn. Mater. 324 (2012) 3899-3905. https://doi.org/10.1016/j.jmmm.2012.06.042

Powder Metallurgy and Advanced Materials – RoPM&AM 2017   Materials Research Forum LLC
Materials Research Proceedings **8** (2018) 18-27   doi: http://dx.doi.org/10.21741/9781945291999-3

[3] L.A. Dobrzanski, M. Drak, B. Ziebowicz, New possibilities of composite materials application-materials of specific magnetic properties, J. Mater. Process. Technol., 191 (2007) 352–355. https://doi.org/10.1016/j.jmatprotec.2007.03.029

[4] T. Caillot, G. Pourroy, D. Stuerga, Novel metallic iron/manganeseezinc ferrite nanocomposites prepared by microwave hydrothermal flash synthesis, J. Alloys Compd. 509 (2011) 3493-3496. https://doi.org/10.1016/j.jallcom.2010.12.096

[5] B.D. Cullity, C.D. Graham, Introduction to magnetic materials, second ed., Press&Wiley, New Jersey, USA, 2009.

[6] A. Goldman, Modern ferrite technology, second ed., Springer, Pittsburgh, USA, 2006.

[7] K. Hirota, M. Obatal, M. Kato, H. Taguchi, Fabrication of full-density Mg ferrite/Fe-Ni permalloy nanocomposites with a high-saturation magnetization density of 1 T, Int. J. Appl. Ceram. Technol. 8 (2011) 1-13. https://doi.org/10.1111/j.1744-7402.2009.02477.x

[8] T.F. Marinca, I. Chicinaş, O. Isnard, Structural and magnetic properties of the copper ferrite obtained by reactive milling and heat treatment, Ceram. Int., 39 (2013) 4179-4186. https://doi.org/10.1016/j.ceramint.2012.10.274

[9] T.F. Marinca, H.F. Chicinas, B.V. Neamtu, O. Isnard, A. Mesaros, I. Chicinas, Composite magnetic powder of Ni_3Fe/Fe_3O_4 type obtained from $Fe/NiO/Fe_2O_3$ mixtures by mechanosynthesis and annealing, J. Alloys Compd. 714 (2017) 484-492. https://doi.org/10.1016/j.jallcom.2017.04.263

[10] G. Couderchon, Alliages fer-nickel et fer-cobalt e Proprietes magnetiques, Publisher Techniques de l'ingenieur, Traite Genie Electrique, 1994, pp. 1e24. D2 130.

[11] I. Chicinas, Soft magnetic nanocrystalline powders produced by mechanical, alloying routes, J. Optoelectron. Adv. Mater. 8 (2006) 439-448.

[12] S. Chikazumi, Physics of Ferromagnetism, second ed., Oxford University Press, New York, USA, 1997.

[13] A.L. Patterson, The Scherrer formula for X-ray particle size determination, Phys. Rev. 56 (1939) 978-982. https://doi.org/10.1103/PhysRev.56.978

[14] T.F. Marinca, H.F. Chicinas, B.V. Neamtu, O. Isnard, I. Chicinas, Structural, thermal and magnetic characteristics of Fe_3O_4/Ni_3Fe composite powder obtained by mechanosynthesis-annealing route, J. Alloys Compd. 652 (2015) 313-321. https://doi.org/10.1016/j.jallcom.2015.08.249

[15] C.N. Chinnasamy, A. Narayanasamy, N. Ponpandian, K. Chattopadhyay, H. Guerault, J-M Greneche, Magnetic properties of nanostructured ferrimagnetic zinc ferrite, J. Phys.: Condens. Matter. 12 (2000) 7795–7805. https://doi.org/10.1088/0953-8984/12/35/314

[16] C. Suryanarayana, Mechanical Alloying and Milling, Marcel Dekker, 2004. New York, USA. https://doi.org/10.1201/9780203020647

Powder Metallurgy and Advanced Materials – RoPM&AM 2017   Materials Research Forum LLC
Materials Research Proceedings 8 (2018) 28-34   doi: http://dx.doi.org/10.21741/9781945291999-4

Green synthesis of silver chloride nanoparticles using Rhodotorula Mucilaginosa

Ioana GHIUȚĂ [1, a *], Daniel CRISTEA [1, b], Rodica WENKERT [2,c],
Daniel MUNTEANU [1,d]

[1]Transilvania University, Eroilor 29, Brasov, Romania

[2] Yitzhack I. Rager Blvd 151, Beersheba, Israel

[a*]ioana.ghiuta@unitbv.ro, [b]daniel.cristea@unitbv.ro [c]RodicaWe@clalit.org.il

[d]danielmunteanu@unitbv.ro

Keywords: Silver chloride nanoparticles, *Rhodotorula Mucilaginosa*, Green synthesis.

Abstract. The biosynthesis of silver chloride nanoparticles (AgCl NPs) is presented in this paper. Silver chloride nanoparticles were synthesized using fungi culture from *Rhodotorula Mucilaginosa* and aqueous AgNO$_3$ solution, as precursor. The plasmon resonance of the nanoparticles containing solution has shown through UV-visible spectrophotometry an absorbance peak at about 437 nm. Scanning Electron Microscopy, Energy Dispersive Spectroscopy, and X-ray Diffraction analyses confirmed the presence of spherical silver chloride nanoparticles with a face centered cubic crystal structure and an average particle size of 25 nm. Silver chloride nanoparticles have been shown to be able to inhibit the growth of different microorganisms, including bacteria and fungi, which would make them suitable for antimicrobial applications.

Introduction

The development of materials and structures at nanoscale dimensions has gained a huge interest in the nanomaterials and nano-technology research fields. One of the most important properties of metallic nanoparticles is their antimicrobial activity.

Eco-friendly methods concerning nanomaterials synthesis present a substantial importance for biological applications, mainly due to nontoxic substances and environmentally friendly procedures employed [1].

Green synthesis of silver chloride nanoparticles has been reported to be mediated by different kinds of organisms, from bacteria to plant extracts. Cell-free culture supernatant of *Streptomyces* strain, *Klebsiella planticola*, biomass of *Bacillus subtilis*, leaf extract of *Cissus quadrangularis*, aqueous extract of *Sargassum plagiophyllum*, extract from needles of *Pinus densiflora*, *Prunus persica L.* outer peel extract are just a few examples of organisms able to synthesize AgCl NPs [2]-[8].

Weili Hu et al. have presented the synthesis of silver chloride nanoparticles under ambient conditions in nanoporous bacterial cellulose membranes as nanoreactors. It has been demonstrated that the synthesized silver chloride nanoparticles exhibited high hydrophilic ability and a strong antimicrobial activity against *Staphylococcus aureus* and *Escherichia coli* bacteria [9]. The antibacterial effect of biosynthesized AgCl NPs investigated against *Escherichia coli* was found to be dose-dependent [6]. The biosynthesized silver chloride nanoparticles exhibited besides the antimicrobial activity, cytotoxicity activity against HeLa and SiHa cancer cell lines [2].

Powder Metallurgy and Advanced Materials – RoPM&AM 2017   Materials Research Forum LLC
Materials Research Proceedings 8 (2018) 28-34   doi: http://dx.doi.org/10.21741/9781945291999-4

M. Sophocleous and J. K. Atkinson have described in their review the significant development of Ag/AgCl screen printed sensors [10]. Moreover, Ag/AgCl NPs are examined for further applications of nanoparticles as a plasmonic photocatalyst [6].

The reports which have shown the importance of the AgCl NPs application, from sensors, catalysts, to antimicrobial activity, have led to the research results presented herein. Even though several methods to obtain nanoparticles are currently developed, the green methods have captured the interest of researchers due to their lack of toxicity. In the present work, the green synthesis of silver chloride nanoparticles is described. This is, to the best of our knowledge, the first report for green synthesis of AgCl NPs mediated by *Rhodotorula Mucilaginosa*.

Materials and Methods
1. Fungi and culture conditions
The fungi *Rhodotorula Mucilaginosa* used in this study were provided from Soroka University Medical Center from Beersheva, Israel. The fungi were cultivated in the solid media, Sabouraud agar supplied by Scharlau Chemicals, and incubated at 35 °C for 48 h.

2. Biosynthesis of AgCl nanoparticles using *Rhodotorula Mucilaginosa*
In order to synthesize the silver chloride nanoparticles, 1µl of bacterial strains were freshly inoculated in test tubes containing 15 ml of growth medium, namely Brain Heart Infusion from Sigma Aldrich. The liquid media contained beef heart (infusion from 250 g), 5 g/L; calf brains (infusion from 200 g), 12.5 g/L; disodium hydrogen phosphate, 2.5 g/L; D(+)-glucose, 2 g/L; peptone, 10 g/L; sodium chloride, 5 g/L. The liquid culture was kept in a thermostat at 35 °C for 24 h, followed by centrifugation at 4000 rpm for 30 min. The supernatant and biomass were tested in parallel. In the first situation 5 ml of supernatant was used, while for the second one the biomass was kept with the addition of 5 ml of distilled water. The next step was similar by adding the 5-ml culture over 40 ml Ag NO_3 precursor solution, at 1 mM, 2 mM and 3 mM concentration, respectively. The culture, supernatant and biomass + distilled water and precursors were kept as control. The samples were kept in a thermostat set at 35 °C, for 48 h.

3. Characterization of AgNPs
Ultraviolet-visible spectral analysis was carried-out by using Jasco V-630 spectro-photometer. The UV-visible spectra were measured in the range 200-600 nm with a wavelength step size of 1.5 nm.

For morphological characteristics and chemical composition, a JSM 7400f scanning electron microscope (SEM) with a platform for Energy Dispersive Spectroscopy (EDS) was used. The silver chloride nanoparticles colloid was dropped on a copper grid and coated with a platinum thin film. After this the samples were mounted on a double sided adhesive carbon-tape. The acceleration voltage was fixed to 10kV.

The crystalline nature of silver chloride nanoparticles was analyzed by XRD using a Philips PW 1050/70 X-ray powder diffractometer with graphite monochromator using $CuK_{\alpha1}$ ($\lambda = 1,54\text{Å}$), at a voltage of 40 kV, a current of 28 mA, in the scan range 10÷80 °, in Bragg-Brentano geometry.

Results and Discussion
After incubation in the thermostat for 48 h at 35 °C, the final color of the colloid, containing biomass of *Rhodotorula Mucilaginosa*, changed from light yellow to light brown. The change in color is an indication for the formation of nanoparticles. In Fig.1 the obvious difference in color can be observed, depending on the precursor concentration. The samples with supernatant and the control have remained unchanged. Furthermore, considering the optical indication (changing the color), the UV-visible absorption spectra of the colloidal solutions with nanoparticles was measured (shown in Fig. 2).

29

Powder Metallurgy and Advanced Materials – RoPM&AM 2017   Materials Research Forum LLC
Materials Research Proceedings 8 (2018) 28-34   doi: http://dx.doi.org/10.21741/9781945291999-4

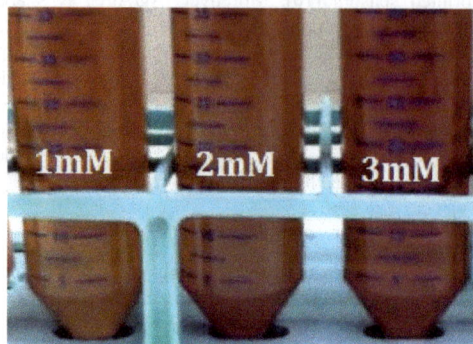

Fig. 1. Image of colloidal nanoparticles synthesized in the presence of
Rhodotorula Mucilaginosa.

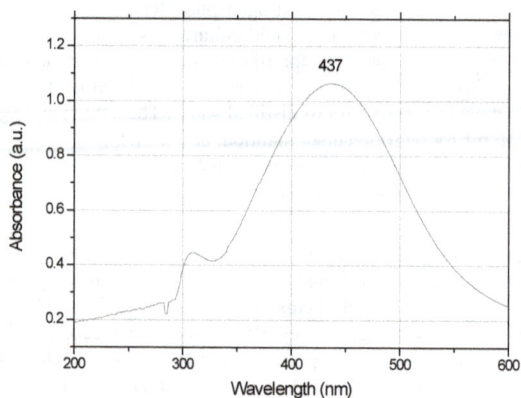

Fig. 2. UV-visible spectrum of the AgCl nanoparticle containing colloids3.

Fig. 2 corresponds to the UV-visible absorption spectrum of the solution containing AgCl NPs synthesized using biomass of *Rhodotorula Mucilaginosa* fungi. The spectra showed the maximum absorbance at 437 nm for samples with concentration 3 mM $AgNO_3$. Several reports have provided similar results concerning the absorption spectrum. The reports have confirmed that peaks around 440 nm coincide to the plasmon resonance of silver chloride nanoparticles [4], [5].

The crystalline nature of synthesized nanoparticles, obtained from X-ray diffraction, is confirmed by the diffraction peaks shown in Fig. 3, which correspond to the (111), (200), and (220) planes of the face centered cubic structure of AgCl crystal (Fig. 3) [11].

Powder Metallurgy and Advanced Materials – RoPM&AM 2017   Materials Research Forum LLC
Materials Research Proceedings 8 (2018) 28-34   doi: http://dx.doi.org/10.21741/9781945291999-4

An explanation for the presence of silver chloride nanoparticles can be based on the interaction between silver nitrate and bacteria, which was previously grown in Brain Heart Infusion Broth media containing sodium chloride.

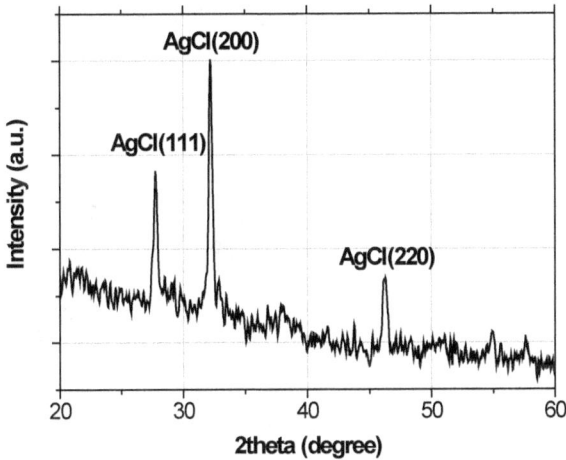

Fig. 3. X-Ray Diffraction pattern for AgCl nanoparticles.

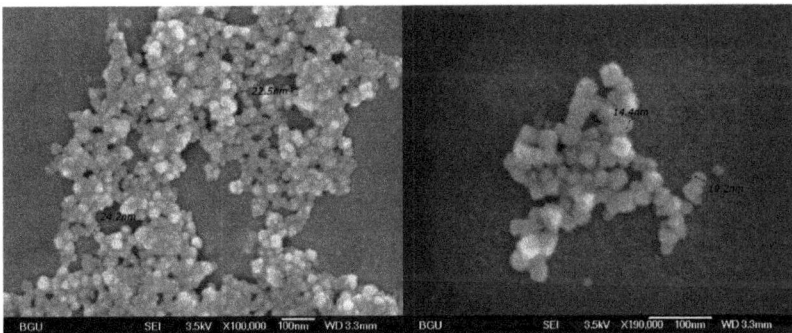

Fig. 4. SEM images of silver chloride nanoparticles
at 1mM, and 3Mm AgNO₃, respectively.

Powder Metallurgy and Advanced Materials – RoPM&AM 2017   Materials Research Forum LLC
Materials Research Proceedings 8 (2018) 28-34   doi: http://dx.doi.org/10.21741/9781945291999-4

The SEM analysis (Fig. 4) confirmed the presence of nano-scaled particles and also showed the spherical shape of them. As can be seen on the SEM micrographs, the silver chloride nanoparticles have dimensions smaller than 25 nm. The particle size/antimicrobial efficacy relation was reported previously, smaller particles, due to their much larger surface area, are expected to exhibit a much higher antibacterial efficacy.

The EDS spectra presented in Fig. 5 show the presence of the principal elements, namely Ag and Cl. The EDS analysis also revealed others elements which can be found on the samples due to the preparation stages (copper grid, carbon tape, thin film of platinum).

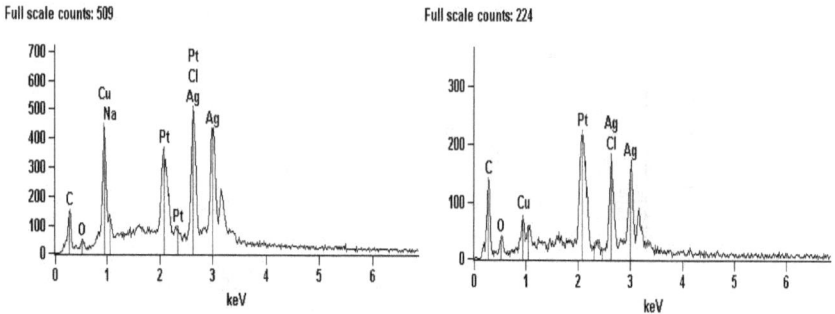

Fig. 5. EDS spectra of the biosynthesized AgCl NPs at 1mM, and 3Mm AgNO$_3$, respectively.

In table 1, one can observe that silver chloride nanoparticles can be synthesized by different approaches, most of them starting from the AgNO$_3$ like precursor. The difference consists in the type of stabilizer and surfactant.

The spherical shape is predominant in the case of silver chloride nanoparticles. Very closely to silver nanoparticles, as shown by the authors in [13], AgCl NPs present significant anti-microbial activity against *Escherichia coli, Candida albicans, Staphylococcus Aureus*.

The antimicrobial effect of colloidal silver chloride nanoparticles has been tested against different microorganisms, including fungi and bacteria. The results will be used for further investigation to determine the synergism of the samples in the presence of different antibiotics, how is shown in the paper *"Characterization and antimicrobial activity of silver nanoparticles, biosynthesized using Bacillus species"* [13].It is important to emphasize that the precursor used was AgNO$_3$, a reason for which we have to take into account that the recipe for growth media used for the microorganisms is very important for the final results. For example, the Brain Heart Infusion used for growing the present fungi, because it contains sodium chloride, may be the main factor which has led to the silver chloride nanoparticles formation by reducing the precursor AgNO$_3$ in the presence of biomass of *Rhodotorula Mucilaginosa*.

Table 1. Characteristics and applications of AgCl NPs synthesized through different methods.

Synthesis method	Size [nm]	Shape	Wavelength max [nm]	Application	References
Green Synthesis *Streptomyces exfoliatus*	10 -40	Spherical rod	410	Antibacterial Cytotoxicity: HeLa and SiHa cell lines	[2]
Green Synthesis *Bacillus subtilis*	20-60	spherical	400	Antifungal	[4]
Green Synthesis *Cissus quadrangularis Linn*	15–23	spherical	440	Antibacterial	[5]
Green Synthesis marine alga S. plagiophyllum	18–42	spherical	417	Antibacterial	[6]
Chemical synthesis Precipitation	6–7	hexagonal	-	Catalysis	[9]
Green Synthesis Prunus persica L. outer peel extract	15-50	spherical	440	Antibacterial, Anticandidal, Antioxidant	[8]
Chemical synthesis One pot method	Aroun d 100	Spherical ellipsoidal	268	Antibacterial	[12]

Summary
In the present work, silver chloride nanoparticles were synthesized via bio-reduction. It has been proved that the aqueous enzymatic extract of *Rhodotorula Mucilaginosa* fungi is able to synthesize silver chloride nanoparticles. It has been found that the silver chloride nanoparticles bioreduced by *Rhodotorula Mucilaginosa* decrease in size with increasing precursor concentration. The low-cost synthesis method is strengthened by the environmentally friendly steps of the procedure.

References
[1]   X. Li, H. Xu, Z-S. Chen, G., Chen, Biosynthesis of nanoparticles by microorganisms and their applications, J. Nanomater. 2011 (2011), Article ID 270974, 16 pages.
[2]   A.M. Iniyan et al., In vivo safety evaluation of antibacterial silver chloride nanoparticles from Streptomyces exfoliatus ICN25 in zebrafish embryos, Microb.Pathog., 112 (2017) 76 - 82. https://doi.org/10.1016/j.micpath.2017.07.054
[3]   K. Paulkumar, et al., Eco-friendly synthesis of silver chloride nanoparticles using Klebsiella planticola (MTCC 2277), IJGCB, 3:1 (2013) 12 - 16.
[4]   K. Paulkumar, et al., Biosynthesis of silver chloride nanoparticles using bacillus subtilis MTCC 3053 and assessment of its antifungal activity, 2013 (2013), Article ID 317963, 8 pages.
[5]   V. Gopinath, et al., Biogenic synthesis of antibacterial silver chloride nanoparticles using leaf extracts of Cissus quadrangularis Linn, Mater. Lett. 91 (2013) 224 – 227. https://doi.org/10.1016/j.matlet.2012.09.102

Powder Metallurgy and Advanced Materials – RoPM&AM 2017　　　　Materials Research Forum LLC
Materials Research Proceedings 8 (2018) 28-34　　　　doi: http://dx.doi.org/10.21741/9781945291999-4

[6]　　T. Stalin Dhas, et al., Facile synthesis of silver chloride nanoparticles using marine alga and its antibacterial efficacy, Spectrochim. Acta Mol. Biomol.Spectrosc. 120 (2014) 416 – 420. https://doi.org/10.1016/j.saa.2013.10.044

[7]　　V.A. Kumar, et al., Synthesis of nanoparticles composed of silver and silver chloride for a plasmonic photocatalyst using an extract from needles of *Pinus densiflora*, Mater. Lett. 176 (2016) 169 – 172. https://doi.org/10.1016/j.matlet.2016.04.077

[8]　　J.K. Patra, K.H. Baek, Green synthesis of silver chloride nanoparticles using Prunus persica L. outer peel extract and investigation of antibacterial, anticandidal, antioxidant potential, Green Chemistry Letters and Reviews, 9:2 (2016) 132 – 142. https://doi.org/10.1080/17518253.2016.1192692

[9]　　W. Hu, et al., In situ synthesis of silver chloride nanoparticles into bacterial cellulose membranes, Mater. Sci. Eng. C. 29 (2009) 1216 – 1219. https://doi.org/10.1016/j.msec.2008.09.017

[10]　　M., Sophocleous, J.K. Atkinson, A review of screen-printed silver/silver chloride (Ag/AgCl) reference electrodes potentially suitable for environmental potentiometric sensors, Sen. Actuators A Phys. https://doi.org/10.1016/j.sna.2017.10.01.

[11]　　S. Majumder, Synthesis and characterization of surfactant stabilized nanocolloidal dispersion of silver chloride in aqueous medium, Colloids Surf. A. 443 (2014) 156 – 163. https://doi.org/10.1016/j.colsurfa.2013.10.064

[12]　　I. Ghiuta, et al., Characterization and antimicrobial activity of silver nanoparticles, biosynthesized using Bacillus species, Appl. Surf. Sci. 438 (2018) 66 – 73. https://doi.org/10.1016/j.apsusc.2017.09.163

[13]　　H. Ha, J. Payer, The effect of silver chloride formation on the kinetics of silver dissolution in chloride solution, Electrochim. Acta 56 (2011) 2781 – 2791. https://doi.org/10.1016/j.electacta.2010.12.050

Powder Metallurgy and Advanced Materials – RoPM&AM 2017   Materials Research Forum LLC
Materials Research Proceedings 8 (2018) 35-51   doi: http://dx.doi.org/10.21741/9781945291999-5

Helical structure of linear homopolymers

Sorana D. BOLBOACĂ [1,a] and Lorentz JÄNTSCHI [2,3,b*]

[1] Department of Medical Informatics and Biostatistics, Iuliu Haţieganu University of Medicine and Pharmacy, 400349 Cluj-Napoca, Romania

[2] Department of Physics and Chemistry, Technical University of Cluj-Napoca, Muncii Blvd., no. 103-105, 400641 Cluj-Napoca, Romania

[3] Chemistry Doctoral School, Babeş-Bolyai University, Arany János Str., no. 11, 400028 Cluj-Napoca, Romania

[a] sbolboaca@umfcluj.ro, [b] lorentz.jantschi@gmail.com

Keywords: Helical structure, Computational study, Polymers.

Abstract. The aim of our research was to conduct a computational study on helical geometries of several homopolymers. Simple helix of polymers with seventeen (poly(lactic acid)) or eighteen (poly(1-chloro-trans-1-butenylene), poly(1-methyl-trans-1-butenylene), poly(1,4,4-trifluoro-trans-1-butenylene), polyacrylonitrile and respectively polychlorotrifluoroethylene) monomers were investigated. The X, Y, and Z coordinates obtained after optimization of the geometry of polymers were used as input data to identify the rotation and translation of the coordinates and respectively the coefficient of the helix. The values of interest were calculated by minimization of residuals using two different protocols. The first protocol investigated the whole polymer by imposing (step 1) or not (step 2) a fixed value of the helix coefficient. The second protocol investigated by minimization of residuals if the monomer (one or two) from each end of the polymer is or not an outlier of the helical geometry of the polymer.

Introduction

The conformation of natural polymers goes back in 1937 when the helical structure of α-amylose was proposed by Hanes [1, 2] and further investigated by Freudenbers [3]. Linus Pauling and Robert B. Corey [4] showed in the 1950s that proteins chain form by monomeric units of proteins represented by amino-acids twists into a helix (α-helix) called secondary structure. These discoveries along with the identification of double-helical structure of DNA by James Watson and Francis Crick [5] were major breakthroughs.

Simulations studies conducted on some synthetic polymers (e.g. m-terphenyl-based π-conjugated polymer [6], poly(ethylene glycol) (PEG) [7], poly(ethylene imine) (PEI) [8], squaraine polymers [9]) shown their tendency to form helixes.

The helical shape translation of biopolymers in synthetic materials could lead to identification of materials with desired properties.

Ten linear chained homopolymers (polymers made by identical units) were subject to the investigation of shaping in a helix form. The aim of the study was to identify the helix parameters where helix shapes appears.

Polymers Generic Structures and Their Main Characteristics

The repeated sequence, abbreviation when existent, name and the feasibility to provide helix [10] are given in Table 1.

Powder Metallurgy and Advanced Materials – RoPM&AM 2017   Materials Research Forum LLC
Materials Research Proceedings 8 (2018) 35-51   doi: http://dx.doi.org/10.21741/9781945291999-5

Table 1. Generic structure of polymers

No	Structure	Formula, abbreviation, name and remarks
#1		$(C_2H_4O)_n$ PEG Name: poly(ethylene glycol) Likely to provide helix: No, is purely linear
#2		$(C_2H_5N)_n$ PEI Name: polyethyleneimine Likely to provide helix: No, is purely linear
#3		$(C_3H_4O_2)_n$ PLA Name: poly(lactic acid) Likely to provide helix: Yes, internal helix
#4		$(C_6H_9NO)_n$ PVP Name: poly(N-vinyl-pyrrolidone) Likely to provide helix: Yes, double helix
#5		$(C_4H_6)_n$ Name: poly(trans-1-butenylene) Likely to provide helix: No, is purely linear
#6		$(C_4H_5Cl)_n$ Name: poly(1-chloro-trans-1-butenylene) Likely to provide helix: Yes

No	Structure	Formula, abbreviation, name and remarks
#7		$(C_5H_8)_n$ Name: poly(1-methyl-trans-1-butenylene) Likely to provide helix: Yes
#8		$(C_4H_3F_3)_n$ Name: poly(1,4,4-trifluoro-trans-1-butenylene) Likely to provide helix: Yes
#9		$(C_3H_3N)_n$ PAN Name: polyacrylonitrile Likely to provide helix: Yes, internal helix
#10		$(CF_2CClF)_n$ PCTFE / PTFCE Name: polychlorotrifluoroethylene Likely to provide helix: Yes, internal helix

Poly(ethylene glycol) or PEG, available in different geometries such as branched (3 to 10 PEG), star (10 to 100 chains) or comb (multiple PEG chains grafted onto a polymer backbone) is a compound with applications in different fields from industry to medicine.

PEGs are used as the basis of laxative [11], excipient in pharmacological products, or delivery systems [12-15]. PEGs are also used as the basis of skin creams [16], toothpaste or as an anti-foaming agent in food [17], and despite the fact that is considered biologically inert and safe can be responsible by contact dermatitis [16].

Polyethyleneimine (PEI) is a synthetic polymer, weakly basic and non-toxic that can take linear, branched or dendrimeric forms [18]. PEIs have been successfully used as drug carriers [19,20], gene delivery system [21,22], or growth inhibitors of microbial species [23] or cancer cells [24].

Poly(lactic acid) (polylactic acid or polylactide, PLA) is a biodegradable polyester obtained from renewable resources [25] and used in 1974 in combination with polyglycolic acid (PGA) as suture materials [26]. Due to biocompatibility and dissolvability in the body, PLA and its co-polymers were used in tissue engineering [27,28], or as carriers for bone morphogenetic proteins [29], in suture materials [30], and delivery systems [31,32].

Powder Metallurgy and Advanced Materials – RoPM&AM 2017   Materials Research Forum LLC
Materials Research Proceedings **8** (2018) 35-51   doi: http://dx.doi.org/10.21741/9781945291999-5

Poly(N-vinyl-pyrrolidone) (polyvidone or povidone, PVP) is a water-soluble synthetic polymer used as a blood plasma substituent due to its hemocompatibility [33]. Poly(N-vinyl-pyrrolidone) nanoparticles have shown potential as drug delivery systems for hydrophobic drugs [34,35].

Poly(trans-1-butenylene), poly(1-chloro-trans-1-butenylene), poly(1-methyl-trans-1-butenylene), and poly(1,4,4-trifluoro-trans-1-butenylene) are polyalkadienes. They have a glass transition temperature (T_g/K) from 207 (poly(1-methyl-trans-1-butenylene)) to 238 (poly(1,4,4-trifluoro-trans-1-butenylene)) [36].

Polyacrylonitrile (PAN) is a synthetic, semicrystalline organic polymer resin, first synthesized in 1930 by Dr. Hans Fikentscher and Dr. Claus Heuck [37]. Because is hard, relatively insoluble, and a high-melting material [38], PAN is used to develop carbon fibers by stabilization, carbonization, and graphitization under controlled conditions [39,40]. PAN is also used in development of nanofibers [41,42], microcapsules/microspheres with diverse application (such as foam targets, carbon molecular sieve, cargo storage, or medical instruments) [43], semiconductor PAN powder (10^{-10} S/cm < conductivity < 10^{-3} S/cm when treated at 285-300°C) [44], or in the treatment of metals [45].

Polychlorotrifluoroethylene (PCTFE or PTFCE) is a thermoplastic chlorofluoropolymer discovered in 1934 by Fritz Schloffer and Otto Scherer [46]. PTFCE is nonflammable, has a high optical transparency, is chemical resistant, and has near-zero moisture absorption and excellent electrical properties [47-49]. The co-deposition of PCTFE (at a concentration less than or equal to 4 g/l) into Ni-W coating proved a feasible manufacturing of fluid lubricant [50]. The conductivity of PTFCE was double by treatment with nano-size copper iodide [51]. In pharmaceutical and medical industry, the PCTFE is used for packaging [52,53]. PCTFE find its application in many sectors such as chemical industry, manufacturing, electronics, architecture, energy, health and domestic sectors [54].

Helix Coefficients by Minimization of Residuals

The problem of finding the helixes determined by the atoms positions was investigated on those structures presented in Table 1 that are likely to provide a helix. Finding the helixes on these polymers is essentially a double problem. The proper rotations (eq1) and translation (eq2) of the coordinate system in such way that the helix to be aligned to the z-coordinate must be identify before and after identification of the helix (eq3).

$$
\begin{bmatrix} x_1 \\ y_1 \\ z_1 \end{bmatrix} = \begin{bmatrix} \cos(a_0) & \sin(a_0) & 0 \\ -\sin(a_0) & \cos(a_0) & 0 \\ 0 & 0 & 1 \end{bmatrix} \cdot \begin{bmatrix} x_0 \\ y_0 \\ z_0 \end{bmatrix}
\tag{1a}
$$

$$
\begin{bmatrix} x_2 \\ y_2 \\ z_2 \end{bmatrix} = \begin{bmatrix} \cos(a_1) & 0 & -\sin(a_1) \\ 0 & 1 & 0 \\ \sin(a_1) & 0 & \cos(a_1) \end{bmatrix} \cdot \begin{bmatrix} x_1 \\ y_1 \\ z_1 \end{bmatrix}
\tag{1b}
$$

$$
\begin{bmatrix} x_3 \\ y_3 \\ z_3 \end{bmatrix} = \begin{bmatrix} 1 & 0 & 0 \\ 0 & \cos(a_2) & \sin(a_2) \\ 0 & -\sin(a_2) & \cos(a_2) \end{bmatrix} \cdot \begin{bmatrix} x_2 \\ y_2 \\ z_2 \end{bmatrix}
\tag{1c}
$$

$$\begin{bmatrix} x_4 \\ y_4 \\ z_4 \end{bmatrix} = \begin{bmatrix} x_3 \\ y_3 \\ z_3 \end{bmatrix} - \begin{bmatrix} a_3 \\ a_4 \\ a_5 \end{bmatrix} \tag{2}$$

$$\begin{cases} x_4 & = & a_7 \cdot \sin(a_6 \cdot t) \\ y_4 & = & a_8 \cdot \cos(a_6 \cdot t) \\ z_4 & = & a_9 \cdot t \end{cases} \tag{3}$$

where (x_0, y_0, z_0) are the raw Cartesian coordinates of the atoms belonging to the helix, (x_1, y_1, z_1), (x_2, y_2, z_2) and (x_3, y_3, z_3) are new Cartesian coordinates obtained through elementary rotations, and (x_4, y_4, z_4) includes a translation. In eq1-eq3, a_i $(0 \leq i \leq 9)$ are unknown coefficients (a_0, a_1 and a_2 are angles of rotation; a_3, a_4 and a_5 are scalars of a translation; a_6, a_7, a_8 and a_9 are the helix parameters; for circular-based helixes $a_7 = a_8$), and t is the changing parameter defining the evolution of the helix.

The helix is rotated and translated such that its xOy projection is a circle when $a_7 = a_8$ while $a_7 \neq a_8$ provides an ellipse. The main difficulty is given that starting roughly from atoms coordinates (x_0, y_0, z_0), all coefficients (from a_0 to a_9) must be identified at once.

The problem defined by the eq1 to eq3 is a problem of optimization. If n represents consecutive points (atoms positions) from the polymer converged as helix, the residual sum $\Sigma_{1 \leq i \leq n} f_0$ should be minimized (f_0 in eq4, $t \leftarrow i$).

$$f_0 = \Sigma\{[a_7\sin(a_6 \cdot i)\text{-}x_4(i)]^2 + [a_8\cos(a_6 \cdot i)\text{-}y_4(i)]^2 + [a_9 i\text{-}z_4(i)]^2\} \tag{4}$$

Since some of the unknowns are inside of periodic functions (especially concerns should be taken for a_6), it is essential that the starting values of the coefficients to be good.

In the case of a double helix, the eq3 can be changed to include a double time-dependence (eq5).

$$\begin{cases} x_4 & = & a_7 \cdot \sin(a_6 \cdot t) + a_{11} \cdot \sin(a_{10} \cdot t) \\ y_4 & = & a_8 \cdot \cos(a_6 \cdot t) + a_{12} \cdot \cos(a_{10} \cdot t) \\ z_4 & = & a_9 \cdot t \end{cases} \tag{5}$$

Going back to the problem of minimization (eq4), for fixed values of the a_0, a_1, a_2, a_6 (and a_{10} for eq5) coefficients, the problem may be simplified (by embedding eq3 or eq5 in eq4) to correspond to triple linear regression (eq6 for embedding eq3 in eq4; for embedding eq5 in eq4 is similar).

$$0 = \frac{\partial f_1}{\partial a_3} = \frac{\partial f_1}{\partial a_7}, \ f_1 = \sum_{i=1}^{n} \left(x_3(i) - a_3 - a_7 \cdot \sin(a_6 \cdot i) \right)^2 \tag{6a}$$

$$0 = \frac{\partial f_2}{\partial a_4} = \frac{\partial f_2}{\partial a_8}, \ f_2 = \sum_{i=1}^{n} \left(y_3(i) - a_4 - a_8 \cdot \cos(a_6 \cdot i) \right)^2 \tag{6b}$$

$$0 = \frac{\partial f_3}{\partial a_5} = \frac{\partial f_3}{\partial a_9}, \; f_3 = \sum_{i=1}^{n} \left(z_3(i) - a_5 - a_9 \cdot i\right)^2$$

(6c)

The solutions of eqs6 are therefore easily obtained (eqs7):

$$a_7 = \frac{n \sum_{i=1}^{n} x_3(i)\sin(a_6 i) - \sum_{i=1}^{n} x_3(i)\sum_{i=1}^{n} \sin(a_6 i)}{n \sum_{i=1}^{n} \sin(a_6 i)\sin(a_6 i) - \sum_{i=1}^{n} \sin(a_6 i)\sum_{i=1}^{n} \sin(a_6 i)}$$

(7a1)

$$a_3 = (\sum_{i=1}^{n} x_3(i) - a_7 \sum_{i=1}^{n} \sin(a_6 i))/n$$

(7a2)

$$a_8 = \frac{n \sum_{i=1}^{n} y_3(i)\cos(a_6 i) - \sum_{i=1}^{n} y_3(i)\sum_{i=1}^{n} \cos(a_6 i)}{n \sum_{i=1}^{n} \cos(a_6 i)\cos(a_6 i) - \sum_{i=1}^{n} \cos(a_6 i)\sum_{i=1}^{n} \cos(a_6 i)}$$

(7b1)

$$a_4 = (\sum_{i=1}^{n} y_3(i) - a_8 \sum_{i=1}^{n} \cos(a_6 i))/n$$

(7b2)

$$a_9 = \frac{n \sum_{i=1}^{n} z_3(i)i - \sum_{i=1}^{n} z_3(i)\sum_{i=1}^{n} i}{n \sum_{i=1}^{n} i^2 - \sum_{i=1}^{n} i\sum_{i=1}^{n} i}$$

(7c1)

$$a_5 = (\sum_{i=1}^{n} z_3(i) - a_9 \cdot \sum_{i=1}^{n} i)/n$$

(7c2)

The complexity of finding the helix was reduced from a problem of optimum with 10 (eq3) or respectively 11 (eq5) unknown coefficients to a problem of optimum with 4 (eq3) or respectively 5 (eq4) unknown coefficients that must be determined.

The study was conducted with polymers that embedded seventeen (#3, Table 1) or eighteen monomers (#4, #7, #8, #9, #10, Table 2) to assure the reliability of the calculation with the available calculation power. The polymers of interest were drawn with HyperChem and their geometries were optimized with the Spartan program at Hartree-Fock [55] 6-31G* level of the theory [56].

X, Y, and Z coordinates resulted after optimization of the geometry were used to shape the helixes. These data were used as input data to identify the rotation and translation of the coordinates and respectively the coefficients of the helix. The graphical representation of these input data is presented in Fig. 1 where each triangle is made from the coordinates of a monomer.

40

The above-presented equations were used to identify the optimum solution by using a guess value for the a_6 coefficient and the same initial starting values for all other coefficients. The guess value for a_6 coefficient was obtained by inspection of the graphical representations of the molecules.

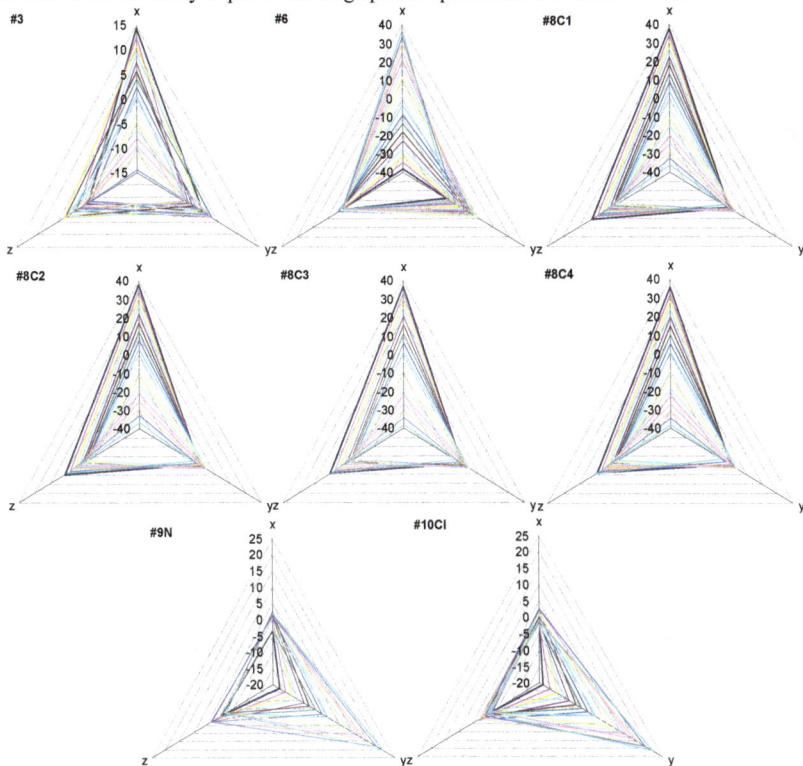

Fig. 1. Distribution of coordinates as resulted from geometry optimization.

Two different protocols were applied to calculate the values of interest:

- *Protocol I* was dedicated to the evaluation of the whole polymer. In the first step, the helix coefficient (a_6) was kept fixed and the values of the remaining coefficients were iterated to minimize the value of residuals until convergence. In the second step, the algorithm was applied again, starting with the final values obtained in the first step and the values of all coefficients (including helix coefficient a_6) were iterated to minimize the residuals until convergence.
- Since usually at the end the geometrical constrains are different than in the middle of the polymer, the *Protocol II* was dedicated to the evaluation of a smaller inside portion of the polymer and was done by exclusion of one or two monomers from each end. The starting values in the calculation of

Powder Metallurgy and Advanced Materials – RoPM&AM 2017   Materials Research Forum LLC
Materials Research Proceedings **8** (2018) 35-51   doi: http://dx.doi.org/10.21741/9781945291999-5

the values of coefficients were the values obtained as optimum values at the end of the first protocol. The coefficients were iterated until convergence.

The input data were the atom coordinates. Thus, for #3 the checkpoint was the double bonded oxygen, for #6 the chlorine, for #7 the carbon from methyl position. Four different checkpoints (C1 - carbon connected with one fluorine atom, and C2, C3, and C4 - the other carbon atoms) were used for #8. The checkpoint for #9 was the nitrogen atom, while for the #10 was chlorine atom.

Convergence Analysis

The pattern of convergence to the minimum value of residuals was different for the applied steps in the Protocol I. The starting values for the angular coefficients (a_0-a_2) were 0.1 in all cases. The number of iterations until the convergence ranged between 26 and 41 (Fig. 2) when the value of a_6 was fixed and was between 5 and 46 when all coefficients were calculated (Fig. 3).

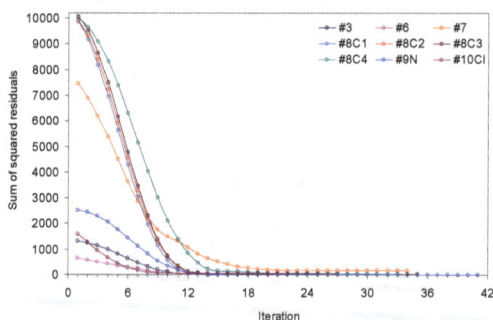

Fig. 2. Residuals convergence for helix coefficient fixed.

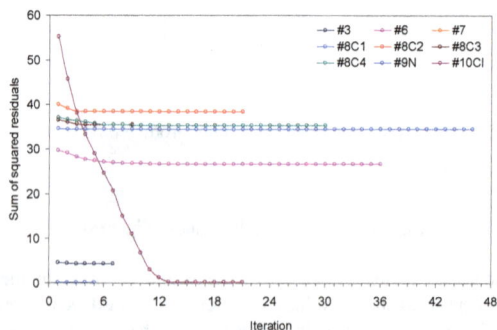

Fig. 3. Residuals convergence for helix coefficient also optimized.

Powder Metallurgy and Advanced Materials – RoPM&AM 2017   Materials Research Forum LLC
Materials Research Proceedings 8 (2018) 35-51   doi: http://dx.doi.org/10.21741/9781945291999-5

In the case with fixed value of a_6, a similar (#8C4 similar with all other #8s, Fig. 2) and sometimes overlapping (#8C1, #8C2, and #8C3) behavior was observed for polymer #8 when different carbon atoms were imposed as a checkpoint. The residuals have the tendency to significant decrease until around the 12th iteration and after this iteration, the changes in the values of residuals are small (Fig. 2).

The minimization of residuals in the second step mimic a change in a plateau with small differences between iteration on the same polymer with one exception represented by #10Cl (Fig. 3). The residuals of the #10Cl follow a change similar to the one observed in the first step with significant decrease until the 12th iteration follows by a plateau where the changes are very small.

With no exception, the values of residuals reduced significantly at the end of the first step (Table 2). The highest reduction was observed for the #9 polymer (its monomer contain a triple bond -C≡N) when the value of residual was 20,731 times lower at the end of the first step compared with the start value. The value of residuals at the end of the second step is less than 190 times as compared with the value of residuals at the end of the first step.

Table 2. The values of residuals and coefficients: first protocol

		residuals	a_0	a_1	a_2	a_3	a_4	a_5	a_6	a_7	a_8	a_9	Iteration
#3	start	1316.371	0.100	0.100	0.100	0.450	0.435	1.480	1.867	1.372	2.818	-0.177	0
	end-1	4.720	1.538	0.923	1.521	0.059	0.061	16.081		3.276	3.091	-1.736	27
	end-2	4.410	1.545	0.843	1.527	0.026	0.028	16.082	1.876	3.264	3.103	-1.736	7
#6	start	10562.800	Identical with #3			-18.547	-2.735	-2.541	0.155	25.095	2.612	0.319	0
	end-1	29.772	1.536	-0.064	1.587	-16.419	0.479	-44.933		25.049	1.022	4.624	37
	end-2	26.670	1.600	-0.073	1.580	-13.788	0.535	-44.939	0.165	22.452	1.088	4.633	36
#7	start	7444.306	Identical with #3			15.736	-0.745	-3.299	0.194	-27.304	2.934	0.374	0
	end-1	177.895	1.431	-0.037	1.637	11.725	-0.222	-37.635		-22.925	-1.005	4.074	34
	end-2	124.517	1.622	-0.041	1.639	19.548	0.019	-38.459	0.165	-30.481	-1.455	4.143	18
#8C1	start	9846.996	Identical with #3			-8.205	-1.086	6.253	-0.172	-14.596	-3.541	-0.662	0
	end-1	34.633	1.363	1.214	1.361	-11.097	-0.186	44.230		-17.638	-0.694	-4.533	27
	end-2	34.452	1.420	1.214	1.417	-11.884	-0.142	44.256	-0.168	-18.362	-0.577	-4.535	46
#8C2	start	9879.200	Identical with #3			-11.256	-1.619	4.242	-0.176	-18.787	-3.461	-0.461	0
	end-1	40.091	1.463	1.143	1.490	-10.456	0.551	43.714		-17.387	0.118	-4.556	27
	end-2	38.377	1.589	1.142	1.607	-12.553	0.598	43.790	-0.165	-19.356	0.214	-4.562	21
#8C3	start	10042.268	Identical with #3			-12.599	-2.999	3.398	-0.176	-18.747	-3.476	-0.417	0
	end-1	36.508	1.487	1.139	1.483	-10.503	-0.109	42.449		-17.505	-0.973	-4.565	26

	end-2	35.542	1.585	1.131	1.575	-12.112	-0.024	42.526	-0.167	-19.058	-0.850	-4.571	9
#8C4	start	10025.977	Identical with #3			-12.690	-3.337	2.918	-0.175	-17.411	-2.597	-0.323	0
	end-1	37.168	-1.578	2.051	-1.573	-10.648	0.127	41.182		-16.974	0.093	-4.518	33
	end-2	35.331	-1.464	2.057	-1.475	-12.656	0.175	41.234	-0.164	-18.981	0.019	-4.523	30
#9N	start	2510.015	0.100	0.100	0.100	0.132	0.113	1.910	2.103	-2.119	1.882	-0.215	0
	end-1	0.121	0.000	-0.782	1.571	0.002	0.002	19.676		-3.221	-3.212	-2.220	41
	end-2	0.121	0.000	-0.785	1.571	0.002	0.002	19.676	2.104	-3.220	-3.212	-2.220	5
#10CL	start	2680.676	Identical with #3			0.117	0.102	1.668	2.103	-1.120	2.215	-0.196	0
	end-1	55.205	-0.008	-0.792	1.582	-0.072	-0.073	20.803		-1.723	-1.564	-2.328	30
	end-2	0.290	-0.010	-2.342	1.564	0.036	0.035	20.805	2.268	-2.415	-2.382	-2.328	21

end-1: the values obtained by minimizing the residuals when a_6 is fixed;

end-2: the values obtained by minimizing the residuals when all coefficients are calculated

The lowest value of residuals is obtained at the end of the second step while the values of the coefficients of the helixes have generally the tendency to increase (see Table 2). However, for the #7 and #8 polymers, the coefficient of the helix decrease at the end of the second step compared with the value obtained at the end of the first step.

The smallest absolute value of the helix coefficient a_6 when the first protocol was applied was of 0.155 (#6, poly(1-chloro-trans-1-butenylene)) at the end of the first step and 0.164 (#8C4, poly(1,4,4-trifluoro-trans-1-butenylene)) at the end of the second step. The highest absolute value of the helix coefficient a_6 was of 2.103 (#9N - polyacrylonitrile and #10Cl - polychlorotrifluoroethylene) at the end of the first step and 2.268 (#10Cl - polychlorotrifluoroethylene) at the end of second step in this first protocol.

For the second protocol, the exclusion of the monomers from the ends of the polymer was investigated to identify the outliers of helix on the investigated polymers.

As expected, the behavior to the convergence to the minimum value of residuals was different when the second protocol was applied compared with the first protocol.

The number of iterations ranges from 7 to 80 when one monomer from each end of the polymer was excluded (the number of monomers in the polymers decreased with two, Fig. 4) and from 7 to 168 when two monomers from each end of the polymer were excluded (the number of monomers in the polymers decreased with four, Fig. 5).

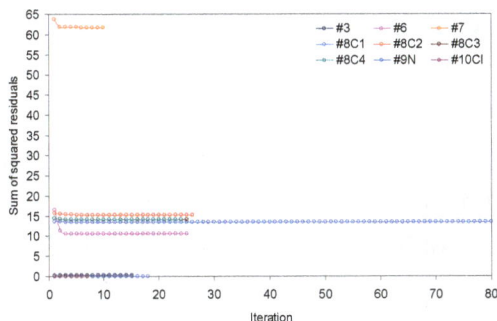

Fig. 4. Convergence of residuals for exclusion of one monomer from each end.

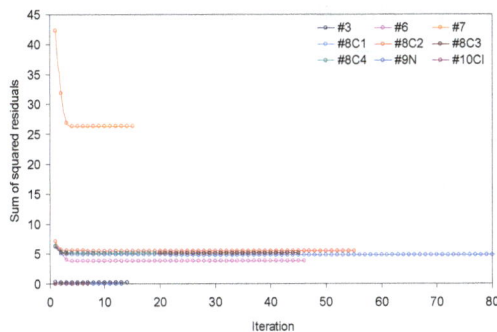

Fig. 5. Convergence of residuals for exclusion of two monomers from each end.

The changes in the residuals were smaller in the case of the second protocol compared with the first protocol. The minimization of residuals led to different values when one monomer is excluded compared with the case when two monomers are excluded (Table 3). The analysis of the values of residuals could let to the identification of the existence of outliers of the helix on the investigated polymers.

The values of residuals are similar after the exclusion of a monomer from each end of the polymer; as compared with the initial value, this indicates that the ended monomers are not outliers.

In two cases (#9N & #10Cl) the residual data were lower than 0.1 and the values of the helix coefficients were different at a value of 0.001 when compared the exclusion of a monomer relative to the baseline value and identical for exclusion of one monomer or two monomers (see Table 3).

For #9N, the exclusion of two monomers in each end of the polymer decrease the residuals 4 times (from 0.030 to 0.008) compared with the exclusion of one monomer, but in this case the ended monomers could not be considered outliers since the residuals are in all investigated cases very small.

In three investigated polymers (#3, #6, and #7), one monomer from each end of the polymers could be considered as outlier of the helix since the value of residuals were three times lower compared with the values of residuals without any exclusion. Furthermore, the values of helix coefficient (a_6) are very closed for #6 and #7 to the start values and identical if one or two monomers at the end of each polymer were excluded (see Table 3).

Table 3. The values of residuals and coefficients: second protocol

		residuals	a_0	a_1	a_2	a_3	a_4	a_5	a_6	a_7	a_8	a_9	Iteration
#3	start	0.711	1.545	0.843	1.527	1.876	0.010	15.932	0.012	3.255	3.220	-1.711	0
	excl-1	0.240	1.548	0.895	1.545	0.012	0.011	15.935	1.868	3.249	3.233	-1.711	15
	excl-2	0.205	1.544	0.897	1.545	0.020	0.019	15.939	1.868	3.238	3.245	-1.713	14
#6	start	28.438	1.600	-0.073	1.580	-13.525	1.626	-45.839	0.165	22.133	1.481	4.728	0
	excl-1	10.547	1.603	-0.075	1.539	-13.363	0.615	-45.687	0.166	21.978	2.873	4.716	25
	excl-2	3.764	1.605	-0.075	1.506	-13.094	0.666	-46.281	0.166	21.673	4.258	4.782	46
#7	start	167.269	1.622	-0.041	1.639	19.883	-2.457	-40.294	0.165	-30.886	-0.518	4.337	0
	excl-1	61.771	1.625	-0.041	1.756	20.026	0.253	-39.945	0.164	-31.030	-4.198	4.296	10
	excl-2	26.278	1.623	-0.046	1.850	20.329	0.411	-41.187	0.164	-31.383	-6.314	4.423	15
#8C1	start	19.043	1.420	1.214	1.417	-11.900	0.509	44.871	-0.168	-18.382	-1.269	-4.602	0
	excl-1	13.498	1.352	1.216	1.321	-11.015	-0.201	44.738	-0.173	-17.562	-2.331	-4.591	80
	excl-2	4.790	1.348	1.220	1.292	-11.016	-0.240	45.138	-0.173	-17.563	-3.620	-4.638	168
#8C2	start	21.448	1.589	1.142	1.607	-12.600	1.281	44.478	-0.165	-19.412	-0.527	-4.634	0
	excl-1	15.234	1.538	1.143	1.531	-11.758	0.607	44.421	-0.169	-18.597	-1.481	-4.629	26
	excl-2	5.441	1.524	1.147	1.494	-11.574	0.582	44.913	-0.170	-18.423	-2.844	-4.683	55
#8C3	start	19.882	1.585	1.131	1.575	-12.140	0.632	43.175	-0.167	-19.092	-1.553	-4.640	0
	excl-1	14.102	1.544	1.135	1.511	-11.500	-0.003	43.066	-0.170	-18.458	-2.438	-4.631	25
	excl-2	5.059	1.538	1.138	1.484	-11.451	0.007	43.508	-0.171	-18.411	-3.689	-4.681	45
#8C4	start	19.798	-1.464	2.057	-1.475	-12.699	0.825	41.890	-0.164	-19.033	-0.700	-4.591	0
	excl-1	14.166	-1.506	2.057	-1.536	-11.971	0.241	41.839	-0.168	-18.282	-1.468	-4.587	24
	excl-2	5.094	-1.521	2.054	-1.570	-11.753	0.265	42.312	-0.169	-18.061	-2.711	-4.639	19
#9N	start	0.037	0.000	-0.785	1.571	0.007	0.007	19.630	2.104	-3.218	-3.223	-2.215	0

	excl-1	0.030	-0.002	-0.785	1.571	0.006	0.006	19.629	2.103	-3.219	-3.224	-2.215	18
	excl-2	0.008	-0.002	-0.794	1.570	0.007	0.006	19.632	2.104	-3.224	-3.217	-2.215	13
#10Cl	start	0.076	-0.010	-2.342	1.564	0.029	0.029	20.740	2.268	-2.422	-2.416	-2.320	0
	excl-1	0.056	-0.010	-2.337	1.568	0.028	0.028	20.740	2.267	-2.422	-2.418	-2.320	7
	excl-2	0.048	-0.009	-2.350	1.567	0.027	0.027	20.756	2.268	-2.416	-2.417	-2.322	7

Summary

Helical geometries are natural in biological polymers structures as seen in DNA [5], RNA [57], or globular proteins [58]. The understanding of how this natural helical geometry occurs could guide the design of material at macro and nano-scale levels with specific desired uses and properties [59].

Our computational study showed that some of the investigated homopolymers have a good chance to stabilize as helical geometries and for these structures, several coefficients were calculated by minimization of residuals applying two different protocols. In our analysis, the polymer with the highest helix coefficient was the one formed by 18 polychlorotrifluoroethylene monomers with chlorine as checkpoint of the helix. Opposite, the polymer with the lowest helix coefficient was the one formed by 18 poly(1-chloro-trans-1-butenylene) monomers. The second protocol was applied to identify if the monomers (one or two) from the ends of the polymers are or not outliers of the helical geometry. In other words, this second protocol analyzed if helical geometry of investigated polymers are of 15/16 monomers instead of 17/18 monomers (a monomer in each end of the polymer was excluded) or respectively 13/14 (two monomers from each end of the polymer). The investigation of the helical geometry pattern with more monomers in the polymer could bring more sounds in the knowledge of their behavior.

Acknowledgements

This work was supported by a grant from the Romanian National Authority for Scientific Research and Innovation, CCCDI – UEFISCDI, project number 8/2015, acronym GEMNS (under the frame of the ERA-NET EuroNanoMed II European Innovative Research and Technological Development Projects in Nanomedicine).

References

[1]   C.S. Hanes, The action of amylases in relation to the structure of starch and its metabolism in the plant. Parts IV-VII, New Phytologist 36 (1937) 101-141. https://doi.org/10.1111/j.1469-8137.1937.tb06906.x

[2]   C.S. Hanes, The action of amylases in relation to the structure of starch and its metabolism in the plant. Parts I-III, New Phytologist 36 (1937) 189-239. https://doi.org/10.1111/j.1469-8137.1937.tb06913.x

[3]   K. Freudenberg, E. Schaaf, G. Dumpert, T. Ploetz, Neue Ansichten über die Stärke, Naturwissenschaften 27 (1939) 850-853. https://doi.org/10.1007/BF01489430

[4]   L. Pauling, R.B. Corey, Configuration of polypeptide chains, Nature, 168 (1951) 550-551. https://doi.org/10.1038/171737a0

[5]   J.D. Watson, F.H.C. Crick, Molecular structure of nucleic acids: a structure for deoxyribose nucleic acid, Nature 171 (1953) 737-738. https://doi.org/10.1038/171737a0

[6]   H. Dong, J.Y. Shu, N. Dube, Y. Ma, M.V. Tirrell, K.H. Downing, T. Xu, 3-Helix micelles stabilized by polymer springs, J. Am. Chem. Soc. 134 (2012) 11807-11814. https://doi.org/10.1021/ja3048128

[7]   Y. Chatani, T. Kobatake, H. Tadokoro, R. Tanaka, Structural studies of poly(ethylenimine). 2. Double-stranded helical chains in the anhydrate, Macromolecules 15 (1982) 170-176. https://doi.org/10.1021/ma00229a034

[8]   C. Lambert, F. Koch, S.F. Völker, A. Schmiedel, M. Holzapfel, A. Humeniuk, M.I.S. Röhr, R. Mitric, T. Brixner, Energy Transfer Between Squaraine Polymer Sections: From Helix to Zigzag and All the Way Back, J. Am. Chem. Soc. 137 (2015) 7851-7861. https://doi.org/10.1021/jacs.5b03644

[9]   R.W. Newberry, R.T. Raines, n→π* interactions in poly(lactic acid) suggest a role in protein folding, Chemical Communications 49 (2013) 7699-7701. https://doi.org/10.1039/c3cc44317e

[10]   L. Jäntschi, S.D. Bolboacă, Study of geometrical shaping of linear chained polymers stabilized as helixes, Studia Universitatis Babes-Bolyai Chemia LXI (2016) 131-145.

[11]   J.A. Di Palma, M.B. Cleveland, J. McGowan, J.L. Herrera, A randomized, multicenter comparison of polyethylene glycol laxative and tegaserod in treatment of patients with chronic constipation, The American Journal of Gastroenterology 102 (2007) 1964-1971. https://doi.org/10.1111/j.1572-0241.2007.01365.x

[12]   A. Kolate, D. Baradia, S. Patil, I. Vhora, G. Kore, A. Misra, PEG - A versatile conjugating ligand for drugs and drug delivery systems, J. Controlled Release 192 (2014) 67-81. https://doi.org/10.1016/j.jconrel.2014.06.046

[13]   Z. Zhang, Y. Zhao, X. Meng, D. Zhao, D. Zhang, L. Wang, C. Liu, A Simple Zn^{2+} complex-based composite system for efficient gene delivery, PLoS One 11 (2016) e0158766. https://doi.org/10.1371/journal.pone.0158766

[14]   X. Ping, K. Jiang, S.-Y. Lee, J.-X. Cheng, X. Jin, PEG-PDLLA micelle treatment improves axonal function of the corpus callosum following traumatic brain injury, J Neurotrauma 31 (2014) 1172-1179. https://doi.org/10.1089/neu.2013.3147

[15]   V.N. Bharadwaj, J. Lifshitz, P.D. Adelson, V.D. Kodibagkar, S.E. Stabenfeldt, Temporal assessment of nanoparticle accumulation after experimental brain injury: Effect of particle size, Sci. Rep. 6 (2016) 29988. https://doi.org/10.1038/srep29988

[16]   A. Lasek-Duriez, M.C. Castelain, P. Modiano, Allergic contact dermatitis due to methoxy PEG-22 dodecyl glycol present in a cosmetic cold cream. Annales de Dermatologie et de Vénéréologie 140 (2013) 528-530. https://doi.org/10.1016/j.annder.2013.01.427

[17]   Current EU approved additives and their E Numbers. UK Government – Food Standards Agency. Retrieved 21 October 2016.

[18]   O. Yemul, T. Imae, Synthesis and characterization of poly(ethyleneimine) dendrimers, Colloid. Polym. Sci. 286 (2008) 747-752. https://doi.org/10.1007/s00396-007-1830-6

[19]   K. Kim, B. Bae, Y.J. Kang, J.-M. Nam, S. Kang, J.-H. Ryu, Natural polypeptide-based supramolecular nanogels for stable noncovalent encapsulation, Biomacromolecules 14 (2013) 3515-3522. https://doi.org/10.1021/bm400846h

Powder Metallurgy and Advanced Materials – RoPM&AM 2017   Materials Research Forum LLC
Materials Research Proceedings 8 (2018) 35-51   doi: http://dx.doi.org/10.21741/9781945291999-5

[20]   Y. Li, M. Guo, Z. Lin, M. Zhao, M. Xiao, C. Wang, T. Xu, T. Chen, B. Zhu, Polyethylenimine-functionalized silver nanoparticle-based co-delivery of paclitaxel to induce HepG2   cell   apoptosis.   Int.   J   Nanomedicine   11   (2016)   6693-6702. https://doi.org/10.2147/IJN.S122666

[21]   X. Zhao, H. Cui, W. Chen, Y. Wang, B. Cui, C. Sun, Z. Meng, G. Liu, Morphology, structure and function characterization of PEI modified magnetic nanoparticles gene delivery system, PLoS One 9 (2014) e98919. https://doi.org/10.1371/journal.pone.0098919

[22]   J.D. Ziebarth, D.R. Kennetz, N.J. Walker, Y. Wang, Structural comparisons of PEI/DNA and PEI/siRNA complexes revealed with molecular dynamics simulations, J Phys. Chem. B 121 (2017) 1941-1952. https://doi.org/10.1021/acs.jpcb.6b10775

[23]   M.M. Azevedo, P. Ramalho, A.P. Silva, R. Teixeira-Santos, C. Pina-Vaz, A.G. Rodrigues, Polyethyleneimine and polyethyleneimine-based nanoparticles: novel bacterial and yeast biofilm inhibitors, J Medical Microbiology 63 (2014) 1167-1173. https://doi.org/10.1099/jmm.0.069609-0

[24]   N. Sahiner, S. Sagbas, M. Sahiner, R.S. Ayyala, Polyethyleneimine modified poly(Hyaluronic acid) particles with controllable antimicrobial and anticancer effects. Carbohydr. Polym. 159 (2017) 29-38. https://doi.org/10.1016/j.carbpol.2016.12.024

[25]   O. Martin, L. Avérous, Poly(lactic acid): plasticization and properties of biodegradable multiphase   systems,   Polymer   42   (2001)   6209-6219.   https://doi.org/10.1016/S0032-3861(01)00086-6

[26]   R. Mehta, V. Kumar, H. Bhunia, S.N. Upadhyay, Synthesis of poly(lactic acid): a review. J Macromolecular   Sci.:   Part   C: Polymer   Rev.   45   (2005)   325-349. https://doi.org/10.1080/15321790500304148

[27]   P. Gentile, V. Chiono, I. Carmagnola, P.V. Hatton, An overview of Poly(lactic-co-glycolic) Acid (PLGA)-based biomaterials for bone tissue engineering, Int. J Mol. Sci. 15 (2014) 3640-3659. https://doi.org/10.3390/ijms15033640

[28]   J.S. Fernandes, R.L. Reis, R.A. Pires, Wetspun poly-L-(lactic acid)-borosilicate bioactive glass scaffolds for guided bone regeneration. Mater. Sci. Eng.: C Mater. Biological Appl. 71 (2017) 252-259. https://doi.org/10.1016/j.msec.2016.10.007

[29]   N. Saito, K. Takaoka, New synthetic biodegradable polymers as BMP carriers for bone tissue   engineering,   Biomaterials   24   (2003)   2287-2293.   https://doi.org/10.1016/S0142-9612(03)00040-1

[30]   R. Mehta, V. Kumar, H. Bhunia, S.N. Upadhyay, Synthesis of poly(lactic acid): a review, J Macromolecular   Sci.   Polymer   Rev.   45   (2005)   325-349. https://doi.org/10.1080/15321790500304148

[31]   C. Chen, G. Lv, C. Pan, M. Song, C. Wu, D. Guo, X. Wang, B. Chen, Z. Gu, Poly(lactic acid) (PLA)-based nanocomposites-a novel way of drug-releasing, Biomedical Materials 2 (2007) L1-4. https://doi.org/10.1088/1748-6041/2/4/L01

[32]   J. Li, C. Chen, X. Wang, Z. Gu, B. Chen, Novel Strategy to Fabricate PLA/Au nanocomposites as an efficient drug carrier for human leukemia Cells in Vitro, Nanoscale Research Letters 6 (2011) 29.

Powder Metallurgy and Advanced Materials – RoPM&AM 2017   Materials Research Forum LLC
Materials Research Proceedings 8 (2018) 35-51   doi: http://dx.doi.org/10.21741/9781945291999-5

[33]   X. Liu, Y. Xu, Z. Wu, H. Chen, Poly(N-vinylpyrrolidone)-modified surfaces for biomedical   applications,   Macromolecular   Bioscience   13   (2013)   147-154. https://doi.org/10.1002/mabi.201200269

[34]   A.N. Kuskov, P.P. Kulikov, A.V. Goryachaya, M.N. Tzatzarakis, A.O. Docea, K. Velonia, M.I. Shtilman, A.M. Tsatsakis, Amphiphilic poly-N-vinylpyrrolidone nanoparticles as carriers for non-steroidal, anti-inflammatory drugs: In vitro cytotoxicity and in vivo acute toxicity study. Nanomedicine: Nanotechnology,   Biology   and   Medicine   13   (2017)   1021-1030. https://doi.org/10.1016/j.nano.2016.11.006

[35]   X. Yao, C. Xie, W. Chen, C. Yang, W. Wu, X. Jiang, Platinum-incorporating Poly(N-vinylpyrrolidone)-poly(aspartic acid) pseudoblock copolymer nanoparticles for drug delivery, Biomacromolecules 16 (2015) 2059-2071. https://doi.org/10.1021/acs.biomac.5b00479

[36]   R.B. Fox, Glass transition temperature for selected polymers. In: Haynes W.M. (Ed). CRC Handbook of Chemistry and Physics, 97th Edition, 13-10, 2017 CRC Press Taylor & Francis Group, Boca Raton, FL, USA.

[37]   H. Finkentscher, C. Heuck, DE Patent 654989, Verfahren zur Herstellung von Polymerisationprodukten, Anmeldetag 18.2.1930

[38]   M. Schwartz, Encyclopedia of materials, parts, and finishes. (2nd ed.) CRC Press, Boca Raton, Florida, 2002.

[39]   M.S.A. Rahaman, A.F. Ismail, A. Mustafa, A review of heat treatment on polyacrylonitrile fiber,   Polym.   Degrad.   Stab.   92   (2007)   1421-1432. https://doi.org/10.1016/j.polymdegradstab.2007.03.023

[40]   S. Nunna, M. Naebe, N. Hameed, B.L. Fox, C. Creighton, Evolution of radial heterogeneity in polyacrylonitrile fibres during thermal stabilization: An overview, Polym. Degrad. Stab. 136 (2017) 20-30. https://doi.org/10.1016/j.polymdegradstab.2016.12.007

[41]   L. Zhang, A. Aboagye, A. Kelkar, C. Lai, H. Fong, A review: carbon nanofibers from electrospun polyacrylonitrile and their applications, J Mater. Sci. 49 (2014) 463-480. https://doi.org/10.1007/s10853-013-7705-y

[42]   M. Bourourou, M. Holzinger, F. Bossard, F. Hugenell, A. Maaref, S. Cosnier, Chemically reduced electrospun polyacrilonitrile-carbon nanotube nanofibers hydrogels as electrode material for   bioelectrochemical   applications,   Carbon   87   (2016)   233-238. https://doi.org/10.1016/j.carbon.2015.02.026

[43]   W. Han, S. Dong, B. Li, L. Ge, Preparation of polyacrylonitrile- based porous hollow carbon microspheres, Colloids and Surfaces A: Physicochemical and Engineering Aspects 520 (2017) 467-476. https://doi.org/10.1016/j.colsurfa.2017.02.009

[44]   S. Koutsonas, Electrical conductivity of degraded polyacrylonitrile powder by microwave irradiation for supercapacitor devices or other mobile applications, Mater. Letter. 193 (2017) 203-205. https://doi.org/10.1016/j.matlet.2017.02.001

[45]   D. Liu, H. Chen, P. Yin, N. Ji, G. Zong, R. Qu, Synthesis of polyacrolonitrile by single-electron transfer-living radical polymerization using Fe(0) as catalyst and its absorption properties after   modification,   J Polymer   Sci.   Part   A: Polymer   Chem.   49   (2011)   2916-2923. https://doi.org/10.1002/pola.24727

Powder Metallurgy and Advanced Materials – RoPM&AM 2017   Materials Research Forum LLC
Materials Research Proceedings 8 (2018) 35-51   doi: http://dx.doi.org/10.21741/9781945291999-5

[46]   T. Nakajima, H. Groult, Fluorinated materials for energy conversion. Elsevier. 2005, pp. 472. ISBN 978-0-08-044472-7.

[47]   R. Winter, A Consumer's Dictionary of Household, Yard and Office Chemicals: Complete Information About Harmful and Desirable Chemicals Found in Everyday Home Products, Yard Poisons, and Office Polluters. iUniverse. 2007, p. 255.

[48]   F. Cardarelli, Materials handbook: a concise desktop reference. Springer. 2008, pp. 708-709.

[49]   T. Hongxiang, Overview of the development of the fluoropolymer indus. Applied Sciences 2 (2012) 496-512. https://doi.org/10.3390/app2020496

[50]   M.G. Hosseini, H. Teymourinia, A. Farzaneh, S. Khameneh-asl, Evaluation of corrosion, mechanical and structural properties of new Ni-W-PCTFE nanocomposite coating, Surf. Coat. Technol. 298 (2016) 114-120. https://doi.org/10.1016/j.surfcoat.2016.04.060

[51]   R.V. Mazurenko, S.M. Makhno, G.M. Gunya, P.P. Gorbyk, Effect of dispersion of copper-iodide particles on the electrical properties of composites based on polychlortrifluoroethylene, Metallofizika   i   Noveishie   Tekhnologii   38   (2016)   647-656. https://doi.org/10.15407/mfint.38.05.0647

[52]   T.P. O'Brien, G.M. McNally, W.R. Murphy, B.G. Millar, G.S. Garrett, A.H. Clarke, R.P. McGinley, Assessing the thermoformability of high-performance polymers for use in medical packaging applications. 66th Annual Technical Conference of the Society of Plastics Engineers, Plastics Encounter at ANTEC 2008; Milwaukee, WI; United States; 4 May 2008 through 8 May 2008; Technical Papers, Regional Technical Conference - Society of Plastics Engineers Volume 2, 2008, 963-96.

[53]   R.M.V. Alves, S.B.M. Jaime, M.R. Goncalves, R.W. Suzuki, Plastic and glass packages for pharmaceutical products: Evaluation of light barrier properties, Revista de Ciencias Farmaceuticas Basica e Aplicada 29 (2008) 167-178.

[54]   J. Gardiner, Fluoropolymers: origin, production, and industrial and commercial applications, Aust. J. Chem. 68 (2015) 13-22. https://doi.org/10.1071/CH14165

[55]   J.C. Slater, The electronic structure of atoms – the hartree-fock methods and correlations, Reviews of Modern Physics 35 (1963) 484-487. https://doi.org/10.1103/RevModPhys.35.484

[56]   R. Ditchfield, W.J. Hehre, J.A. Pople, Self-consistent molecular-orbital methods. IX. An extended gaussian-type basis for molecular-orbital studies of organic molecules, J Chem. Phys. 54 (1971) 724-728. https://doi.org/10.1063/1.1674902

[57]   N. Safaee, A.M. Noronha, D. Rodionov, G. Kozlov, C.J. Wilds, G.M. Sheldrick, K. Gehring, Structure of the parallel duplex of poly(A) RNA: evaluation of a 50 year-old prediction, Angewandte   Chemie   International   Edition   52   (2013)   10370-10373. https://doi.org/10.1002/anie.201303461

[58]   J.W. Prothero, Correlation between the distribution of amino acids and alpha helices. Biophysical Journal 6 (1966) 367-370. https://doi.org/10.1016/S0006-3495(66)86662-6

[59]   M. Bassas-Galia, S. Follonier, M. Pusnik, M. Zinn, 2 – Natural polymers: a source of inspiration. bioresorbable polymers for biomedical applications. From fundamentals to translational medicine, Woodhead Publishing, Elsevier Ltd. 2017, pp. 31-64. https://doi.org/10.1016/B978-0-08-100262-9.00002-1

Powder Metallurgy and Advanced Materials – RoPM&AM 2017
Materials Research Proceedings 8 (2018) 52-60

Materials Research Forum LLC
doi: http://dx.doi.org/10.21741/9781945291999-6

Synthesis and characterization of various surfactants for stabilized CuO powder

Alina MATEI [1,a *], Vasilica TUCUREANU [1,2,b], Marian POPESCU [1], Cosmin ROMANIȚAN [1,3], Bogdan BIȚĂ [1,3], Ileana CERNICA [1]

[1]National Institute for Research and Development in Microtehnologies IMT-Bucharest, Bucharest, Romania

[2]Department of Materials Science, Transilvania University of Brasov, Brasov, Romania

[3]Faculty of Physics, University of Bucharest, Magurele, Romania

[a*]alina.matei@imt.ro,[b]vasilica.tucureanu@imt.ro

Keywords: Copper oxide, Nanoparticles, Surfactants, Chemical method

Abstract: In the present work, CuO nanoparticles were successfully prepared by the coprecipitation method using copper acetate ($Cu(CH_3COO)_2$) as a basic precursor, sodium hydroxide (NaOH) as a precipitator material, sodium dodecyl sulfate (SDS) and cetyltrimethylammonium bromide (CTAB) as anionic and cationic surfactants, respectively. The synthesized powders samples were characterized by Fourier transform infrared spectrometry (FT-IR), field emission scanning electron microscopy (FESEM) and X-ray diffraction (XRD). The investigation showed that the added types of surfactants have effects on the decrease of the crystallite size, on the CuO particles morphology, shape and uniform distribution as it is noticed in the XRD and SEM characterizations. Additionally, the FTIR spectra for all the powders samples showed the same Cu-O stretching vibration mode which indicates the presence of a crystalline CuO monoclinic structure. The obtained results create premises for further advanced the applications of CuO powders in various domains.

Introduction

Over the years, the interests in developing nanoparticles metal oxides have considerably increased due to the necessity of obtaining materials with outstanding physical and chemical properties. Various methods of metal oxide synthesis have been know so far, researches continue to development a new approaches with a strict control over nanoparticles morphology, size and composition for several technological applications.

Copper oxide (CuO), belonging to the nanomaterials class, which has attracted recent research because of its excellent properties, cost effectiveness and wide spectrum of practical applications (solar and electrochemical cells, gas sensors, field emitters, active catalyst and antimicrobial activity, etc.). Also, CuO as nanostructured oxide being classified as a *p*-type monoclinically structured semiconductor material with a direct band-gap value of 1.85 eV presents a particular attention. This type of material has a special concern because it extends the use in a board range of applications, such as electronics and optoelectronics, catalysts, sensors and biosensors, chemical sensing devices, nanofluids and field emitters, desinfection, cosmetic pigments, antibacterial agent, etc. [1–4]. In order for this material to exhibit viable properties in the desired field of applicability, it is intended to establish its method of obtaining and its synthesis parameters; there

Powder Metallurgy and Advanced Materials – RoPM&AM 2017   Materials Research Forum LLC
Materials Research Proceedings 8 (2018) 52-60   doi: http://dx.doi.org/10.21741/9781945291999-6

have been elaborated and known various physical and chemical methods so far, such as sol-gel, coprecipitation, hydrothermal synthesis, mechanical mixing, solid state reaction, thermal descomposition of precursors, microemulsion, microwave irraditiaon, physical vapor deposition, ablation, etc. [3, 5, 6].

From the bottom-up type methods, precipitation is a cheap one, with applicability on a broad-scale, and which does not involve the addition of secondary reaction products [7–9]. Particularly, in order to obtain CuO, the method involves precipitation of various soluble copper salts (nitrates, chlorides, sulphates and acetates) in aqueous solutions, followed by their thermal decomposition with oxide formation [4, 5, 10].

Inorganic salts are mixed in an aqueous medium and under a rigorous pH control by using solutions of NaOH, KOH, $(NH_4)_2CO_3$ or NH_4OH, finally the precipitates obtained are subject to characteristic thermal treatment. In the synthesis process the reaction parameters (pH, rate of addition of reactants and speed of stirring, solution concentration, reaction temperature, etc.) have a determining role in the particle size, morphology and granulometry [11 – 13].

In the literature various thermal treatments at relatively high temperature (between 700 °C and 1100 °C) are presented, leading to the phenomenon of agglomeration and increase of particles average size. To prevent the tendency of agglomeration and to favour the formation of nanostructured materials, the use surfactants have been proposed to be used. The use of different surfactants allows the improvement of particles structural, physico-chemical and morphological properties due to the electrostatic and stearic stabilizing mechanisms that reduce the solutions surface tension and improve the nanocrystalline material properties [6, 14 – 18].

Also, the addition of surfactants in the oxide materials precipitation process leads to increase nanoparticle stability and interaction between surfactant molecules and metal ions. The surfactants are chemical substances, focusing on the surface and solubilized materials with low affinity relative to each other. They have an asymmetrical molecular structure, consisting of a non-polar (hydrocarbon) and a polar (ionisable or non-ionisable) part. The role of surfactants is to provide an effective and efficient coating to induce electrostatic or stearic repulsions that can counterbalance van der Waals attractions [19 – 22].

The specialized literature shows research studies to improve the properties and formation of CuO nanoparticles by using various surfactants of the type of of oleic acid (OA), polyethylene glycol (PEG), cetyltrimethylammonium bromide (CTAB), hypochlorite dodecyl sulphate (SDS), polyvinylpyrrolidone (PVP), tetraoctylammonium bromide (TOAB), playing an important role in the synthesis process steps since the incipient phase. The reason for selecting the two types of surfactants (CTAB and SDS) in the experiments carried out in the present study is based on their remarkable effects on particle stability, size changing, morphology and the surface properties of the precipitated particles, but also because they have low price, can be found relatively easy on the market and have low toxicity [7, 14, 23 – 25].

This paper presents the study of obtaining CuO nanoparticles by the coprecipitation method, in the absence and in the presence of two types of surfactants, anionic (SDS) and cationic (CTAB). It is presented the effect of surfactants on CuO nanoparticles morphology, average size and crystalline structure. The morphological and structural properties are highlighted by using field emission scanning electron microscopy (FESEM), X-ray diffraction (XRD) and Fourier transform infrared spectrometry (FT-IR).

Powder Metallurgy and Advanced Materials – RoPM&AM 2017   Materials Research Forum LLC
Materials Research Proceedings 8 (2018) 52-60   doi: http://dx.doi.org/10.21741/9781945291999-6

Experimental details

For the synthesis of CuO powders: copper acetate monohydrate [Cu(CH$_3$COO)$_2$], sodium hydroxide [NaOH] and surfactants of type anionic [C$_{12}$H$_{25}$NaO$_4$S, SDS 99%] and [C$_{19}$H$_{42}$BrN, CTAB > 99 %], were used as starting materials. All reagents were purchased from the company Sigma-Aldrich without any previous purification.

For the synthesis of CuO in the absence and in the presence of surfactants the stock solutions of 1M [Cu(CH$_3$COO)$_2$] and 1M [NaOH] were prepared, respectively. From the stock solution of acetate one part is taken and the pH is adjusted in basic medium by adding NaOH solution in dropwise until the formation of a greenish-blue precipitate was observed.

In the case of samples with surfactant, over the source of copper the afferent surfactants (SDS and CTAB concentration 0.1 M) was added and under continuous stirring the precipitating agent (NaOH) was adding in dropwise until the precipitate formed and the pH was adjusted to about 10-11. After the precipitation formation, the stirring continues up to a temperature of 80 °C. The introduction of SDS anionic surfactant induces a homogeneous nuclear process due to considerable size effect of counter-ions on the crystal facets. By adding CTAB as a cationic surfactant there is a complete ionization and cation formation in the tetrahedric structure, but it also determines a control of the growth rates of different faces of the CuO nanoparticles [14, 26].

For all synthesized samples the same synthetic conditions (time, temperature and pH) were maintained. The precipitates thus formed are left in the rest position, then have been filtered under vacuum using a Buchner funnel, and following they were washed with a water-ethanol mixture for purification and removal of the secondary compounds. After washing, the samples were subjected to the drying step in the oven at a temperature of 80 °C, preceded by calcination sintering at a temperature of 550 °C for 3 hours in normal atmosphere. Due to the sintering temperature of the dry samples changed colour from green to blue to black.

The functional groups and the chemical bonds of the synthesized oxide samples were analyzed by Fourier Transform Infrared spectrometry (Bruker Optics, Vertex 80V) using the KBr pellet method in the wavenumber range of 4000-400 cm^{-1} by averaging 64 scans. In the processing of all spectra, the bands attributable to the vibration mode of the C=O bond in CO$_2$ were extracted.

In order to investigate the morphology and particle size a Field Emission Scanning electron microscope (FE-SEM), obtained at an operating voltage at 10 kV and a magnitude of 30 000x has been used.

X-ray diffraction measurements of the synthesized CuO particles were recorded using a Rigaku Smartlab diffractometer with the radiation CuKa=1.540593 Å, indicating the limit of variation for the current between 150 mA and 190 mA. Data were collected at a scan rate of 12°/min. in the range $2\theta = 20$-95°.

Results and Discussion

FTIR Spectra

The FTIR spectra of the different samples are presented as a comparison between CuO powders obtained in the absence of surfactants (Figure 1a) and in the presence of surfactants (Fig. 1b and c).

For all samples treated at 550 °C, bands attributed to both the vibration mode of the Cu-O linkages in the precursor and to the surfactants vibration mode can be seen, thus suggesting the binding of surfactants to particle surface. Bands centred at wavelengths less than 600 cm^{-1}, can be attributed to the vibration mode of Cu-O bond from samples, indicating a temperature and time sufficient for total synthesis to obtain CuO.

Thus, bands that can be attributed to the Cu-O bond at: a) 599, 511 and 430 cm^{-1} for the sample of CuO obtained in the absence of surfactants; b) 590, 488 and 423 cm^{-1} for the sample of CuO obtained in the presence of anionic surfactant (SDS); c) 608, 495 and 411 cm^{-1} for the sample of CuO obtained in the presence of the cationic surfactant (CTAB), confirming the formation of a CuO monoclinic crystalline structure.

In the case of CuO samples in the presence of surfactants (Fig. 1b and c) is observed the spectral bands centred in the region of 3000-800 cm^{-1} which can be attributed to the type of surfactant used. The FTIR spectra for the two samples presents the intense bands at 2920 cm^{-1}, 2855 cm^{-1} and 1405 cm^{-1} for the sample with anionic surfactant and 2917 cm^{-1}, 2852 cm^{-1} and 1415 cm^{-1} respectively for the sample with cationic surfactant which may be due to vibration of the C-H from organic materials. Also, the peaks from 1700-800 cm^{-1} may be assigned to specific bonds from surfactants. Based on the appearance of bands corresponding to the surfactants used in synthesized samples, reveals the interaction and binding of the surfactant molecules on the surface of CuO particles.

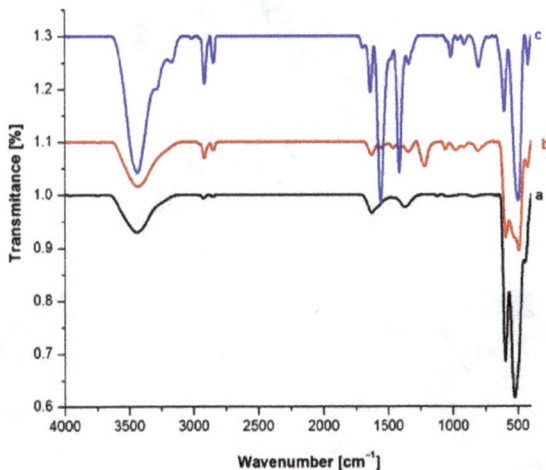

Fig. 1. FTIR spectra of CuO powders in the absence of surfactant (a) and in the presence of surfactants (b), SDS, (c) CTAB.

Morphological Analysis
Fig. 2 shows micrographs of CuO samples in the absence and the presence of surfactants. Analysis of samples by SEM was carried out for the correct estimation of their morphology, nanoparticle size, nanostructure formation and the agglomeration degree.

The CuO sample synthesized in the absence of surfactants (Fig. 2a) revealed agglomeration of particles, with dimensions in the range 40-70 nm, showing an polyhedral shape with rounded edges for defining the particles shape. The use of different types of surfactants has been derived from the necessity to control the size and morphology of the nanostructures, as well as to reduce the agglomeration tendency of particles [14,23].

Powder Metallurgy and Advanced Materials – RoPM&AM 2017   Materials Research Forum LLC
Materials Research Proceedings 8 (2018) 52-60   doi: http://dx.doi.org/10.21741/9781945291999-6

Thus, in the case of the CuO samples with anionic surfactant - SDS (Fig. 2b), from SEM micrograph we observed a slight decrease of the tendency of agglomeration concomitantly with particle size reduction, but the particle size diminishes, the particle size being estimated between 20 and 60 nm. As a result of this observation, the necessity to find another surfactant in order to decrease furthermore the tendency of agglomeration arises.

The micrograph of CuO with surfactant cation - CTAB sample (Fig. 2b) shows the formation of particles with a well-defined shape, polyhedral with rounded edges, but with larger particle size due to the type of surfactant used and the high sintering temperature. This increase of particles size can be determined by the global tension forces that minimize free surface energy by controlling particle interaction, nucleation and particle size modification by keeping the particles at a nanometric level. It is well known that the speed of the cationic surfactants adsorption is rapid, indicating that the adsorption is faster compared with anionic surfactants, showing that the stabilization and the properties of the nanoparticules differ from the other surfactants.

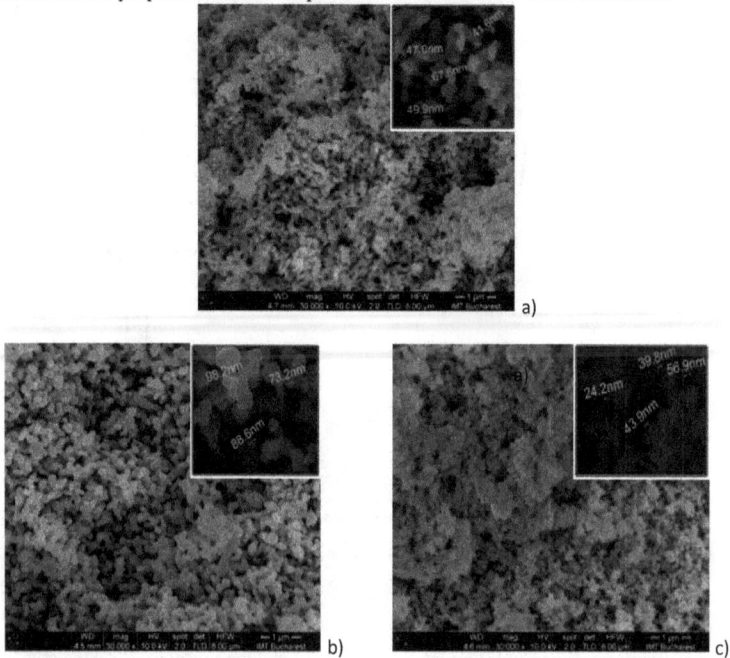

Fig. 2. SEM micrograph of CuO powder in the absence of surfactant (a) and in the presence of surfactants (b), SDS, (c) CTAB.

XRD Analysis

XRD diffraction spectra for CuO powders in the absence and in the presence of surfactants are highlighted in Fig. 3 (a-c). By using this characterization method the crystallite mean size, crystallographic structure and lattice parameters are estimate. In all three samples the diffraction

Powder Metallurgy and Advanced Materials – RoPM&AM 2017   Materials Research Forum LLC
Materials Research Proceedings 8 (2018) 52-60   doi: http://dx.doi.org/10.21741/9781945291999-6

peaks were indexed in accordance with the standard data of the No 00-101-1194, being in good accordance with the existing research literature [5,25,27].

From the analyzed samples it was found that the XRD diffractograms are similar and typical with the CuO single phase monoclinic structure, the surfactants do not influence the samples crystalline structure and no other intermediate phases characteristic of them have been observed. Also, the characteristic diffraction peaks centered at about 20 (hkl) = 32°, 35°, 38°, 48°, 53°, 58°, 61°, 68°, 72°, 75° are shown, corresponding to the crystalline planes characteristic of CuO with Miller indices at (110), (002), (111), ($\overline{2}02$), (020), (202), ($\overline{1}13$), (220), (311), (004).

Depending on the sample type, the two main peaks of the crystalline planes with indices (002) and (111) can be found at 2θ = 35,474 ° and 38,714 ° for the CuO sample in the absence of surfactants, and at 2θ = 35,426 ° and 38,617 ° for CuO samples in the presence of anionic surfactant (SDS) and at 2θ = 35,377 ° and 38,665 ° respectively for CuO samples in the presence of CTAB. As a result of the use of surfactants, changes in the average size of crystallites and lattice parameters can be observed. The details of these parameters and the variation of the size of the crystals are shown in Table 1.

Calculation of the average crystallite size for all three samples was based on the Debye Scherer's formula (1):

$$D=K\lambda/(\beta\cos\theta) \tag{1}$$

Where: K is the shape coefficient (0.94), λ is the radiation wavelength (1.54 A), β is the peak FWHM (full width half maximum width in radians), and θ is the Bragg diffraction angle obtained from 2θ value corresponding to maximum intensity peak (in radians).

Based on this formula the results indicate that the average size of crystallites varied from 16.7 nm for CuO and 12.5 nm for CuO in the presence of SDS and 20.02 nm for CuO in the presence of CTAB respectively, thus it was found that depending on the type of surfactant used, a control is performed on the degree of crystallites growth.

Table 1. Crystallite sizes, lattice parameters and range particle size of CuO powders in the absence and in the presence of surfactants.

Sample number	Crystallite sizes (nm)	Lattice parameters		Range particle sizes (nm)
		a (A)	c (A)	
1 *CuO*	16.7	4.6859	5.1327	50
2 *CuO in the presence of SDS*	12.5	4.6857	5.1320	40
3 *CuO in the presence of CTAB*	20.02	4.6923	5.1394	70

Powder Metallurgy and Advanced Materials – RoPM&AM 2017   Materials Research Forum LLC
Materials Research Proceedings 8 (2018) 52-60   doi: http://dx.doi.org/10.21741/9781945291999-6

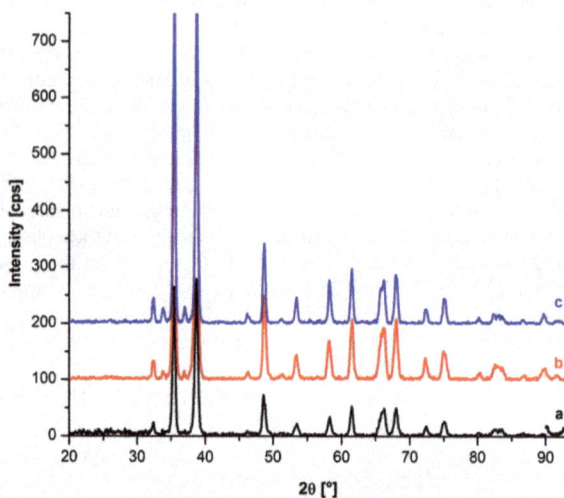

Fig. 3. XRD pattern of CuO powder in the absence of surfactant (a) and in the presence of surfactants (b), SDS, (c) CTAB.

Summary

In the present study, CuO nanoparticles were synthesized by the chemical co-precipitation method, in the absence and in the presence of two types of surfactants (SDS and CTAB), sintered at 550°C. Structural analysis clearly indicates the formation of a single crystalline phase in all synthesized samples.

FTIR spectra reveal bands centered at wavelengths less than 600 cm^{-1} characteristic of the Cu-O metal bond. Also, the presence of peaks in the region 3000-800 cm^{-1} indicates the binding of surfactants to the surface of the particles without affecting the particle crystalline structure, confirmed by XRD.

The XRD analysis indicates that regardless of the type of surfactant used, the powders synthesized had a CuO single phase monoclinic structure.

SEM images indicate that the sample obtained in the absence of surfactants shows an agglomeration tendency with dimensions ranging from 40-70 nm.

In the case of samples synthesized depending on the type of surfactant, it is found that there is a control over the morphology and particle sizes ranging between 20-80 nm and reduces the agglomeration of the CuO particles.

The results showed that function of the type of surfactant used an improvement in the dispersion of CuO nanoparticles, a control of the morphology and particle size was exerted. By optimizing the conditions of processing and the establishment of surfactants, this method may be accessible for the synthesis of the various types of oxide materials with effect in a wide range of domains.

Acknowledgments

This work was supported by a grant of the Ministry of National Education and Scientific Research, RDI Program for Space Technology and Advanced Research—STAR, project number 639/2017. Also, this work was supported by National Basic Funding Programme TEHNOSPEC – Project No. PN1632/2016.

References

[1]   R. R. Sandupatla, P. Veerasomaiah, Synthesis, characterization and photoluminescence study of CuO nanoparticles using aqueous solution method, *Int. J Nanomater. Biostruct.* 6 (2016) 30–33.

[2]   A. El-Trass, H. ElShamy, I. El-Mehasseb et al., CuO nanoparticles : synthesis, characterization, optical properties and interaction with amino acids, Appl. Surf. Sci. 258 (2012) 2997–3001. https://doi.org/10.1016/j.apsusc.2011.11.025

[3]   E. Darezereshki, F. Bakhtiari, Synthesis and characterization of tenorite (CuO) nanoparticles from smelting furnace dust (SFD), J Min. Metall. Sect. B-Metall. 49 (2013) 21–26. https://doi.org/10.2298/JMMB120411033D

[4]   A. Rahnama, M. Gharagozlou, Preparation and properties of semiconductor CuO nanoparticles via a simple precipitation method at different reaction temperatures, Opt. Quant. Electron. 44 (2012) 313–322. https://doi.org/10.1007/s11082-011-9540-1

[5]   S.J. Davarpanah, R. Karimian, V. Goodarzi et al., Synthesis of Copper (II) Oxide (CuO) nanoparticles and its application as gas sensor, J Appl. Biotechnol. Rep. 2 (2015) 329–332.

[6]   M.E. Grigore, E.R. Biscu, A.M. Holban, et al., Methods of synthesis, properties and biomedical applications of CuO nanoparticles, Pharmaceuticals 9 (2016) 1–14. https://doi.org/10.3390/ph9040075

[7]   J. Singh, G. Kaur, M. Rawat, A brief review on synthesis and characterization of copper oxide nanoparticles and its applications, J Bioelectron. Nanotechnol. 1 (2016) 1–9.

[8]   A. Matei, V. Tucureanu, L. Dumitrescu, Aspects regarding synthesis and applications of ZnO. *Bull Transilv Univ Brasov*, Ser.I: Eng. Sci. 7 (2014) 45-52.

[9]   T. Iqbal, A. Hassan, S. Ghazal, Synthesis of iron oxide, cobalt oxide and silver nanoparticles by different techniques: a review, Int. J Sci. Eng. Res. 7 (2016) 1178–1221.

[10]   K. Phiwdang, S. Suphankij, W. Mekprasart, Synthesis of CuO nanoparticles by precipitation method using different precursors, Energy Procedia. 34 (2013) 740–745. https://doi.org/10.1016/j.egypro.2013.06.808

[11]   M.F. Romadhan, N.E. Suyatma, F.M. Taqi, Synthesis of ZnO nanoparticles by precipitation method with their antibacterial effect, Indones. J Chem. 16 (2016) 117–123. https://doi.org/10.22146/ijc.21153

[12]   D. Raoufi, Synthesis and microstructural properties of ZnO nanoparticles prepared by precipitation method, Renew. Energy. 50 (2014) 932–937. https://doi.org/10.1016/j.renene.2012.08.076

[13]   M.C. Mascolo, Y. Pei, T.A. Ring, Nanoparticles in a large pH window with different bases, Materials 6 (2013) 5549–5567. https://doi.org/10.3390/ma6125549

[14]   H. Siddiqui, M.S. Qureshi, F.Z. Haque, Optik surfactant assisted wet chemical synthesis of copper oxide (CuO) nanostructures and their spectroscopic analysis, Optik 127 (2016) 2740–2747. https://doi.org/10.1016/j.ijleo.2015.11.220

[15]   A.K. Arora, Synthesis of nanosized CuO particles: a simple and effective method, Int. J Chem. Sci. 11 (2013) 1270–1276.

[16]   T.H. Tran, V.T. Nguyen, Copper oxide nanomaterials prepared by solution methods, some properties, and potential applications:a brief review, *Int. Sch. Res. Notices* 2014 (2014) 1-14. https://doi.org/10.1155/2014/856592

[17]   R. Eview, Methods of preparation of nanoparticles- a review, Int. J Adv. Eng. Techol. 7 (2015) 1806–1811.

[18]   P. Derakhshi, R. Lotfi, Synthesis and surfactant effect on structural analysis of nickel doped cobalt ferrite nanoparticles by c-precipitation method, J Appl. Chem. Res. 6 (2012) 60–65.

[19]   S.K. Mehta, S. Kumar, S. Chaudhary et al., Evolution of ZnS nanoparticles via facile CTAB aqueous micellar solution route: a study on controlling parameters, Nanoscale Res. Lett. 4 (2009) 17–28. https://doi.org/10.1007/s11671-008-9196-3

[20]   L.P. Singh, S.K. Bhattacharyya, G. Mishra, S. Ahalawat, Functional role of cationic surfactant to control the nano size of silica powder, Appl. Nanosci. 1 (2011) 117–122. https://doi.org/10.1007/s13204-011-0016-1

[21]   M. Manisha, P. Muthuprasanna, Surya prabha K et al., Basics and potential applications of surfactants - a review, Int. J Pharm. Tech. Res. 1 (2009) 1354–1365.

[22] R. Azarmi, A. Ashjaran, Type and application of some common surfactants, J Chem. Pharmacal. Res. 7 (2015) 632-640.

[23]   M. Saterlie, H. Sahin, B. Kavlicoglu et al., Particle size effects in the thermal conductivity enhancement of copper-based nanofluids, Nanoscale Res. Lett. 6 (2011) 1–7. https://doi.org/10.1186/1556-276X-6-217

[24]   A. Ananth, S. Dharaneedharan, M.S. Heo et al., Copper oxide nanomaterials: synthesis, characterization and structure-specific antibacterial performance, Chem. Eng. J. 262 (2015) 179-188. https://doi.org/10.1016/j.cej.2014.09.083

[25]   X. Wang, J. Yang, L. Shi et al., Surfactant-free synthesis of CuO with controllable morphologies and enhanced photocatalytic property, Nanoscale Res. Lett. 11 (2016) 125. https://doi.org/10.1016/j.cej.2014.09.083

[26]   L.S. Cavalcante, J.C. Sczancoski, M.S. Li et al., Aspects method: growth process and photoluminescence properties, Colloids Surf A: Physicochem Eng. Asp. 396 (2012) 346–351. https://doi.org/10.1016/j.colsurfa.2011.12.021

[27]   B. Shaabani, E. Alizadeh-Gheshlaghi, E. Azizian-Kalandaragh et al., Preparation of CuO nanopowders and their catalytic activity in photodegradation of Rhodamine-B, Adv. Powder Technol. 25 (2014) 1043–1052. https://doi.org/10.1016/j.apt.2014.02.005

Powder Metallurgy and Advanced Materials – RoPM&AM 2017   Materials Research Forum LLC
Materials Research Proceedings 8 (2018) 61-69   doi: http://dx.doi.org/10.21741/9781945291999-7

Analysis of plasma jet depositions on a C45 steel used in crankshaft manufacturing

G. MAHU [1], C. MUNTEANU[1],*, B. ISTRATE [1], M. BENCHEA[1]

[1]"Gheorghe Asachi" Technical University of Iasi, Faculty of Mechanical Engineering, 43 "D. Mangeron" Street, 700050, Iasi, Romania

[a]cornelmun@gmail.com

Keywords: C45 steel, Plasma coatings, Crankshaft, SEM, XRD, Mechanical properties

Abstract. Thermal Spray coatings are used in the automotive industry, with multiple applications, offering effective protection solutions in the suspension system, transmission and engine components. Plasma jet deposition may be a solution to classical methods of enhancing wear resistance by chemical or thermochemical treatments in main and crankpin journals. Through this study the deposition of layers on the surface of a C45 steel was analyzed, using the following powders: Cr_3C_2- $(Ni_{20}Cr)$, Al_2O_3-13TiO_2, Cr_2O_3-SiO_2-TiO_2. Powders were deposited with the 9MCE Spraywizard at atmospheric pressure. Microstructural and morphological analyzes were performed by means of optical, electronic and X-ray diffraction microscopy. Microstretch and indentation tests were performed to determine the adhesion of the deposited layer to the base material. As a result of the tests, the layer deposited using the Al_2O_3 13TiO_2 powder has higher hardness characteristics than the layers of Cr_3C_2- $(Ni_{20}Cr)$ and Cr_2O_3-SiO_2-TiO_2 powders and that of the base material. This study confirms that plasma jet deposition is an effective solution to classic methods used to increase crankshaft wear resistance, at least in limited production series.

Introduction

To improve the properties of crankshafts, such as wear resistance and increased durability, various heat treatments such as carburizing, nitration, chrome plating, carbonitriding, etc. are applied [1 - 4]. The plasma jet deposition process is one of the most common thermal spray deposition processes [5]. Studies has been carried out on plasma deposition on the surface of a C45 steel, the most widespread plasma jet deposition method being nitride according to the studies of T.Bell and H.Liu [6, 7]. Plasma jet spraying is an effective method for improving the wear, corrosion, and high temperature resistance of layers deposited on the surface of a material [8]. F. Hakami et al. performed a study on duplex treatment of C45 steel resulting from chromium followed by plasma nitrating; the study found that the resistance of the deposited layer was significantly increased compared to the initial chrome treatment reaching 1540 HV compared to 1270 HV in the case of simple chrome plating [9, 10]. The purpose of this research is to highlight the characteristics of the layers deposited by the plasma jet spray method on the surface of a C45 steel used in the construction of crankshafts in terms of microstructure, X-ray diffraction, micro scratch test and module Young's. For plasma jet spraying are very important the working parameters. In the case of depositing a layer of $Al_2O_3TiO_2$, nanostructural analysis and tests performed reveal that important changes can occur, so values of layer microhardness may vary from 611 HV to 772 HV in in the case of a current variation from 550 A to 650 A [11 - 13].

Experimental

The steel used in this study has the components in Table 1:

Powder Metallurgy and Advanced Materials – RoPM&AM 2017   Materials Research Forum LLC
Materials Research Proceedings **8** (2018) 61-69   doi: http://dx.doi.org/10.21741/9781945291999-7

Table 1. C45 composition.

C45	C	Si	Mn	S	P
%	0,45	0,17	0,52	0,031	0,032

Using the SPRAYWIZARD 9MCE equipment, produced by Sulzer & Metco, a plasma jet deposition process was performed using 3 powders: *Metco 130-Al$_2$O$_3$-13TiO$_2$, Metco 81NS - Cr$_3$C$_2$-25(Ni$_{20}$Cr), Metco 136F - Cr$_2$O$_3$-SiO$_2$-TiO$_2$*. The powders used for the spraying has the components in Table 2:

Table 2. Powders composition.

Product	Chemical composition (nominal wt. %)							
	TiO$_2$	Al$_2$O$_3$	Cr$_3$C$_2$	Ni	Cr	Cr$_2$O$_3$	SiO$_2$	TiO$_2$
Metco 130	13	87	-	-	-	-	-	-
Metco 81NS	-	-	75	20	5	-	-	-
Metco136F	-	-	-	-	-	92	5	3

The spray parameters used for all 3 powders are shown in Table 3.

Table 3. Spray parameters.

Powder	**Metco 130**	**Metco 81NS**	**Metco 136F**
Spray distance, (mm)	100	100	100
Injector	1,8	1,8	1,8
Curent Intensity, (A)	600	600	600
Electric arc voltage (U)	75	75	75
Rotation speed (rpm)	55	55	55
Argon flow (m3 / h)	50	50	50
Hydrogen flow (m3 / h)	14	14	14

To perform the study, 9 samples were processed, having the dimensions 30 mm X 10 mm X 2 mm.

Powder Metallurgy and Advanced Materials – RoPM&AM 2017   Materials Research Forum LLC
Materials Research Proceedings **8** (2018) 61-69   doi: http://dx.doi.org/10.21741/9781945291999-7

For the study of the microstructure, the samples were prepared metallographically using 200, 500, 800 and 1200 abrasive paper. After preparation, the samples were cleaned with ethyl alcohol and chemically attacked using the 2% nital reagent.

For the microstructure and samples morphology analysis, the SEM QUANTA 3D Dual Beam microscope, produced by the Dutch FEI, was used. The settings were on the High Vacuum mode using the Large Field Detector (LFG) detector, resulting in magnified 500X, 1000X, and 5000X images with a working distance of about 15 mm.

X-ray diffraction was performed using X-ray diffractometer, X 'Pert Pro MRD, with an X-ray tube with Cu kα, $\lambda = 1,54$ Å, using a voltage of 45 KV with an intensity of 40 mA, the diffraction angle (2 theta) ranging between 25 and 130°. The adhesion testing of the deposited layers was done by the micro scratch method and the indentation method by means of the CETR UMT-2 tribometer equipped with a DFH-20 Dual Friction / Load Sensor type duralumin, on which a blade whose top has been mounted the radius of 0.4 mm. The indenting speed was 10 mm / min.

Results and discussion
Microstructural analysis
In Fig. 1 we can see the SEM images of the layer obtained by spraying the Metco 130 powder. Variations in layer thickness can be seen in Figure 1b.

Values vary in the 279-306 μm range, indicating that the spray had a uniform distribution on the surface of the base material, respecting the manufacturer's spraying specifications. We can see a porous structure with few microcracks ,unmelted compounds and cavities (Figure 1c).

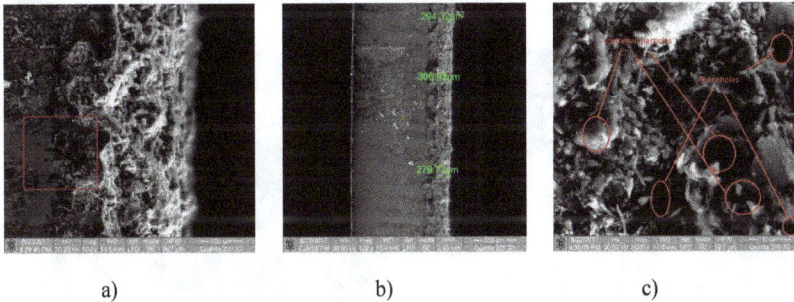

a)   b)   c)

Fig. 1. SEM images of the Metco 130 layer deposited on the surface of the specimens: a) layer separation zone (X5000), b) thickness of the layer (X5000), c) layer structure (X5000).

In Fig. 2a, we can see SEM images in the cross section of the layer obtained after Metco 81NS spraying. The morphology of the Metco 81NS powder reveals a more compact structure (Figure 2c), with splats without unmelted particles, just with isolated areas with segregated unmelted particles matter and cracks. The thickness of the deposited layer (Fig. 2b) reaching 107 μm, being inferior to the Metco 130 powder spray.

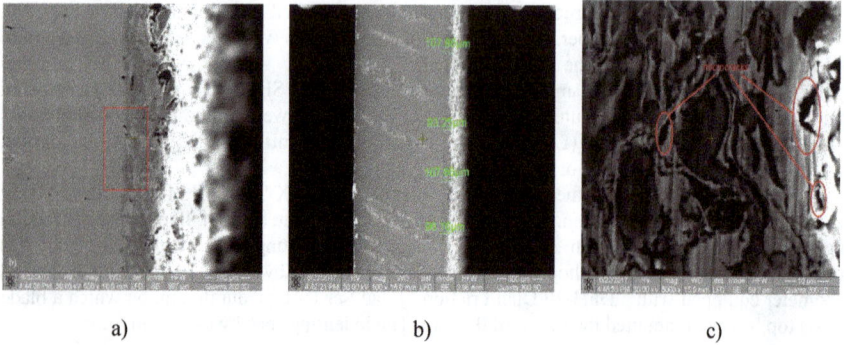

a)   b)   c)

Fig. 2. SEM images of the Metco 81NS layer deposited on the surface of the specimens:
a) layer separation zone (X5000), b) thickness of the layer (X5000), c) structure of the layer (X5000).

The structure obtained from the Metco 136 F powder spray has a non-uniform accentuated character with few isolated splats and unmelted particles areas segregation, as can be seen from the Fig. 3c. Also high accentuated porosity and isolated microcracks are highlighted. The minimum layer thickness is 96 µm, reaching maximum values of 110 µm (Fig. 3b).

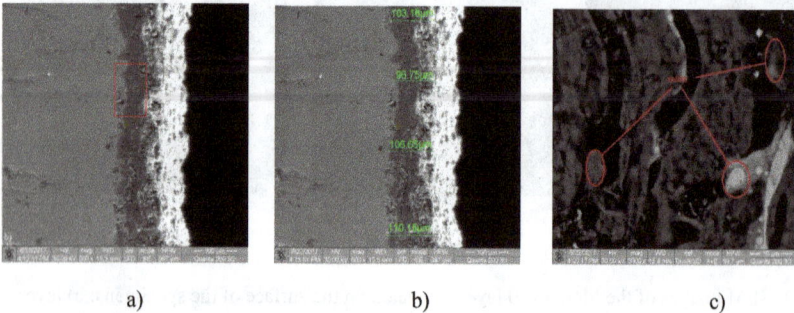

a)   b)   c)

Fig. 3. SEM images of the Metco 136F layer deposited on the surface of the specimens: a) layer separation zone (X5000), b) layer thickness (X5000), c) layer structure (X5000).

Powder Metallurgy and Advanced Materials – RoPM&AM 2017   Materials Research Forum LLC
Materials Research Proceedings 8 (2018) 61-69   doi: http://dx.doi.org/10.21741/9781945291999-7

XRD Analysis

The predominant phases of the Al_2O_3 identified in Fig. 4, for the Metco130 powder, present characteristic peaks at: 25,59°; 35,168°; 43,39°; 57,53° and 68,28°. Al_2O_3 is identified by a rhombohedral crystalline structure. TiO_2 is identified through 2 phases: rutile and anatase, with peaks at: 25,27°, 27,47°, 36,08°, 48,03° and 54,37°. Rutile and anatase display a tetragonal structure.

Fig. 4b, presents Metco 81NS powder diffraction. Crystalline cubic type structure is revealed for Cr, having peaks at: 44,58°, 64,65°, 82,31°. The predominant phases of Cr_7C_3 powder display peaks at: 39,21°, 39,35°, 42,58° and 44,20°. Cr_7C_3 powder presents an orthorhombic type crystalline structure. Metco 136F powder's diffraction, shown in figure 4c, has in it's component Cr, displaying a cubic type structure with characteristic peaks at 44,3° and 81,68°. Cr_2O_3 has peaks at 33,62°, 36,23° and 54, 89°.

a)   b)

c)

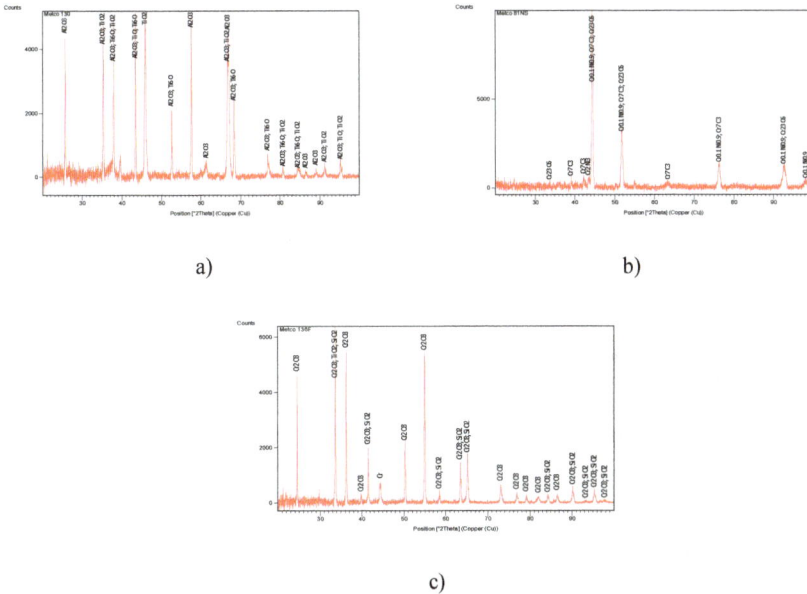

Fig. 4. X-ray diffraction, diffraction interval: 2 theta = 20…100°: a) Metco 130, b) Metco 81NS, c) Metco 136F.

MICROINDENT Analysis

In the Fig. 5 there are presented the variation curves of the penetration force relative to the indentation depth following the micro indentation test for each deposited powder. It can be noticed that the indentation depth for the METCO 136F has the most reduced values compared to the METCO 130 and METCO 81 NS.

METCO 130 is the powder with maximum values for microhardness. The applied force value for the indentation was 9N for each sample. Table 4 presents the obtained values for each of the 3 powders used following micro indentation.

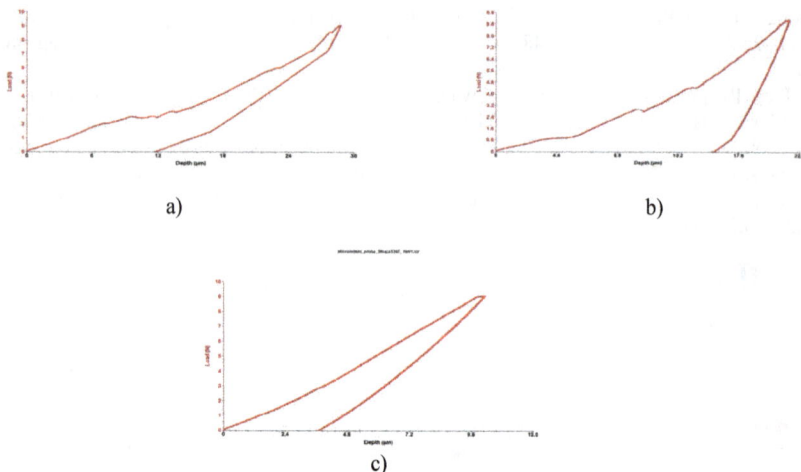

a)

b)

c)

Fig. 5. Variation curves of the penetration force relative to the indentation depth:

a) Metco 130, b) Metco 81NS, c) Metco 136F.

Table 4. Mechanical properties of coatings.

Powder	Microhardness (GPa)	COF	Stiffness(N/μm)	Young Modulus (GPa)
Metco 130	0.4728	0.58	0.5866	3.5586
Metco 81NS	0.5684	0.79	2.4043	16.1916
Metco 136F	1.0573	0.44	1.6833	15.387

The highest value of the apparent coefficient of friction is obtained for the Metco 81NS powder (Fig. 6). The Metco 136F coated products exhibit superior values of microhardness compared with other Metco 130 powders with the lowest value.

Powder Metallurgy and Advanced Materials – RoPM&AM 2017   Materials Research Forum LLC
Materials Research Proceedings **8** (2018) 61-69   doi: http://dx.doi.org/10.21741/9781945291999-7

Elastic modulus values for the three deposited powders range from 4 to 16 GPa, with Metco 130 having the lowest modulus of Young's 4 GPa module (Fig. 7). In Figs. 7 and 8 are represented comparatively the Young modulus parameters for the 3 deposited powders and respectively a comparative graph for the 3 used powders.

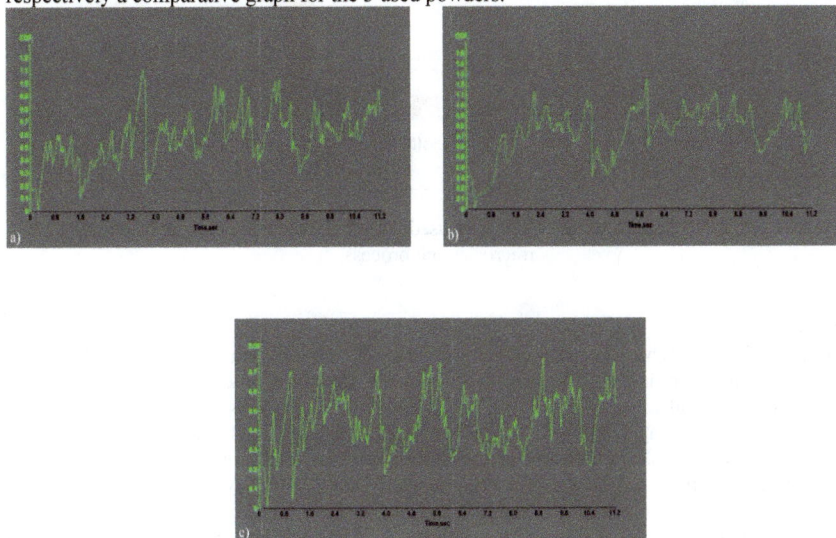

Fig. 6. Time variation diagram for apparent coefficient of friction : a) Metco 130;

b) Metco 81 NS, c) Metco 136 F.

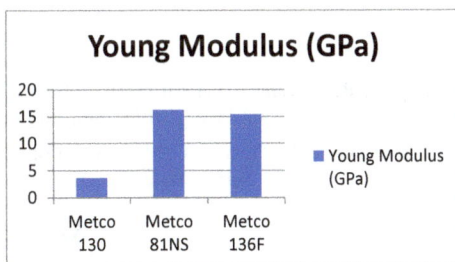

Fig. 7. Determination of the Young modulus parameters for the 3 deposited powders.

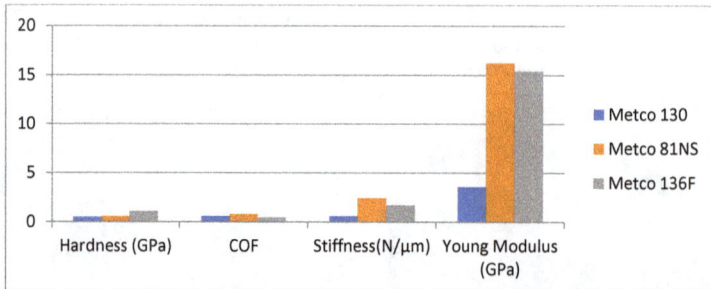

Fig. 8. Comparative graph for the 3 powders used with the obtained parameters following the micro scratch process.

Summary
Although all 3 powders have been developed for intensive wear environments, notable differences are revealed following deposited layers analysis. In the deposition process the same parameters were taken into consideration, but different layers had different thickness. Samples with METCO 130 layer had 306 μm layer, METCO 81 NS had 107 μm and METCO 136F had 110 μm. XRD Analysis revealed complex microstructural characteristics for all 3 powders. According to the microstructural analysis, the layers have been deposited uniformly, having low porosity and no major structural defects could be observed.
   The highest value of the apparent coefficient of friction is obtained for the Metco 81NS powder and METCO 136F has the lowest value. The highest values for the Young modulus are obtained by METCO 81 NS powder, while METCO 136F has the highest microhardness values.
Micro scratch analysis revealed that the obtained layers had a good adhesion and do not require intermediate layer that would require increased expenses and more time for the deposition process. Plasma jet deposition would represent an interesting alternative for classic treatment solutions of the crankshaft.

References

[1] H. Hermann, S. Sampath, R. McCune, Thermal spray: current status and future trends, Thermal Spay Coatings 25 (2000) 17-25. https://doi.org/10.1557/mrs2000.119

[2] C. Paulin, D.L. Chicet, B. Istrate, M. Panțuru, C. Munteanu, Corrosion behavior aspects of Ni-base self-fluxing coatings, IOP Conference Series: Mater. Sci. Eng. 147 (2016), Article number 012347.

[3] C.A. Tugui, C. Nejneru, D.G. Galusca, M.C. Perju, M. Axinte, N. Cimpoesu, P. Vizureanu, The influence of the Al deposition by MOC-CVD method on stainless steel thermal conductivity depending on the substrate roughness, J Optoelectron. Adv. Mater. 17 (2015) 1855-1861.

[4] I. B. Roman, M.H. Tierean, J. L. Ocaña, Effects of laser shock processing on 316L stainless steel welds, J Optoelectron. Adv. Mater. 15 (2013) 121-124.

Powder Metallurgy and Advanced Materials – RoPM&AM 2017   Materials Research Forum LLC
Materials Research Proceedings 8 (2018) 61-69   doi: http://dx.doi.org/10.21741/9781945291999-7

[5]   M.B. Beardsley, P.G. Happoldt, K.C. Kelley, Thermal barrier coatings for low emission, high efficiency diesel engine applications- SAE International 1999-01-2255.April 26-28, 1999.

[6]   T. Bell, Y. Sun, A. Suhadi, Environmental and technical aspects of plasma nitrocarburizing, Vacuum 59 (2000) 14-23. https://doi.org/10.1016/S0042-207X(00)00250-5

[7]   H. Liu, J.C. Li, F. Sun, J. Hu, Characterization and effect of pre-oxidation on D.C. plasma nitriding   for   AISI4140   steel,   Vacuum   109   (2014)   170-174. https://doi.org/10.1016/j.vacuum.2014.07.017

[8]   J.R. Davis (Ed.), Surface hardening of steels: understanding the basics, ASM International, Materials Park, OH, 2002.

[9]   F. Hakami, M.H. Sohi, J. R. Ghani, Duplex surface treatment of AISI 1045 steel via plasma   nitriding   of   chromized   layer,   Thin   solid   Films   519   (2011)   6792-6796. https://doi.org/10.1016/j.tsf.2011.04.054

[10]   Sh. Ahangarani, F. Mahboubi, A.R. Sabour, Effects of various nitriding parameters on active screen plasma nitriding behavior of a low-alloy steel, Vacuum 80 (2006) 1032. https://doi.org/10.1016/j.vacuum.2006.01.013

[11] A. Rahim, M. Sahaba, N. H. Saadb, S. Kasolangb, J. Saedonb, Impact of plasma spray variables parameters on mechanical and wear behaviour of plasma sprayed Al_2O_3 3%wt TiO_2 coating in abrasion and erosion application, Procedia Engineering 41 (2012) 1689-1695. https://doi.org/10.1016/j.proeng.2012.07.369

[12] W. Żórawski, A. Góral, O. Bokuvka, K. Berent, Microstructure and mechanical properties of plasma sprayed nanostructured and   conventional Al2O3–13TiO2 coatings, Proceedings of the International Thermal Spray (ITSC) Conference. Busan, Republic of Korea, 2013.

[13] Kim G E, Thermal Sprayed, Nanostructured coatings: applications and developments, Perpetual Technologies, Inc., Ile des Soeurs, Quebec, Canada (2011) 91-92.

Powder Metallurgy and Advanced Materials – RoPM&AM 2017
Materials Research Proceedings 8 (2018) 70-79

Materials Research Forum LLC
doi: http://dx.doi.org/10.21741/9781945291999-8

Mössbauer spectroscopic analysis of (Nd,Pr,Dy)$_2$(Fe,Co,Ga)$_{14}$B/α-Fe permanent magnetic nanocomposites

Božidar CEKIĆ[1,a], Valentin IVANOVSKI[1,b], Mirela Maria CODESCU[2,c*], Ana UMIĆEVIĆ[1,d], Katarina ĆIRIĆ[1,e], Eugen MANTA[2,f]

[1]Institute of Nuclear Sciences Vinča, University of Belgrade, Belgrade, Serbia

[2]Research and Development National Institute for Electrical Engineering ICPE-CA
313 Splaiul Unirii, Bucharest - 3, Romania

[a]cekic@vinca.rs; [b]valiva@vinca.rs; [c]mirela.codescu@icpe-ca.ro; [d]umicev@vinca.rs; [e]kciric@vinca.rs; [f]eugen.manta@icpe-ca.ro

Keywords: NdFeB nanocomposites, XRD analysis, Mössbauer spectroscopy, Internal magnetic field, Quadrupole splitting

Abstract. In this paper, it is reported the structural and magnetic properties of $Nd_{13.7}Pr_{0.7}Dy_{0.2}Fe_{73.1}Co_{6.3}Ga_{0.4}B_{5.6}$ and $Nd_{7.7}Pr_{0.7}Dy_{0.2}Fe_{79.1}Co_{6.3}Ga_{0.4}B_{5.6}$ magnetic nanocomposites, synthesized by melt-spinning and annealing methods. The Nd-Fe-B ribbons are melt-spun at v=30 m/s in high vacuum and annealed at 715°C for 4 min. in argon. Furthermore, X-ray diffraction and transmission [57]Fe Mössbauer spectra at RT are used to investigate the effects of substituent elements: Dy, Pr, Co, Ga on the hard magnetic properties and microstructure of both nanocomposites. Analysis of Mössbauer spectra for $Nd_{13.7}Pr_{0.7}Dy_{0.2}Fe_{73.1}Co_{6.3}Ga_{0.4}B_{5.6}$ is done in terms of ten Zeeman sextets, one paramagnetic doublet related to $Nd_{1.1}Fe_4B_4$ phase and two hyperfine magnetic fields distributions extracted from spectrum. Similar result of analysis of the second nanocomposite is obtained with eleven sextets, one doublet and one distribution. One sextet corresponds to α-Fe phase, while we have identified six iron sextets corresponding to the six distinct iron sites in the $Nd_2Fe_{14}B$ structure: $16k_1$, $16k_2$, $8j_1$, $8j_2$, 4c and 4e. The three remaining sextets belong to Fe$_3$B structure with three inequivalent Fe sites: Fe$_I$(8g), Fe$_{II}$(8g) and Fe$_{III}$(8g). The eleventh sextet of $Nd_{7.7}Pr_{0.7}Dy_{0.2}Fe_{79.1}Co_{6.3}Ga_{0.4}B_{5.6}$ belongs to FeB. All relevant parameters for both nanocomposites: magnetic hyperfine field, isomer shift and quadrupole splitting are determined for each of these sites. To highlight the thermally induced structural transformations, the quenched samples have been analysed by differential scanning calorimetry and thermo-magnetic measurements. The magnetic properties, measured at RT on the quenched and annealed ribbons, revealed the relationship between the alloy chemical composition and processing.

1. Introduction

Hard magnetic materials were divided into the group of conventional metallic and oxide magnets and the group of modern magnets based on intermetallic compounds of rare earth elements with Co and/or Fe. The importance of advanced permanent magnetic materials in many electro-, magneto-chemical and electronic applications is depending on the significant improvement of the magnetic energy density and a high coercivity or "magnetic hardness" of the new hard magnetic materials. The basis for these high-performance magnets are rare earth intermetallic phases SmCo$_5$ [1] and Nd$_2$Fe$_{14}$B [2-4]. Discovered in '80, the NdFeB magnets imposed due to their remarkable magnetic performance at room temperature, but they present some disadvantages, related to their thermal and corrosion stability [5-7], disadvantages that limit their operating regime in various applications field.

High performance $Nd_2Fe_{14}B$-based permanent magnets are produced with different composition and processing techniques, the magnetic properties of rare earths based permanent magnets being strongly dependent on their chemical composition. In this respect, any alloy contamination, especially with oxygen, occurred during processing, depletes the alloy of the rare earth elements. The first, negative consequence is the shift of the composition on the side of the phase diagram, rich in transition metals, that means a disadvantageous phase distribution. Depending on the alloy composition, the nanocomposite structures can be $Nd_2Fe_{14}B/Fe_3B$ or $Nd_2Fe_{14}B/\alpha$-Fe or mixture of both. Many efforts are continuously devoted to improve the performance of $Nd_2Fe_{14}B/\alpha$-Fe nanocomposite magnets. One of the most effective methods was through the compositional modification of $Nd_2Fe_{14}B/\alpha$-Fe system [8-10]. Two main types of elements can be added to $Nd_2Fe_{14}B/\alpha Fe$ composites: (a) substitutional elements such as Pr, Dy, Co, Ga and La [11,12] and (b) two different types of dopants such as Al, Cu, Zn, Ge, Sn and Ti, Zr, V, Mo, Nb and W [13-15]. Two different types of dopants, influencing microstructure in different ways, can be distinguished independently of the processing route; sintering, melt-spinning, mechanical alloying and hot worked magnets: for example, dopants M1 type as Al, Cu, Ga form binary M1-Nd or ternary M1-Fe-Nd phases and dopants M2 type as Ti, Zr, V, Mo, Nb, W form binary M2-Nd or ternary M2-Fe-B phases. Substituent and dopant elements influence the microstructure, coercivity and corrosion resistance of advanced (Nd,Pr,Dy)-(Fe,Co,Ga)-B magnets. Generally, two types of substituent elements, which replace the rare earth element or transition element sites in the hard-magnetic phase, and two types of dopant elements are distinguished for the highest value of energy density product, obtained so far [16]. Selected elements substitute the Nd-atoms (Pr and Dy) and the Fe-atoms (Co and Ga), in the hard magnetic Φ-phase. Their introduction changes the intrinsic properties; the spontaneous polarization, the Curie temperature and the magnetocrystalline anisotropy according to their solubility range within the $Nd_2Fe_{14}B$ phase.

2. Experimental
2.1 Preparation and XRD
In the Nd-Fe-B system, our studied multicomponent alloys: $Nd_{13.7}Pr_{0.7}Dy_{0.2}Fe_{73.1}Co_{6.3}Ga_{0.4}B_{5.6}$ and $Nd_{7.7}Pr_{0.7}Dy_{0.2}Fe_{79.1}Co_{6.3}Ga_{0.4}B_{5.6}$ are all based on the stoichiometric composition of $Nd_2Fe_{14}B$ hard magnetic phase. The alloys are melted using the induction furnace, Leybold - Heraeus type, in an Al_2O_3 crucible, starting from elements or masteralloys (B20-Fe, Nd84-Fe, Dy80-Fe, in wt. %). In order to avoid the oxidation of the rare earths alloys, their processing is performed in vacuum ($10^{-3} – 10^{-4}$ mbar) and then inert atmosphere (argon, 99.95% purity), and the iron was previously deoxidized, through annealing in reducing atmosphere (hydrogen), at $800 – 850°C$, for 3 hours. The purity of all elements was higher than 99.8%. The ingots are remelted in high vacuum ($5·10^{-4} – 10^{-5}$ mbar), by the melt-spinning technique, using a quartz crucible with the diameter nozzle of $\Phi = 0.4$ mm and the rare earths-based ribbons are prepared under argon atmosphere (900 mbar). The technological parameters of melt-spinning process were: wheel speed v = 30 m/s, and for melt alloy ejection, an argon overpressure $\Delta p = 0.5$ bar. The obtained Nd-Fe-B based ribbon, 15-50 μm thick, 5-8 mm long and 1.5-2 mm wide, have been annealed in vacuum ($2·10^{-4}$ mbar)/argon at 715°C for 4 min. to improve the microstructure and the magnetic properties.

The X-ray diffraction measurements were performed on Philips PW 1050 powder diffractometer with Ni filtered Cu Kα radiation ($\lambda = 1.5418$ Å) and scintillation detector within $10–120°$ 2θ range in steps of 0.02°, and scanning time of 12 s per step.

Powder Metallurgy and Advanced Materials – RoPM&AM 2017   Materials Research Forum LLC
Materials Research Proceedings 8 (2018) 70-79   doi: http://dx.doi.org/10.21741/9781945291999-8

Mössbauer absorption spectra were obtained in a standard transmission geometry using a source of [57]Co in Rh (920 MBq) at room temperature (RT). A calibration was done with laser and isomer shifts values are referred to α-Fe. The measurements were made on the powder sample contained in a Plexiglas holder; the absorbers surface densities the were 39.9 mg/cm^2 for Nd$_{13.7}$Pr$_{0.7}$Dy$_{0.2}$Fe$_{73.1}$Co$_{6.3}$Ga$_{0.4}$B$_{5.6}$ and 31.6 mg/cm^2 for Nd$_{7.7}$Pr$_{0.7}$Dy$_{0.2}$Fe$_{79.1}$Co$_{6.3}$Ga$_{0.4}$B$_{5.6}$. The data were stored in 1024 multichannel analyzer. Laser spectrums were recorded and fitted in order to recalculate channels in mm/s. Sample thickness corrections were carried out by transmission integral. The spectra have been examined by fitting data with WinNormos-Dist software that enables distribution of hyperfine parameters by histogram method and allows for Lorentz sextets and doublets on well-defined ion sites [17].

To highlight the thermally induced structural transformations, the melt-spun samples from the prepared alloys have been analysed with a DSC 204 F1 Phoenix instrument from Netzsch, Germany, at 10 K/min heating rate in the temperature range (25 - 590)°C. The studied samples were placed in open crucible and heated in Ar (gas purity higher than 99.996%). Also have been performed thermo-magnetic measurements on the same samples, through vibrating sample magnetometry (VSM). The measurements were performed with a VSM 7300 Lake Shore in the temperature range (25-830)°C and 800 kA/m magnetic field.

3. Results and Data Analysis
The XRD patterns of the Nd$_{13.7}$Pr$_{0.7}$Dy$_{0.2}$Fe$_{73.1}$Co$_{6.3}$Ga$_{0.4}$B$_{5.6}$ and Nd$_{7.7}$Pr$_{0.7}$Dy$_{0.2}$Fe$_{79.1}$Co$_{6.3}$Ga$_{0.4}$B$_{5.6}$ nanocomposites contain four phases and are presented on the Fig. 1. All the melt-spun alloys have amorphous structure. XRD phase analysis of both nanocomposites confirmed the presence of main hard magnetic phase Nd$_2$Fe$_{14}$B (space group P4$_2$/mnm), soft magnetic phases Fe$_3$B and partially Fe as well as minor quantities of hard magnetic Nd$_{1.1}$Fe$_4$B$_4$ boride phase.

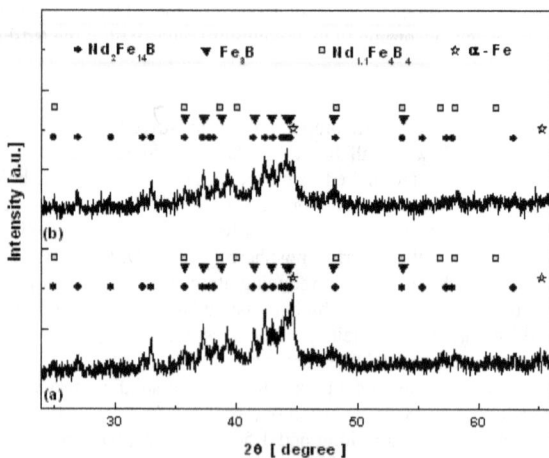

Fig. 1. XRD results of melt-spun Nd$_{13.7}$Pr$_{0.7}$Dy$_{0.2}$Fe$_{73.1}$Co$_{6.3}$Ga$_{0.4}$B$_{5.6}$ a) and Nd$_{7.7}$Pr$_{0.7}$Dy$_{0.2}$Fe$_{79.1}$Co$_{6.3}$Ga$_{0.4}$B$_{5.6}$ b) alloys after annealing treatments.

Powder Metallurgy and Advanced Materials – RoPM&AM 2017   Materials Research Forum LLC
Materials Research Proceedings **8** (2018) 70-79   doi: http://dx.doi.org/10.21741/9781945291999-8

The Mössbauer absorption spectra at 294 K of both alloys are presented in Fig. 2a and Fig.2b. The spectra consist of ten Zeeman sextets and one paramagnetic doublet related to the $Nd_{1.1}Fe_4B_4$ phase. One sextet corresponds to the α-Fe phase, whereas others are attributed to six non-equivalent Fe sites in the $Nd_2Fe_{14}B$ structure, namely $16k_1$, $16k_2$, $8j_1$, $8j_2$, 4c, and 4e. The constraints associated with the computer fit of the spectra were that six subspectra were assumed in intensity ratios 4:4:2:2:1:1 The three remaining sextets belong to the Fe_3B structure with three inequivalent Fe sites $Fe_I(8g)$, $Fe_{II}(8g)$ and $Fe_{III}(8g)$. The relative intensity from crystalline phase in Fe_3B was assumed to have 1:1.4:1.5. The Mössbauer absorption spectrum at 294K of $Nd_{7.7}Pr_{0.7}Dy_{0.2}Fe_{79.1}Co_{6.3}Ga_{0.4}B_{5.6}$ (Fig. 2b), contains still one, the eleventh sextet which belongs to the FeB.

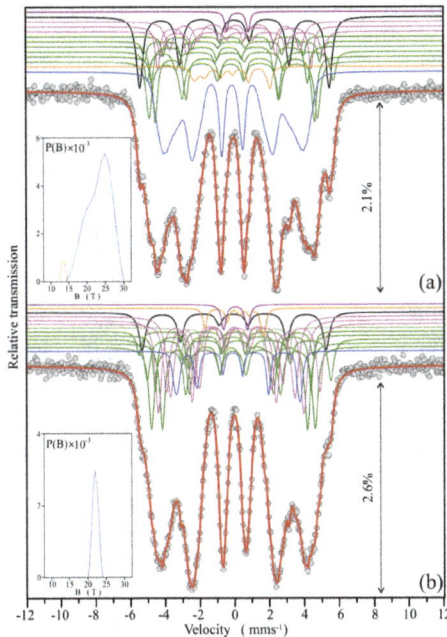

Fig. 2. Mössbauer absorption spectra at 295K in a) $Nd_{13.7}Pr_{0.7}Dy_{0.2}Fe_{73.1}Co_{6.3}Ga_{0.4}B_{5.6}$ and b) $Nd_{7.7}Pr_{0.7}Dy_{0.2}Fe_{79.1}Co_{6.3}Ga_{0.4}B_{5.6}$. Insets represent the hyperfine magnetic fields distributions at both nanocomposites.

The inset in Fig. 2 a represents two hyperfine magnetic field distributions extracted from this spectrum. It can be seen that the first distribution of FeB exhibits one peak with the highest probability of 13.4 T. The average magnetic field for this distribution is 12.64(1.91) T. The next distribution of Fe_2B exhibits two peaks with hyperfine fields of 20.4 T and 25.3 T. The average magnetic field is 23.22 (3.19) T. When the $Nd_{7.7}Pr_{0.7}Dy_{0.2}Fe_{79.1}Co_{6.3}Ga_{0.4}B_{5.6}$ nanocomposite is analyzed, the distribution exhibits one peak, corresponding to hyperfine field of 22.0 T the highest probability. The average hyperfine magnetic field calculated with this distribution is 22.07(0.72

)T. The results of the least squares fit of both nanocomposites are summarized in Table 1; the values of isomer shift (δ), quadrupole splitting ($\Delta = eQV_{ZZ}/2$), hyperfine internal magnetic field (B_{hf}) and area of each component, are reported. The results of quantitative phase XRD analysis of the nanocomposites $Nd_{13.7}Pr_{0.7}Dy_{0.2}Fe_{73.1}Co_{6.3}Ga_{0.4}B_{5.6}$ and $Nd_{7.7}Pr_{0.7}Dy_{0.2}Fe_{79.1}Co_{6.3}Ga_{0.4}B_{5.6}$ in the optimized magnetic state (Fig. 1, Table 1) suggest that the alloys are characterized by the dominant fractional amount of hard magnetic $Nd_2Fe_{14}B$ phase (60.2 % and 63.0 %). The presence of other soft and paramagnetic phases such as α-Fe, Fe_3B and $Nd_{1.1}Fe_4B_4$ were determined as well.

In addition, according to the ternary Nd-Fe-B phase diagram, this magnet contains a certain amount of $Nd_{1.1}Fe_4B_4$ and an Nd-rich phase which is essential for further hot processing.

Table 1. Selected fit of Mössbauer parameters for $Nd_{13.7}Pr_{0.7}Dy_{0.2}Fe_{73.1}Co_{6.3}Ga_{0.4}B_{5.6}$ and $Nd_{7.7}Pr_{0.7}Dy_{0.2}Fe_{79.1}Co_{6.3}Ga_{0.4}B_{5.6}$ nanocomposites.

Sample	Phase	Site	Site area [%]	Phase area [%]	I.S. [mm/s]	Magnitude of splitting Δ [mm/s]	B_{hf} [T]	Γ [mm/s]
$Nd_{13.7}Pr_{0.7}Dy_{0.2}$ $Fe_{73.1}Co_{6.3}Ga_{0.4}B_{5.6}$	$(Nd,Pr,Dy)_2$ $(Fe,Co,Ga)_{14}$ B	16k2	9.1(8)		-0.10(1)	0.14(2)	30.0(1)	0.38(5)
		16k1	9.1(8)		-0.02(1)	0.14(2)	28.4(2)	0.33(2)
		8j2	4.6(4)		0.07(3)	0.59(4)	33.2(3)	0.37(3)
		8j1	4.6(4)		-0.26(2)	0.01(2)	27.8(1)	0.29(6)
		4c	2.3(2)		-0.11(3)	0.03(5)	31.2(2)	0.26(5)
		4e	2.3(2)	32.0	-0.09(4)	0.14(5)	25.0(2)	0.28(4)
	$(Fe,Co,Ga)_3B$	8g1	4.1(7)		0.41(5)	0.15(7)	28.9(3)	0.39(6)
		8g2	5.8(9)		-0.05(2)	0.17(6)	27.3(3)	0.32(9)
		8g3	6.2(9)	16.1	0.10(2)	-0.06(4)	23.3(1)	0.52(7)
	Fe – α	2a	13.1(9)	13.1	0.039(7)	-0.01(2)	33.77(3)	0.43(2)
			1.1(2)	1.1	0.23(2)	1.24(5)		0.26(3)
					Distributions:			
						< B_{hf} >	SD [T]	< I.S. >
	$(Fe,Co,Ga)_2B$	8h	34.6	34.6	23.22	3.19	-0.06(3)	
	$(Fe,Co,Ga)B$	4c	3.1	3.1	12.64	1.91	-0.14(9)	
$Nd_{7.7}Pr_{0.7}Dy_{0.2}$ $Fe_{79.1}Co_{6.3}Ga_{0.4}B_{5.6}$	$(Nd,Pr,Dy)_2$ $(Fe,Co,Ga)_{14}$ B	16k2	11.2(7)		-0.08(1)	0.00(1)	29.18(7)	0.38(2)
		16k1	11.2(7)		-0.080(6)	0.19(2)	25.91(7)	0.36(2)
		8j2	5.6(4)		0.20(1)	0.10(3)	32.78(8)	0.37(3)
		8j1	5.6(4)		-0.086(9)	-0.01(1)	27.77(8)	0.27(2)
		4c	2.8(2)		-0.24(2)	-0.51(4)	31.1(1)	0.33(4)
		4e	2.8(2)	39.2	-0.04(1)	0.11(3)	19.31(8)	0.24(2)
	$(Fe,Co,Ga)_3B$	8g1	12(1)		0.361(9)	0.25(2)	27.36(8)	0.41(3)
		8g2	16(2)		-0.111(7)	-0.18(1)	26.00(8)	0.42(3)
		8g3	17(2)	45.0	-0.028(6)	-0.16(2)	22.23(8)	0.50(2)
	Fe – α	2a	6.5(1)	6.5	-0.03(2)	0.02(3)	32.8(1)	0.44(6)
	$(Fe,Co,Ga)B$	4c	2.0(3)	2.0	0.30(1)	-0.56(3)	10.6(1)	0.25(4)
			1.5(2)	1.5	0.07(2)	1.37(4)		0.32(6)
					Distribution:			
						< B_{hf} >	SD [T]	< I.S. >
	$(Fe,Co,Ga)_2B$	8h	5.8	5.8	22.07	0.72	0.06(3)	

Powder Metallurgy and Advanced Materials – RoPM&AM 2017   Materials Research Forum LLC
Materials Research Proceedings **8** (2018) 70-79   doi: http://dx.doi.org/10.21741/9781945291999-8

The appearance of non-ferromagnetic boron rich $Nd_{1.1}Fe_4B_4$ phase, can be explained as a consequence of high boron content in the investigated alloys (above 4.2 at. %) [14]. It was found that $Nd_{1.1}Fe_4B_4$ phase forms in non-uniformly distributed heavily faulted grains of approximately the same dimensions as grains of $Nd_2Fe_{14}B$ phase [15]. It has been observed that, during recrystallization of the amorphous studied alloys after annealing at different temperatures, appears the phenomena of separation and decomposition of the B-rich phase, $Nd_{1.1}Fe_4B_4$, and the addition of Co enhanced these phenomena, increasing the content of the B-rich phase. Further, the B-rich phase dilutes the inter-grain exchange interaction resulting in a decrease of the coercivity for magnets. The $Nd_{1.1}Fe_4B_4$ phase has a very low Curie temperature ($T_C = 13$ K) and the magnetic properties of the magnets are drastically damaged [16]. For example, the presence of this B-rich phase in the composition of NdFeB magnetic nanocomposites has as result the decreasing of coercivity, through the dilution exchange interaction between the grains. The obtained results are in accordance with data reported by the literature [18]. The relative contents of the paramagnetic phase in both nanocomposites are 1.6% and 0.9 % (Table 1).

According to both MS and XRD results, the appearance of soft magnetic phase α-Fe of 8.5% in $Nd_{13.7}Pr_{0.7}Dy_{0.2}Fe_{73.1}Co_{6.3}Ga_{0.4}B_{5.6}$ and 5.8% in $Nd_{7.7}Pr_{0.7}Dy_{0.2}Fe_{79.1}Co_{6.3}Ga_{0.4}B_{5.6}$ is accompanied predominantly of the body centered tetragonal Fe_3B structure. These ribbons consist of a remarkable fractional amount of Fe_3B, 29.7 % and 30.3 % respectively.

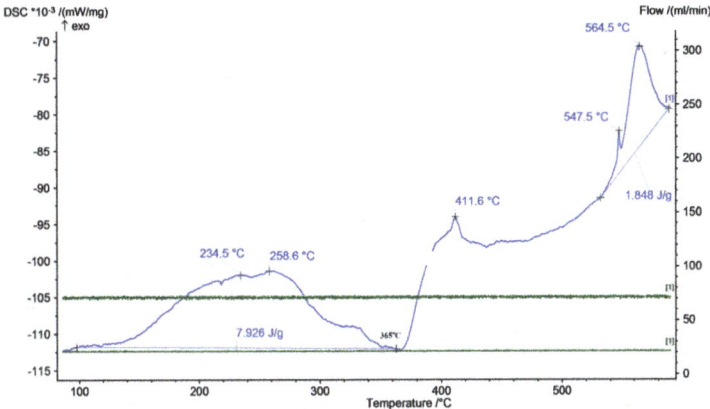

Fig. 3. The DSC curve of the $Nd_{13.7}Pr_{0.7}Dy_{0.2}Fe_{73.1}Co_{6.3}Ga_{0.4}B_{5.6}$ sample.

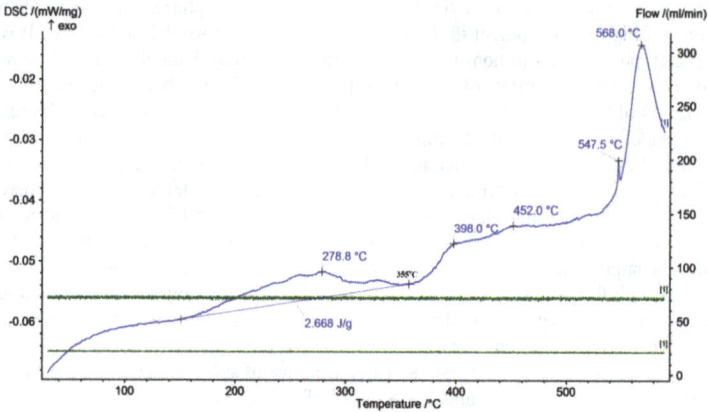

Fig. 4. The DSC curve of the $Nd_{7.7}Pr_{0.7}Dy_{0.2}Fe_{79.1}Co_{6.3}Ga_{0.4}B_{5.6}$ sample.

Samples from the ribbons prepared by melt-spinning have been investigated through DSC to monitor the thermal induced structural changes, the plotted curves being illustrated in Fig. 3 and 4. The experimental data obtained from DSC curves are in accordance with those extracted from the thermomagnetic measurements (see Fig. 5 and 6). The temperature increasing has as main effect in samples the ordering of atoms in the crystalline structure on intermetallic compounds, the process being marked by ascending curve, both in DSC investigation and in thermomagnetic measurements. The Curie point of the hard magnetic $Nd_2Fe_{14}B$ phase is represented on DSC curve trough minimum and on the thermomagnetic variations through sharp decreasing of the magnetisation.

The substitution of Fe by Co in the hard-magnetic phase is highlighted by the values of the Curie temperatures determined for the $Nd_{13.7}Pr_{0.7}Dy_{0.2}Fe_{73.1}Co_{6.3}Ga_{0.4}B_{5.6}$ and $Nd_{7.7}Pr_{0.7}Dy_{0.2}Fe_{79.1}Co_{6.3}Ga_{0.4}B_5$ samples: 365°C respectively 355°C, greater than 312°C, the Curie temperature of the $Nd_2Fe_{14}B$ compound. The crystallisation temperatures of soft magnetic phases existing in the nanocomposites are situated at 564.5°C, respectively 568°C (Fig. 3 and 4). To assure the recrystallisation also for hard magnetic phases (which take place at temperatures greater than 590°C), the annealing process was conducted at 715°C.

In order to highlight the occurrence of the exchange interaction between the soft and the hard magnetic phases from the (Nd,Pr,Dy)-(Fe,Co,Ga)-B studied nanocomposites, was determined the value of the remanent and saturation magnetisation ratio M_r/M_S, whose values, extracted from the plotted hysteresis curves, are presented in Table 2.

It can be seen that in the annealed state the soft and the magnetic phases are exchange coupled, the value of the M_r/M_S ratio being greater than 0.5.

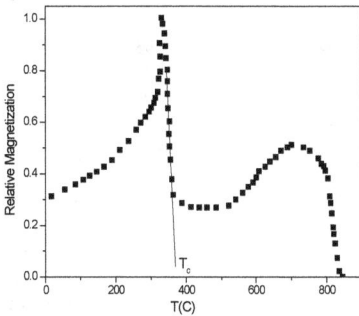

Fig. 5. Relative magnetization *vs* temperature for $Nd_{13.7}Pr_{0.7}Dy_{0.2}Fe_{73.1}Co_{6.3}Ga_{0.4}B_{5.6}$ sample.

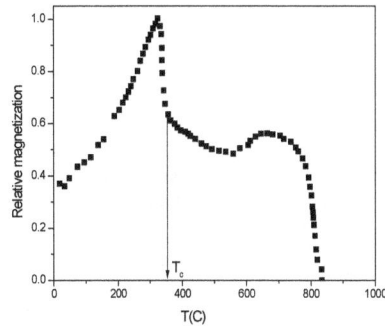

Fig. 6. Relative magnetization *vs* temperature for $Nd_{7.7}Pr_{0.7}Dy_{0.2}Fe_{79.1}Co_{6.3}Ga_{0.4}B_{5.6}$ sample.

Table 2. The main magnetic characteristics of (Nd,Pr,Dy)-(Fe,Co,Ga)-B samples.

Sample	M_s [emu/g]	M_r [emu/g]	H_c [kA/m]	M_r/M_S ratio
$Nd_{13.7}Pr_{0.7}Dy_{0.2}Fe_{73.1}Co_{6.3}Ga_{0.4}B_{5.6}$ – melt spun	55.55	104.23	207	0.52
$Nd_{13.7}Pr_{0.7}Dy_{0.2}Fe_{73.1}Co_{6.3}Ga_{0.4}B_{5.6}$ - annealed	45.43	76.35	557	0.59
$Nd_{7.7}Pr_{0.7}Dy_{0.2}Fe_{79.1}Co_{6.3}Ga_{0.4}B_{5.6}$ – melt spun	82.36	187.25	175	0.44
$Nd_{7.7}Pr_{0.7}Dy_{0.2}Fe_{79.1}Co_{6.3}Ga_{0.4}B_{5.6}$ – annealed	85.23	136.56	233	0.62

In the case of $Nd_{13.7}Pr_{0.7}Dy_{0.2}Fe_{73.1}Co_{6.3}Ga_{0.4}B_{5.6}$ nanocomposites, the coercitive field is developed after the annealing treatment at 715°C for 4 min., increasing from 207 kA/m to 557 kA/m, but this coercivity increasing is accompanied by a remanent magnetization decreasing, from 55.55 emu/g to 45.43 emu/g. In the case of $Nd_{7.7}Pr_{0.7}Dy_{0.2}Fe_{79.1}Co_{6.3}Ga_{0.4}B_{5.6}$ nanocomposites, the annealing at 715°C for 4 min leads to a slight increase, both in remanent magnetisation and in coercivity, but the increased value of 0.62 for M_r/M_S ratio proves that a greater fraction of soft and hard magnetic phases is exchanged coupled.

Summary
Magnetic microstructure of melt-spun (Nd,Pr,Dy)-(Fe,Co,Ga)-B ribbons is very sensitive to variable Nd and Fe contents in comparison with previously published Mössbauer spectroscopic studies of $Nd_2Fe_{14}B$, where average internal magnetic field $<B_{hf}>$ at 295K is estimated as 30.4 T [18,19]. There is a close agreement of $<B_{hf}>$ at 294 K, estimated as 29.4 T and 28.0 T for $Nd_2Fe_{14}B$, dominantly present in $Nd_{13.7}Pr_{0.7}Dy_{0.2}Fe_{73.1}Co_{6.3}Ga_{0.4}B_{5.6}$ and $Nd_{7.7}Pr_{0.7}Dy_{0.2}Fe_{79.1}Co_{6.3}Ga_{0.4}B_{5.6}$ nanocomposites, respectively. It is worthy to note that the volume fraction of Fe₃B remains almost same and the formation of mostly formed tetragonal Fe_3B contributes to the enhancement of total magnetization of the nanocomposite magnets [20]. The $Nd_{1.1}Fe_4B_4$ doublets are with an evident change of quadruple splitting 1.24(5) mm/s and 1.37(4) mm/s, FWHM of about 0.26(3) mm/s ($Nd_{13.7}Pr_{0.7}Dy_{0.2}Fe_{73.1}Co_{6.3}Ga_{0.4}B_{5.6}$) and 0.32(6) mm/s ($Nd_{7.7}Pr_{0.7}Dy_{0.2}Fe_{79.1}Co_{6.3}Ga_{0.4}B_{5.6}$) and an almost constant relative concentration for both nanocomposites, respectively. So, they correspond to a distinct phase which becomes

Powder Metallurgy and Advanced Materials – RoPM&AM 2017
Materials Research Forum LLC
Materials Research Proceedings 8 (2018) 70-79
doi: http://dx.doi.org/10.21741/9781945291999-8

paramagnetic at room temperature. The α-Fe fraction decrease from 13.1(9)% in $Nd_{13.7}Pr_{0.7}Dy_{0.2}Fe_{73.1}Co_{6.3}Ga_{0.4}B_{5.6}$ to 6.5(1)% in $Nd_{7.7}Pr_{0.7}Dy_{0.2}Fe_{79.1}Co_{6.3}Ga_{0.4}B_5$ has influence on the sharp decrease of the magnetic performances: reduced remanence and intrinsic coercitivity [21,22]. The substitution of Fe by Co in the hard magnetic phase is highlighted by the values of the Curie temperatures determined for the $Nd_{13.7}Pr_{0.7}Dy_{0.2}Fe_{73.1}Co_{6.3}Ga_{0.4}B_{5.6}$ and $Nd_{7.7}Pr_{0.7}Dy_{0.2}Fe_{79.1}Co_{6.3}Ga_{0.4}B_5$ samples: 365°C respectively 355°C, greater than 312°C, the Curie temperature of the $Nd_2Fe_{14}B$ compound. The crystallisation temperatures of the soft magnetic phases existing in the nanocomposites are situated at 564.5°C, respectively 568°C $Nd_{13.7}Pr_{0.7}Dy_{0.2}Fe_{73.1}Co_{6.3}Ga_{0.4}B_{5.6}$ and $Nd_{7.7}Pr_{0.7}Dy_{0.2}Fe_{79.1}Co_{6.3}Ga_{0.4}B_{5.6}$ melt spun samples. The magnetic properties, measured at room temperature on the quenched and annealed ribbons, revealed the relationship between the alloy chemical composition and processing. The parameters of the recrystallization process could be carefully chosen in the case of the studied nanocomposites, in order to lead to an improvement of the magnetic performances of the as-quenched NdFeB-based alloys. Analysis of experimental results enabled better insight in relationship between microstructure and magnetic properties of nanocomposites, function of variable percentage fraction of iron.

Acknowledgement

This work has been supported by grant No. 171001G from Serbian Ministry of Science and Technological Development, by Romanian grant 14N 5201/2016 and Romania–JINR Dubna project No. 47/2015, from Romanian Ministry of Research-Development and Innovation.

References

[1]   K. J. Strnat, G. Hoffer, J. Oson, W. Ostertag, A family of new cobalt-base permanent magnet materials, J. Appl. Phys. 38 (1967) 1001-1002. https://doi.org/10.1063/1.1709459

[2]   M. Sagawa, S. Fujimura, N. Togawa, H. Yamamoto, Y. Matsuura, New material for permanent magnets on a base of Nd and Fe, J. Appl. Phys. 55 (1984) 2083-2087. https://doi.org/10.1063/1.333572

[3]   J.J. Croat, J.F. Herbst, R.W. Lee, F.E. Pinkerton, Pr-Fe and Nd-Fe-based materials: A new class of high-performance permanent magnets, J. Appl. Phys. 55 (1984) 2078–2080. https://doi.org/10.1063/1.333571

[4]   F.E. Pinkerton, W.R. Dunham, Mössbauer effect in $R_2Fe_{14}B$ compounds, J. Appl. Phys. 57 (1985) 4121-4123. https://doi.org/10.1063/1.334638

[5]   W. Kappel, M.M. Codescu, D. Popa, Losses in sintered NdFeB magnets, Rom. Rep. Phys. 56 (2004) 391–398.

[6]   W. Kappel, M.M.Codescu, N. Stancu, D. Popa, Evaluation of the corrosion behavior of the permanent magnets based on rare earths, used in aeronautical industry, J. Optoelectron. Adv. Mat. 8 (2006) 523–526.

[7]   M.M. Codescu, W. Kappel, D. Popa, Corrosion tests on alloys and permanent magnets based on NdFeB, used in aerospace industry, J. Optoelectron. Adv. Mat. 10 (2008) 790-793.

[8]   H.A. Davies, A. Manaf, P.Z. Zhang, Nanocrystallinity and magnetic property enhancement in melt-spun iron-rare earth-base hard magnetic alloys, J. Mater. Eng. Perform. 2 (1993) 579-587. https://doi.org/10.1007/BF02661744

Powder Metallurgy and Advanced Materials – RoPM&AM 2017   Materials Research Forum LLC
Materials Research Proceedings 8 (2018) 70-79   doi: http://dx.doi.org/10.21741/9781945291999-8

[9]   G.C. Hadjipanayis, Nanophase hard magnets, J. Magn. Magn. Mater. 200 (1999) 373-391. https://doi.org/10.1016/S0304-8853(99)00430-8

[10]   J.F. Herbst, $R_2Fe_{14}B$ materials: Intrinsic properties and technological aspects, Rev. Mod. Phys. 63 (1991) 819-898. https://doi.org/10.1103/RevModPhys.63.819

[11]   I.R. Betancourt, H.A. Davies, Magnetic properties of nanocrystalline didymium (Nd-Pr)-Fe-B alloys, J. Appl. Phys. 85 (1999) 5911-5913. https://doi.org/10.1063/1.369911

[12]   Z.C. Wang, M.C. Zhang, F.B. Li, S.Z. Zhou, R. Wang, W. Gong, High-coercivity $(NdDy)_2(FeNb)_{14}B$–α–Fe nanocrystalline alloys, J. Appl. Phys. 81 (1997) 5097-5099. https://doi.org/10.1063/1.365187

[13]   J.M. Yao, T.S. Chin, Coercivity of Ti-modified (α-Fe)-$Nd_2Fe_{14}B$ nanocrystalline alloys, J. Appl. Phys. 76 (1994) 7071-7073. https://doi.org/10.1063/1.358030

[14]   A. Inoue, A. Kojima, A. Takeuchi, T. Masumoto, A. Makino, Hard and soft magnetic properties of nanocrystalline Fe–Nd–Zr–B alloys containing intergranular amorphous phase, J. Appl. Phys. 79 (1996) 4836-4836. https://doi.org/10.1063/1.361624

[15]   I. Panagiatopoulos, L. Withanawasam, A. S. Murthy, G.S. Hadjipanayis, E.W. Singleton, D.J. Sellmyer, Magnetic hardening of melt-spun nanocomposite Nd2Fe14B/Fe magnets, J. Appl. Phys. 79 (1996) 4827-4829. https://doi.org/10.1063/1.361621

[16]   T. Schrefl, J. Fidler, Födermayr, Modelling of exchange-spring permanent magnets, J. Magn. Magn. Mater. 177-181 (1998) 970-975. https://doi.org/10.1016/S0304-8853(97)00653-7

[17]   R.A. Brand, (2008) WinNormos Mössbauer fitting program, Universität Duisburg.

[18]   C. You, X.K. Sun, W. Liu, B. Cui, X. Zhao, D. Geng, Z. Zhang, Effects of W and Co additions on the phase transformation and magnetic properties of nanocomposite $Nd_2Fe_{14}B/\alpha$-Fe magnets, J. Phys. D: Appl. Phys. 35 (2002) 943-950. https://doi.org/10.1088/0022-3727/35/10/301

[19]   B. Cekič, V. Ivanovski, M.M. Codescu, A. Umicevic, T. Barudzija, E.A. Patroi, Mössbauer Spectroscopic Analysis of $Nd_2Fe_{14}B/\alpha$-Fe Hard Magnetic Nanocomposites, Sol. St. Phen. 170 (2011) 154-159. https://doi.org/10.4028/www.scientific.net/SSP.170.154

[20]   R. Kamal, Y. Andersson, Mössbauer spectroscopic studies of $Nd_2Fe_{14}B$, Phys. Rev. B 32 (1985) 1756-1760. https://doi.org/10.1103/PhysRevB.32.1756

[21]   W. Kappel, M.M. Codescu, $Nd_2Fe_{14}B/\alpha$-Fe Hard Magnetic Nanocomposite. Performances and Limits, Rom. J. Phys. 49 (2004) 733-741.

[22]   W. Kappel, M.M. Codescu, M. Valeanu, N. Stancu, J. Pintea, F. Lifei, A. Jianu, D. Patroi, E. Patroi, Influence of the recrystallization processes on the structure and magnetic properties of the $Nd_2Fe_{14}B$/alpha-Fe nanocomposites, J. Optoelectron. Adv. Mat. 9 (2007) 1825-1828.

Powder Metallurgy and Advanced Materials – RoPM&AM 2017   Materials Research Forum LLC
Materials Research Proceedings **8** (2018) 88-88   doi: http://dx.doi.org/10.21741/9781945291999-9

Caracterisation of high manganese silicides prepared by mechanical milling

Victor CEBOTARI[1, a], Florin POPA[1, b], Traian Florin MARINCA[1, c], Violeta POPESCU[2, d], Ionel CHICINAȘ[1, e,*]

[1]Materials Science and Engineering Department, Technical University of Cluj-Napoca, 103-105, Muncii Avenue, 400641 Cluj-Napoca, Romania

[2]Physics and Chemistry Department, Technical University of Cluj-Napoca, 103-105, Muncii Avenue, 400641 Cluj-Napoca, România

[a]Victor.Cebotari@stm.utcluj.ro, [b]Florin.Popa@stm.utcluj.ro, [c]Traian.Marinca@stm.utcluj.ro, [d]Violeta.Popescu@chem.utcluj.ro, [e]Ionel.Chicinas@stm.utcluj.ro

Keywords: Thermoelectric, Higher Manganese Silicide, MnSi phase, Mechanical Alloying.

Abstract. The mechanical milling of manganese and silicon powder in a planetary ball mill up to 18 h was performed. In the X-ray diffraction pattern recorded after 18 hours of milling the MnSi phase and $Mn_{15}Si_{26}$ compound are detected. The agglomeration of powders after complete reaction of the elements was observed by scanning electron microscopy. Heating up at 1000 °C, an unreacted sample, milled 4 hours, has found to have the effect of completing the reaction of elements, but forms oxides. Handling of the powder during sampling, without protective atmosphere was found to form oxides. The oxidation of the samples was evidenced by FTIR analysis.

Introduction

The modern society has the tendency to increase the quantity of hydrocarbons which are transformed into energy, with negative effects on the environment. To reduce this impact alternatives are searched. Thermoelectric materials represent a solution to improve the quality of the environment by reducing the combustion product gases. These materials are able to convert the thermal energy directly into electrical energy and vice versa. The quality of a thermoelectric material can be estimated by the figure of merit $ZT=S^2\sigma T/k$ where: S is the Seebeck coefficient, σ is electrical conductivity, T is temperature and k is thermal conductivity [1]. Thermoelectric materials can convert heat from a different source such as solar heat, geothermal heat or exhaust gases [2]. From the studied thermoelectric materials, those based on silicon, especially High Manganese Silicide (HMS) is friendly with the environment and considered as promising candidates. HMS is chemically stable [3] and are preferred in detriment of those based on Pb-Te which operate in the same range of temperature. The HMS materials are nontoxic as well as their constituent chemical elements [4].

HMS are thermoelectric compounds with p-type conduction, having general formula $MnSi_x$ where the x value ranges from 1.67 up to 1.87 [5] and with an energy gap of 0.77 [eV] [6]. HMS system contains four compounds, Mn_4Si_7, $Mn_{11}Si_{19}$, $Mn_{15}Si_{26}$, and $Mn_{27}Si_{47}$, all with the same electronic structure [7]. Crystallographic structure of HMS compounds belongs to Nowotny chimney ladder (NCL) phases, where manganese is located in the corners of tetragon and silicon are arranged inside in the form of a spiral [8]. $MnSi_{1.75}$ compound presents the largest ZT, while the $MnSi_{1.77}$ compound has the smallest value. The low value for the figure of merit is the effect of

Powder Metallurgy and Advanced Materials – RoPM&AM 2017   Materials Research Forum LLC
Materials Research Proceedings 8 (2018) 88-88   doi: http://dx.doi.org/10.21741/9781945291999-9

a large thermal conductivity [6]. Problem with HMS is that obtaining method influence the final phase. Based on the preparation method it is possible to obtain different compounds: by vacuum levitation melting $Mn_{15}Si_{26}$ is obtained, Mn_4Si_7 may obtain by vacuum levitation-induction melting and by dry milling [9-11]. Preparation by melting leads to an inhomogeneous structure and coarse microstructure [12]. The obtaining by mechanical alloying has the advantage of obtaining a small crystallite size which leads to lower thermal conductivity [11]. Also, dry milling leads to the decrease of the quantity of MnSi secondary phase, which reduces the thermoelectric proprieties.

In the milling experiments, using different process control agents (PCA) it is possible to control the MnSi phase. Hexane conducts to the formation of 38.8% of MnSi phase, acetone to 8.7% and ethanol to 5.3%. Milling without any PCA leads to the formation of 49.5% MnSi/HMS phase [10]. In order to obtain the proper HMS, the conditions can be summarized to be small milling time and high rotation speed according to [6, 13]. Prolonged milling conducts to the decomposition of HMS compound in MnSi phase as a result of the excess energy which is generated by collisions [14].

The increase in the thermoelectric properties can be achieved by doping. Adding Yb, the carrier concentration increases, and MnSi phase quantity decreases [14]. By doping with Co a homogenous microstructure is obtained and the ZT increases proportionally to the concentration of Co [12]. Other chemical elements that are studied for increasing the thermoelectric proprieties are Cr, Ti, Fe, Al, and Ge. The doping increases the thermoelectric proprieties only if the concentration of elements does not exceed the limit of solubility because the doped elements are located at Mn sites [15-18].

The present paper is focused on the synthesis of HMS with the chemical composition MnSi1.75. The formation of this compound by mechanical milling is studied as a function of the milling time. The paper presents the evolution of the powder morphology and the distribution of the chemical elements in the samples after milling. The thermal stability of powders is also presented and discussed.

Experimental

The thermoelectric material has been obtained starting from elemental powders of manganese with purity of 99.3% (-325 mesh, Alfa Aesar) and silicon with purity of 99.9% (-100 mesh, Alfa Aesar), in a stoichiometric ratio corresponding to $MnSi_{1.75}$ compound formula. The powder mixture was loaded into the vial with grinding media after the prior homogenisation of the elemental powders. The mechanical milling was made in a planetary ball mill Fritsch Pulverisette 6 using a ball to powder mass ratio (BPR) of 10:1, and a 400 rpm rotational speed of vial. The vial and balls are of stainless steel with a diameter of 14 mm. For the protection of powders which are subjected to milling process from oxidation, the milling was done under argon atmosphere. The milling was conducted up to 18 hours, and sampling was done after the following milling times: 0, 1, 2, 4, 6, 8, 10, 14 and 18 hours.

The structural study and phases composition of the samples were investigated by X-ray diffraction using an INEL 3000 Equinox diffractometer using Kα radiation of Co ($\lambda = 1.79026$ Å). To study the morphology of the powders and the local chemical homogeneity a JEOL- JSM 5600 LV electron microscope equipped with EDX spectrometer (Oxford Instruments, INCA 2000 soft) was used. The thermal stability of the samples was investigated by differential scanning calorimetry (DSC), using a LabSys-Setaram apparatus. The DSC investigations were performed in an argon atmosphere, up to 1000 °C, with heating/cooling rate of 10 °C/min using alumina as a reference sample. The presence of oxide inside the probe was investigated by the Fourier

Powder Metallurgy and Advanced Materials – RoPM&AM 2017   Materials Research Forum LLC
Materials Research Proceedings **8** (2018) 88-88   doi: http://dx.doi.org/10.21741/9781945291999-9

Transform Infrared (FTIR) technique using Spectrum BX II apparatus. The experiment was realised by embedding the Mn-Si powder into the potassium bromide pellet.

Results and discussion

X-ray diffraction patterns of the mechanically milled powders are presented in Fig. 1. In diffraction pattern of the starting sample are identified the peaks corresponding to the used elemental powders. In the diffraction patterns corresponding to the sample milled for one hour is observed a reduction of the peaks intensity and a pronounced broadening. This is assigned to the reduction of crystallite size and an increase of the internal stresses [19]. The sample milled for 2 hours presents similar behavior. A new MnSi phase appears after 4 hours of milling. The formation of the HMS compound begins after 6 hours of mechanical milling. The complete reaction of the elements is observed after 18 hours.

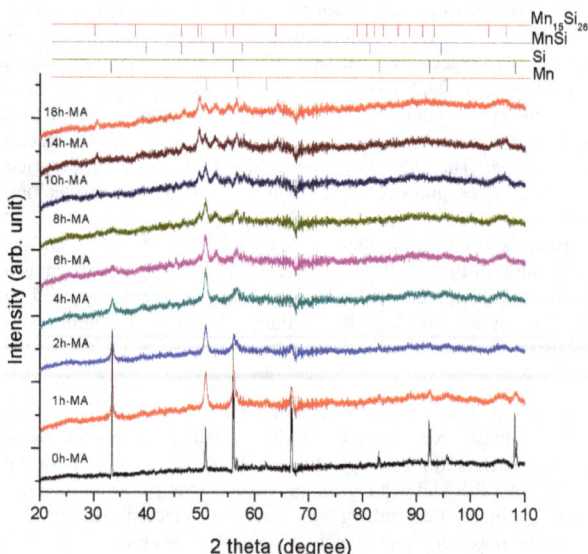

Fig. 1. X-ray diffraction of elemental powder mixture corresponding to chemical composition MnSi1.75 at different milling times.

The evolution of morphology and the distribution map of chemical elements was analyzed and is presented in Fig. 2. The SEM image presented in Fig. 2a is recorded on starting powders mixture. The particles present polyhedral irregular shapes. The distribution map for starting mixture reveals a good homogenisation of particles before loading in the vial for the milling process. A good homogenisation of powders is necessary to reduce as much as possible the silicon deposit on the balls, being more ductile than manganese, this can lead to the increase of the mechanical alloying duration as has been already reported in [20]. The image of powder milled for 4 hours (Figure 2 b), presents agglomeration of powder particles. After 4 hours of milling, manganese is more homogenously distributed as compared with the starting sample, due to initiation of the alloying process by milling. Powder milled 18 hours presents an irregular shape,

Powder Metallurgy and Advanced Materials – RoPM&AM 2017   Materials Research Forum LLC
Materials Research Proceedings **8** (2018) 88-88   doi: http://dx.doi.org/10.21741/9781945291999-9

the dimension of the particles is small than 50 μm and much smaller as compared to particles of
the starting sample. The sample milled for 18 hours shows particles with a size of less than 1 μm
and particles with a size of a few micrometers that are composed of fine particles that are welded
together. After 18 hours of mechanical milling, the distributions maps of manganese and silicon
are uniform.

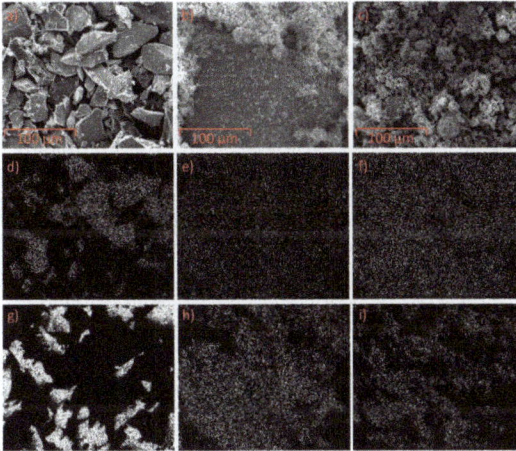

Fig. 2. SEM images for probe milled a) 0h; b) 4h; 18h. Map distribution of manganese is
presented in d, e, f, and for silicon in g, h, i for initial mixture; and 4 hours; and respectively 18
hours.â

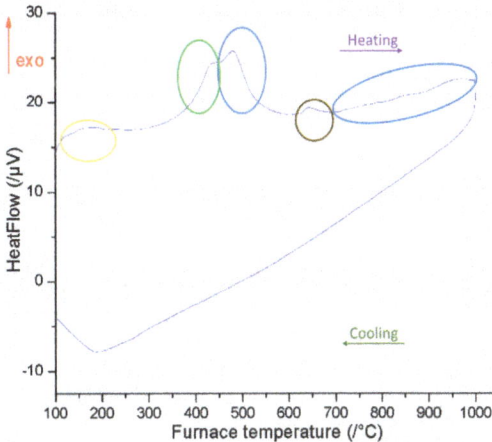

Fig. 3. DSC analysis of the sample milled for 4 hours. Heating was made up to 1000 °C with a
heating rate of 10 °C / min.

Powder Metallurgy and Advanced Materials – RoPM&AM 2017   Materials Research Forum LLC

Materials Research Proceedings **8** (2018) 88-88   doi: http://dx.doi.org/10.21741/9781945291999-9

DSC analyses of the sample milled for 4 hours is shown in Fig. 3. The DSC curve on heating up to 1000 °C present 5 distinct phenomena but the curve on cooling does not present any phenomena. To identify the phase transition after each event, X-ray diffraction was performed after DSC measurements at temperatures corresponding to each event up to 1000 °C. The X-ray diffraction is presented in Fig. 4. The first phenomenon is the stresses release of crystalline structure [21]. The stresses release peak is very broad and has the maximum around 200 °C. The second maximum corresponds to the formation of MnSi phase and has its maximum at 430 °C. The third thermal event corresponds to the formation of the $Mn_{15}Si_{26}$ chemical compound formation. The fourth and the fifth will be explained in the next section.

In the X-ray diffraction pattern corresponding to powders heat treated at 350 °C exhibits in comparison with the as-prepared sample only a peak narrowing. Annealing up to 450 °C conducts to the formation of MnSi phase and a small quantity of $Mn_{15}Si_{26}$ compound. In addition peaks of MnO are observed, indicating the sample oxidation. Heating up the sample at 550 °C, the formation of the $Mn_{15}Si_{26}$ compound is continued. Unfortunately, the oxidation of manganese continues as well, and the MnO further reacts and transforms into Mn_3O_4. By heating up to 700 °C there are no major changes in the samples, compared with the samples heated at 500 °C. While heating at 1000 °C major changes is recorded, the MnO peaks disappear being replaced by Mn_3O_4 peaks. The peaks corresponding to the MnSi phase and that of the $Mn_{15}Si_{26}$ compound grow in intensity and becomes narrower, as the crystallite mean size increases.

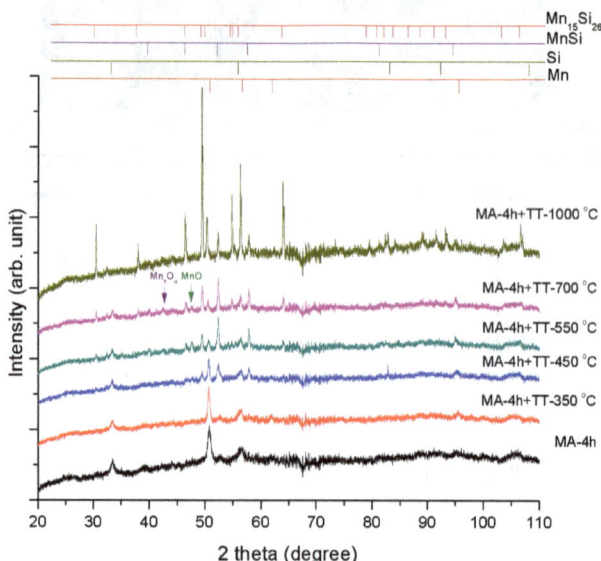

Fig. 4. X-ray diffraction patterns for the 4-hour mechanical alloyed sample and DSC at 350 °C, 450 °C, 550 °C, 700 °C and 1000 °C with a heating rate of 10 °C/min.

The oxidation of the samples can occur during sampling, that was performed in air and powders probably adsorbed oxygen from the atmosphere. To elucidate this assumption the samples were subjected to FTIR analyse. The result of FTIR analyze is presented in Fig. 5. In the FTIR spectrum, several adsorption bands were recorded. It was recorded the adsorption band corresponding to the Mn-O stretching vibration at 655 cm^{-1} [22-25]. At 1029 cm^{-1} was recorded a wide abortion band, attributed to the stretching vibrations of Si-O-Si bond [22, 26], Si-OH [27], Mn-OH. The Mn-OH group has another maximum at 1422 cm^{-1} [28]. The next band, recorded at 1615 cm^{-1}, was attributed to the adsorbed hydroxyl groups (O-H bending mode) [23].

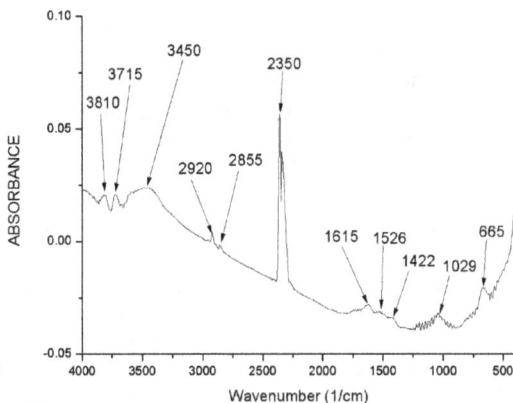

Fig. 5. FTIR spectra for MnSi$_{1.75}$ mechanical alloyed 18 hours.

The bands at 3810 cm^{-1} and 3715 cm^{-1} correspond to asymmetric stretch and symmetric stretch. This fact leads to the idea that molecular water exists in the probe [29]. The most intense maximum (2350 cm^{-1}) correspond to CO_2 [29, 30]. The absorption band in the range of 3400-3600 cm^{-1} corresponds to the O-H stretching vibration, the presence of O-H vibration is possible to appear because the powders absorb the water from the air [22,]. For the upcoming experiments, in order to avoid oxidation, the probes will be drawn in a controlled atmosphere.

Summary
From the preparation of Mn-Si alloy by mechanical milling route in the given condition the following conclusions can be drawn:

1. By solid state reaction of elemental powders, MnSi phase can be obtained, in the first stage, and the Mn$_{15}$Si$_{26}$ compound in the second stage.
2. The time required for complete reaction of the elements on the Fritsch Pulverisette 6 planetary mill with the selected parameters is 18 hours.
3. In the first hours of mechanical alloying, a fragmentation of manganese particles is observed, in which the most fragile component is.
4. After complete reaction, a powder of fine particle size is obtained.
5. The thermal treatment of the milled but unreacted powder leads to the complete reaction of the elements, but for the alloyed powder, the heat treatment only has the effect of increasing the grain size and stress release.

Powder Metallurgy and Advanced Materials – RoPM&AM 2017   Materials Research Forum LLC
Materials Research Proceedings **8** (2018) 88-88   doi: http://dx.doi.org/10.21741/9781945291999-9

References

[1] M. S. Dresselhaus, G. Chen, M. Y. Tang, R. Yang, H. Lee, D. Wang, Z. Ren, J. P. Fleurial, P. Gogna, New directions for low-dimensional thermoelectric materials, Adv. Mater. 19 (2007) 1043–1053. https://doi.org/10.1002/adma.200600527

[2] Z. Du, T. Zhu, X. Zhao, Enhanced thermoelectric properties of $Mg_2Si_{0.58}Sn_{0.42}$ compounds by Bi doping, Mater. Letter. 66 (2012) 76–78. https://doi.org/10.1016/j.matlet.2011.08.031

[3] T. Yamada, Y. Miyazaki, H. Yamane, Preparation of Higher Manganese Silicide (HMS) bulk and Fe-containing HMS bulk using a Na–Si melt and their thermoelectrical properties, Thin Solid Films 519 (2011) 8524–8527. https://doi.org/10.1016/j.tsf.2011.05.032

[4] X. Hu, D. Mayson, M. R. Barnett, Synthesis of Mg_2Si for thermoelectric applications using magnesium alloy and spark plasma sintering, J All. Compd. 589 (2014) 485–490. https://doi.org/10.1016/j.jallcom.2013.11.092

[5] X. Chen, L. Shi, J. Zhou, J. B. Goodenough, Effects of ball milling on microstructures and thermoelectric properties of higher manganese silicides, J All. Compd. 641 (2015) 30–36. https://doi.org/10.1016/j.jallcom.2015.04.048

[6] Z. Zamanipoura, X. Shia, M. Mozafari, J. S. Krasinski, L. Tayebib, D. Vashaee, Synthesis, characterization, and thermoelectric properties of nanostructured bulk p-type $MnSi_{1.73}$, $MnSi_{1.75}$, and $MnSi_{1.77}$, Ceram. Int. 39 (2013) 2353–2358. https://doi.org/10.1016/j.ceramint.2012.08.086

[7] D. B. Migas, V. L. Shaposhnikov, A. B. Filonov, and V. E. Borisenko, Ab initio study of the band structures of different phases of higher manganese silicides, Phys. Rev. B 77 (2008) 075205. https://doi.org/10.1103/PhysRevB.77.075205

[8] H. Lee, G. Kim, B. Lee, J. Kim, S. M. Choi, K. H. Lee, W. Lee, Effect of Si content on the thermoelectric transport properties of Ge-doped higher manganese silicides, Scripta Mater. 135 (2017) 72–75. https://doi.org/10.1016/j.scriptamat.2017.03.011

[9] A.J. Zhou, X.B. Zhao, T.J. Zhu, Y.Q. Cao, C. Stiewe, R. Hassdorf, E. Mueller, Composites of higher manganese silicides and nanostructured secondary phases and their thermoelectric properties, J Electron Mater. 38 (2009) 1072-1077. https://doi.org/10.1007/s11664-009-0774-7

[10]A.J. Zhou, X.B. Zhao, T.J. Zhu, T. Dasgupta, C. Stiewe, R. Hassdorf, E. Mueller, Mechanochemical decomposition of higher manganese silicides in the ball milling process, Intermetallics 18 (2010) 2051-2056. https://doi.org/10.1016/j.intermet.2010.06.008

[11]D.Y.N. Truong, H. Kleinke, F. Gascoin, Preparation of pure Higher Manganese Silicides through wet ball milling and reactive sintering with enhanced thermoelectric properties, Intermetallics 66 (2015) 127-132. https://doi.org/10.1016/j.intermet.2015.07.002

[12]T. Itoh, S. Uebayashi, Cobalt an iron doping effect on thermo-electric properties of higher manganese silicides prepared by mechanical milling and pulse discharge sintering, J Jpn. Soc. Powder Powder Metall. 63 (2016) 491-496. https://doi.org/10.2497/jjspm.63.491

[13]D. K. Shin, K. W. Jang, S. C. Ur, I. H. Kim, Thermoelectric properties of higher manganese silicides prepared by mechanical alloying and hot pressing, J Electron Mater. 42 (2013) 1756-1761. https://doi.org/10.1007/s11664-012-2415-9

[14]M. Saleemi, A. Famengo, S. Fiameni, S. Boldrini, S. Battiston, M. Johnsson, M. Muhammed, M.S. Toprak, Thermoelectric performance of higher manganese silicide nanocomposites Journal of Alloys and Compounds 619 (2015) 31–37. https://doi.org/10.1016/j.jallcom.2014.09.016

Powder Metallurgy and Advanced Materials – RoPM&AM 2017   Materials Research Forum LLC
Materials Research Proceedings 8 (2018) 88-88   doi: http://dx.doi.org/10.21741/9781945291999-9

[15] V. Ponnambalam, D. T. Morelli, S. Bhattacharya, T. M. Tritt The role of simultaneous substitution of Cr and Ru on the thermoelectric properties of defect manganese silicides $MnSi_\delta$ (1.73 < δ < 1.75), J. All. Compd. 580 (2013) 598–603. https://doi.org/10.1016/j.jallcom.2013.07.136

[16] L.D. Ivanova, Preparation of thermoelectric materials based on higher manganese silicide, Inorganic Mater. 47 (2011) 965–970. https://doi.org/10.1134/S002016851109010X

[17] W. Luo, H. Li, F. Fu, W. Hao, X. Tang Improved thermoelectric properties of Al-doped higher manganese silicide prepared by a rapid solidification method, J. Electron. Mater. 40 (2011) 1233-1237. https://doi.org/10.1007/s11664-011-1612-2

[18] A.J. Zhou, T.J. Zhu, X.B. Zhao, S.H. Yang, T. Dasgupta, C. Stiewe, R. Hassdorf, E. Mueller, Improved thermoelectric performance of Higher Manganese Silicides with Ge additions J. Electron. Mater. 39 (2010) 2002-2007. https://doi.org/10.1007/s11664-009-1034-6

[19] T.F. Marinca, H.F. Chicinas, B.V. Neamt, O. Isnard, P. Pascuta, N. Lupu, G. Stoian, I. Chicinas, Mechanosynthesis, structural, thermal and magnetic characteristics of oleic acid coated Fe_3O_4 nanoparticles, Mater. Chem. Phys. 171 (2016) 336-345. https://doi.org/10.1016/j.matchemphys.2016.01.025

[20] T. Itoh, M. Yamada, Synthesis of thermoelectric Manganese Silicide by mechanical alloying and pulse discharge sintering J. Electron. Mater. 38 (2009) 925-929. https://doi.org/10.1007/s11664-009-0697-3

[21] T.F. Marinca, B.V. Neamtu, F. Popa, V.F. Tarta, P. Pascuta, A.F. Takacs, I. Chicinas, Synthesis and characterization of the $NiFe_2O_4/Ni_3Fe$ nanocomposite powder and compacts obtained by mechanical milling and spark plasma sintering, Appl. Surf. Sci. 285P (2013) 2-9. https://doi.org/10.1016/j.apsusc.2013.07.145

[22] A. Belhadi, L. Boudjellal, S. Boumaza, M. Trari. Hydrogen production over the hetero-junction MnO_2/SiO_2. Int. J Hydrogen Energy 43 (2017) 3418-3423. https://doi.org/10.1016/j.ijhydene.2017.06.086

[23] T. K. Ghorai, S. Pramanik, P. Pramanik, Synthesis and photocatalytic oxidation of different organic dyes by using Mn_2O_3/TiO_2 solid solution and visible light, Appl. Surf. Sci. 255 (2009) 9026–9031. https://doi.org/10.1016/j.apsusc.2009.06.086

[24] X. Yang, L. Zhao, L. Zheng, M. Xu, X. Cai, Polyglycerol grafting and RGD peptide conjugation on MnO nanoclusters for enhanced colloidal stability, selective cellular uptakeand cytotoxicity, Colloids and Surfaces B: Biointerfaces 163 (2018) 167–174. https://doi.org/10.1016/j.colsurfb.2017.12.034

[25] M. Zheng, H. Zhang, X. Gong, R. Xu, Y. Xiao, H. Dong, X. Liu, Y. Liu, A simple additive-free approach for the synthesis of uniform manganese monoxide nanorods with large specific surface area, Nanoscale Res. Lett. 1 (2013), 166. https://doi.org/10.1186/1556-276X-8-166

[26] G. Nuyts, S.e Cagno, K. Hellemans, G. Veronesi, M. Cotte, K. Janssens, Study of the early stages of Mn intrusion in corroded glass by means of combined SR FTIR/μXRF imaging and XANES spectroscopy, Procedia Chem. 8 (2013) 239-247. https://doi.org/10.1016/j.proche.2013.03.030

[27] M. Abdelmouleh, S. Boufi, M.N. Belgacem, A.P. Duarte, A. Ben Salah, A. Gandini, Modification of cellulosic fibres with functionalised silanes: development of surface properties Int. J. Adhes. Adhes. 24 (2004) 43–54. https://doi.org/10.1016/S0143-7496(03)00099-X

Powder Metallurgy and Advanced Materials – RoPM&AM 2017   Materials Research Forum LLC
Materials Research Proceedings 8 (2018) 88-88   doi: http://dx.doi.org/10.21741/9781945291999-9

[28] S. A. Moon, B. K. Salunke, B. Alkotaini, E. Sathiyamoorthi, B. S. Kim, Biological synthesis of manganese dioxide nanoparticles by Kalopanax pictus plant extract IET, Nanobiotechnol. 9 (2015) 220-225. https://doi.org/10.1049/iet-nbt.2014.0051

[29] W. Laminack, J. L. Gole, M. G. White, S. Ozdemir, A. G. Ogden, H. J. Martin, Z. Fang, T. H. Wang, D. A. Dixon, Synthesis of nanoscale silicon oxide oxidation state distributions: The transformation from hydrophilicity to hydrophobicity, Chem. Phys. Lett. 653 (2016) 137–143. https://doi.org/10.1016/j.cplett.2016.04.079

[30] B.V. Neamtu, O. Isnard, I. Chicinas, C. Vagner, N. Jumate, P. Plaindoux, Influence of benzene on the Ni_3Fe nanocrystalline compound formation by wet mechanical alloying: An investigation combining DSC, X-ray diffraction, mass and IR spectrometries, Mater. Chem. Phys. 125 (2011) 364–369. https://doi.org/10.1016/j.matchemphys.2010.10.056

[31]   Y. Xu, H. Lin, Y. Li, H. Zhang The mechanism and efficiency of $MnO2$ activated persulfate process coupled with electrolysis Science of the Total Environment 609 (2017) 644–654. https://doi.org/10.1016/j.scitotenv.2017.07.151

Powder Metallurgy and Advanced Materials – RoPM&AM 2017   Materials Research Forum LLC
Materials Research Proceedings 8 (2018) 89-94   doi: http://dx.doi.org/10.21741/9781945291999-10

Influence of the palladium coating on the hydrogen embrittlement of Ni$_{61}$Nb$_{33}$Zr$_6$ amorphous tapes obtained by melt spinning

Gyorgy THALMAIER[1, a *], Ioan VIDA-SIMITI [1,b], Niculina Argentina SECHEL [1,c]

[1] Technical University of Cluj-Napoca, 103-105 Muncii Ave., 400641 Cluj-Napoca, Romania

[a]Gyorgy.Thalmaier@sim.utcluj.ro, [b]Vida.Simiti@stm.utcluj.ro,[c]Niculina.SECHEL@stm.utcluj.ro

Keywords: Amorphous alloys; Hydrogenation; Embrittlement, Palladium coating.

Abstract. The current work is focused towards the properties of Ni$_{61}$Nb$_{33}$Zr$_6$ amorphous alloy for use in hydrogen-related energy applications. The master alloys were prepared by arc melting using high purity metals in a Ti-gettered argon atmosphere. The alloys were melted several times to improve homogeneity. The ingots were induction-melted under a argon atmosphere in a quartz tube and a graphite crucible, injected through a nozzle onto a Cu wheel to produce rapidly solidified amorphous ribbons. The characterization of the amorphous ribbons was done by X-ray diffraction, DSC analysis and hardness tests. The hydrogen charging was done electrochemically for low temperature tests and by heating in a hydrogen atmosphere at different temperatures in the case of the high temperature tests. It was found that the palladium plating reduces the hydrogen embrittlement limit by 50 °C.

Introduction

The amorphous alloys have been proposed for hydrogen separation membranes, because amorphous alloys absorb generally hydrogen without forming metallic hydride and show good mechanical properties. However, since amorphous alloys are thermally unstable, using them as dense, hydrogen permeation membrane at elevated temperatures is very hard. Maintaining an amorphous alloy close to its glass transition temperature will trigger crystallization, decrease of the hydrogen permeability and ultimately its mechanical failure. From this point of view it at utmost importance to have a Tg as high as possible.

Generally, Ni-Nb amorphous alloys have high Tx [1] and according to Inoue [2] it could be further improved by adding more elements to the alloy. Zirconium on the other hand has excellent hydrogen permeability and in general improves the glass forming ability of the alloys [3]. On the other hand, increasing the zirconium content will lead to the reduction of the Tg, so, an optimal balance of these two issues must be found. Different nickel niobium alloys are studied [4, 5] which could be used as a separation membrane.

The studied alloy has a supercooled liquid region of ~ 50K, which would allow it to be shaped by hot-pressing in this temperature range. The purpose of this paper is to evaluate hydrogel embrittlement behavior of the amorphous Ni$_{61}$Nb$_{33}$Zr$_6$ alloy and identifying a temperature range in which the alloy could be used as the hydrogen separation membrane from this point of view.

Experimental

The master alloy (Ni$_{61}$Nb$_{33}$Zr$_6$) was prepared by arc melting using high purity materials in a Ti-gettered argon atmosphere. The alloys were melted several times in order to improve homogeneity. The alloy ingot was induction-melted under a high-purity argon atmosphere in a quartz crucible and injected through a nozzle onto a rotating Cu wheel to produce amorphous

Powder Metallurgy and Advanced Materials – RoPM&AM 2017   Materials Research Forum LLC
Materials Research Proceedings 8 (2018) 89-94   doi: http://dx.doi.org/10.21741/9781945291999-10

tapes. The obtained tapes were 4 mm wide and approximately 50 μm thick. The rotation speed used during the present experiments was 32 m/s. The amorphous nature of the ribbons was investigated by X-ray diffraction using a Shimadzu XRD – 6000 diffractometer and CuKα1 radiation. The samples behavior on heating was investigated by differential scanning calorimetry (SETARAM Labsys system) at the heating rate of 40 K/min. The ultimate tensile strength of the tapes was estimated from the Vickers micro-hardness measurements (40 $gf.$ applied for 15 seconds) as UTS = HV*10/3 [MPa].

The palladium layer was deposited by thermal evaporation in a base pressure of $5*10^{-6}$ torr. The hydrogen embrittlement behavior was studied by heating the palladium coated and uncoated samples in flowing hydrogen to different temperatures (250°C, 300°C, 350°C, 400°C, 450°C, 500°C, 540°C and 580°C). Heating to higher temperatures would result in the crystallization of the tapes.

The critical bending strain was determined by measuring the radius of curvature at which fracture occurs in a bending test between two parallel plates. The strain is then calculated using the following equation: $= \frac{t}{2r-t} \cdot 100$ [%] , where r is the bending radius and t is the sample thickness.

Results and discussions
The amorphous structure of the sample is confined by XRD measurement. The X-ray diffraction pattern shown in Fig. 1a presents a broad maximum (FWHM = 6.3°) characteristic for glassy structures.

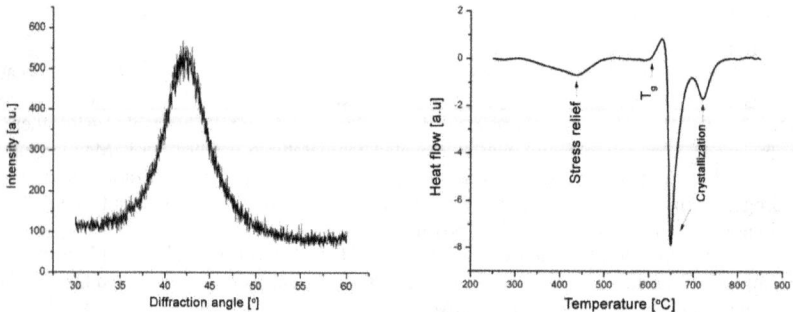

Fig. 1. X-ray diffraction pattern (a) and DSC curve (b) of the as cast tapes.

DSC measurements were performed to determine the thermal transformations that took place in the material and to approximate the thermal stability. The DSC heating curve of an amorphous material presents certain critical temperatures such as: glass transition temperature (T_g), crystallization temperatures (T_X and T_P) and melting temperature (T_S and T_1). The amorphous material remains in vitreous state until the T_X temperature is reached. The crystallization of the amorphous material is indicated by the presence of exothermic peaks, their number depending on the number of crystallization steps through which the material undergoes. The DSC curve presented in Fig. 1b, shows at 420 °C a structural relaxation followed by a glass transition (T_g at 601 ° C and two crystallization steps (T_{x1}= 638 °C and T_{x2}= 702 °C). From the combined analysis we can conclude that these tapes are x-ray amorphous structures.

Powder Metallurgy and Advanced Materials – RoPM&AM 2017   Materials Research Forum LLC
Materials Research Proceedings 8 (2018) 89-94   doi: http://dx.doi.org/10.21741/9781945291999-10

Another advantage of the amorphous structure is the outstanding mechanical properties. Although not as precise, the ultimate tensile strength evaluation from the hardness measurements is a simple and strait forward way to go since even if the samples are prepared by grinding and polishing, there will still remain edges on the margins that act as tension concentrators, leading to an erroneous measurement. In table 1 the microhardness measured using the Vickers method is summarized.

Table 1. Microhardness and estimated UTS of the selected tape.

HV0.04/15 [daN/mm^2]	Rm [MPa]	HV$_{med}$ [daN/mm^2]	Rm$_{med}$ [MPa]
958	3193		
805	2683	835	2783
741	2473		

The obtained values are similar to those presented in the literature for a similar alloy (Ni$_{62}$Nb$_{33}$Zr$_5$), the alloy that has the best glass forming ability in this alloy family but has a smaller supercooled liquid region than the studied composition [6].

To evaluate the hydrogen embrittlement resistance and the Pd coating's influence on the embrittlement coated and uncoated Pd samples were heated in flowing hydrogen atmosphere and then subjected to bending tests.

Table 2. The critical bending parameters at different hydrogen charging temperatures.

Temperature [°C]	Critical bending diameter [mm]		Deformation [%]	
	Pd coated	without Pd	Pd coated	without Pd
250	0.04	0.04	100	100
300	0.04	0.04	100	100
350	0.04	0.04	100	100
400	0.04	0.04	100	100
450	0.04	1.38	100	1.47
500	2.92	2.28	0.68	0.88
540	5.08	4.03	0.39	0.49
580	8	10.14	0.25	0.19

The samples that permitted bending to 180° were considered to have a deformation of 100%. The critical bending diameters were measured with a precision of 0.01 mm and presented in table 2 correlated with the deformation at the testing temperatures. From Fig. 2 is evident that the alloy starts to embrittle at 400 °C for the uncoated alloy and 450 °C for the palladium-coated alloy. As the temperature increases the critical bending radius are also increases suggesting a stronger

Powder Metallurgy and Advanced Materials – RoPM&AM 2017   Materials Research Forum LLC
Materials Research Proceedings 8 (2018) 89-94   doi: http://dx.doi.org/10.21741/9781945291999-10

embrittlement as the temperature rises. Based on these measurements one can speculate that the upper temperature limit of this alloy should be in the 400 - 450 °C area.

X-ray diffraction measurements were performed on both types of hydrogen-loaded alloys at 450 °C to assess the effect of the absorbed hydrogen on the alloy structure. Measurements were made on samples heated to this temperature because the palladium-free alloy had embrittled, and the palladium-coated layer did not.

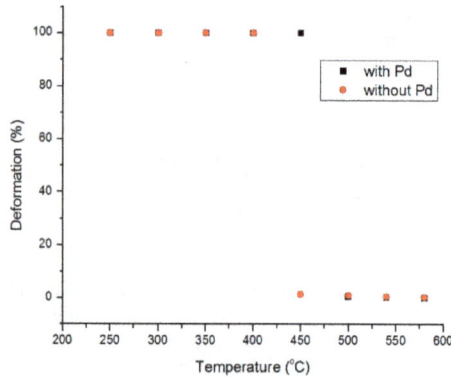

Fig. 2. Embrittlement resistance of coated and uncoated samples.

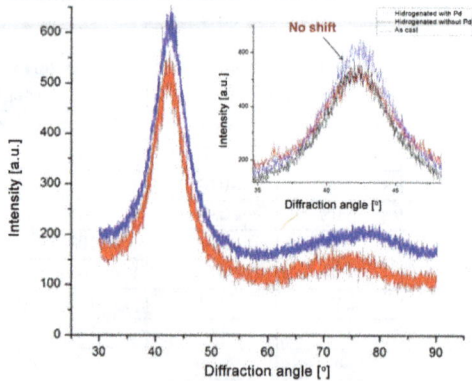

Fig. 3. X-ray diffraction patterns on hydrogen-loaded tapes with and without Pd layer.

The X-ray diffraction patterns presented in Fig. 3 for the samples charged with hydrogen and in the initial state. No shift in the peak position is visible. Also, stable hydrides were not formed during the tests; these observations are in good agreement with the corresponding binary phase diagrams.

Fracture in the case of bending tests is of brittle nature in both cases, specific to metallic glasses; although some localized plastic flows may be observed on the fracture surface of the samples. In the case of metallic glasses their deformation is achieved by two mechanisms: at high temperatures by viscous flow and at ambient temperature (T <0.5Tg) by twinning, the deformation being in the shear bands [7].

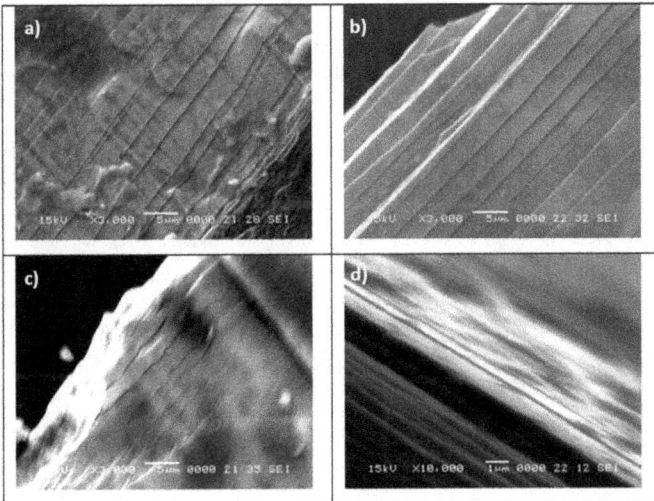

Fig. 4. Fracture surfaces after bending tests.

These materials seem to be fragile, although there are significant local deformations in the rupture. The width of such bands has a width of some micrometers and they appear in the direction of maximum applied tension. The slip occurs in a narrow area (~ 10nm) due to the local temperature increase, the mechanical work performed is proportional to its displacement [8].

The presence of these bands suggests the existence of a plastic deformation zone of thickness s on both surfaces of the strip as shown in Fig. 4. The displacement generated by the shear stresses relaxes the adjacent area, and another slip system cannot be formed close to it. This effect results in the sliding strips being spaced according to their length [8].

The sliding strips for hydrogen heated samples at 400 °C or 450 °C are deep (~ 1μm) however when heated to 540 °C they are significantly shallower. This suggests that the mechanical work consumed to reduce the tensions was lower in this case. This observation is also consistent with the shorter distance between the sliding strips.

If some bands slid more than others during bending, the sliding plane turns into a crack when its displacement exceeds a critical value. According to R. D. Conner et al. [7], in a metallic glass in which the displacement and the distance between the sliding strips is greater, the crack is generated and propagated more easily. This statement is in a slight contradiction with the present observations, but the presence of hydrogen in the structure has an embrittling effect on the material. In samples heated to temperatures below 450 °C, the amount of hydrogen is too low to produce fragility. In the present case the most likely mechanism of hydrogen embrittlement is decohesion.

Summary
The influence of the palladium coating on the hydrogen embrittlement behavior of the $Ni_{61}Nb_{33}Zr_6$ amorphous ribbons was studied in the present paper. X-ray amorphous ribbons 4 mm wide was obtained by rapid solidification. Some of these were Pd coated to improve hydrogen dissociation and recombination. These ribbons presented a slightly higher embrittlement resistance determined by bending tests. This better behavior can be attributed to the lower quantity of residual hydrogen in these samples and a lower decohesion force acting between the atoms.

Fracture in both cases is of brittle nature, although some localized plastic flows are present on the fracture surface of the samples in the form of deformation bands. The final fracture surface is generated when the deformation bands displacement exceeds a critical value.

Further tests are needed to demonstrate the alloys long term stability and an improvement of the structural stability is desired in order to increase the maximum theoretical operating temperature.

References

[1]   Gy. Thalmaier, I. Vida-Simiti, H. Vermesan, C. Codrean, M. Chira, Corrosion resistance measurements of amorphous Ni40Ti40Nb20 bipolar plate material for polymer electrolyte membrane fuel cells, Adv. Eng. For. 8, 335-342.

[2]   A. Inoue, A Stabilization of metallic supercooled liquid and bulk amorphous alloys, Acta Mater., 48 (2000) 279-306. https://doi.org/10.1016/S1359-6454(99)00300-6

[3]   H.J. Chang, E.S. Park, Y.S. Jung, M.K. Kim, D.H. Kim, The effect of Zr addition in glass forming ability of Ni–Nb alloy system, J Alloys Compd. 434–435 (2007) 156-159. https://doi.org/10.1016/j.jallcom.2006.08.292

[4]   T. Lai et al., Hydrogen permeability and mechanical properties of NiNb-M (M = Sn, Ti and Zr) amorphous metallic membranes, J Alloys Compd. 684 (2016) 359-365. https://doi.org/10.1016/j.jallcom.2016.05.100

[5]   O. Palumbo et al., Hydrogen absorption properties of amorphous (Ni $_{0.6}$ Nb $_{0.4-y}$ Ta$_y$) $_{100-x}$ Zr$_x$ membranes, Prog. Nat. Sci.27 (2017) 126-131. https://doi.org/10.1016/j.pnsc.2017.01.002

[6]   Z.W. Zhu, H.F. Zhang, B.Z. Ding, Z.Q. Hu, Synthesis and properties of bulk metallic glasses in the ternary Ni–Nb–Zr alloy system, Mater. Sci. Eng. A 492 (2008) 221-229. https://doi.org/10.1016/j.msea.2008.04.021

[7]   R.D. Conner, W. L. Johnson, Shear bands and cracking of metallic glass plates in bending, J. of Appl. Phys. 94 (2003) 904-911. https://doi.org/10.1063/1.1582555

[8]   Y. Zhang, A.L. Greer, Thickness of shear bands in metallic glasses, Appl. Phys. Lett., 89 (2006) 071907. https://doi.org/10.1063/1.2336598

Powder Metallurgy and Advanced Materials – RoPM&AM 2017   Materials Research Forum LLC
Materials Research Proceedings 8 (2018) 95-104   doi: http://dx.doi.org/10.21741/9781945291999-11

Study on the particle size reduction by milling of quartz sand for magnetic separation

Florin POPA[1, a *], Loredana COPIL[1,b], Victor CEBOTARI[1,c], Traian Florin MARINCA[1,d], Bogdan Viorel NEAMȚU[1,e], Niculina Argentina SECHEL[1,f], and Ionel CHICINAȘ[1,g]

[1]Technical University of Cluj-Napoca, 103-105 Muncii Avenue, 400641, Romania

[a]florin.popa@stm.utcluj.ro, [b]loredana_copil@yahoo.com, [c]Victor.Cebotari@stm.utcluj.ro, [d]traian.marinca@stm.utcluj.ro, [e]bogdan.neamtu@stm.utcluj.ro, [f]niculina.sechel@stm.utcluj.ro, [g]Ionel.Chicinas@stm.utcluj.ro

* corresponding author

Keywords: Quartz sand, Milling, Electron microscopy, Particle size.

Abstract. For being used in crystal glass industry, the iron content of quartz sand must be under 0.09 %. If the reserve contains a higher quantity, methods for iron reduction must be used. Usually the iron phases are present in large quantity in the small particle size fraction. For reducing the sand grain size, milling was performed on a planetary ball mill. Different ball/powders ratio were studied for determining an optimum particle size vs. milling duration. The particle size was determined for each milling experiment. Using Energy Dispersive X-ray spectroscopy (EDX), the elemental distribution for the particle was quantified. By X-ray diffraction, the phase distribution of the sand was analyzed and correlated with the chemical composition. The phases are changing their ratio versus the grain size. The main phase is SiO_2 as quartz, accompanied by minor phases: iron oxides (Fe_3O_4, Fe_2O_3, and $FeTiO_3$) and some oxide of Al, Na, Ca, and K. Testes for magnetic separation were performed for validating the method.

Introduction

The quartz sand is the raw material for glass industry. Unfortunately, as all raw materials, quartz sand purity is the limiting criterion for his usage, since the structure and composition give the properties, the usage and classification criteria for glasses [1]. The most detrimental impurity in the quartz sand is iron, followed by some other metallic oxides (titanium, cobalt, copper, etc.). The effect of metallic impurities in the sand is most commonly observed in color of the resulting glass [2]. The minimum iron quantity in the sand for obtaining a color glass is 0.1 %. The classical way for iron removing is flotation, using toxic reagents as amine, NaOH or H3PO4 [3 - 5]. A cleaner approach is magnetic separation [6]. In magnetic separation experiments, the content in magnetic phase (iron oxides) and particle size represent a key factor for an efficient removing setup [6]. Also, the different types of magnetic separators are considered [7].

A method of controlling the particle size of the sand is by ball milling [8]. In the milling experiments the particle size modification is realized by collision events between balls and sand particles [8]. For our studies is suitable that a high productivity to be achieved, at small milling time and the powder to be produced in a continuous way [9]. For high productivity, the quantity of sand is analyzed and in the milling experiments can be expressed in the form of ball to powder mass ratio (BPR). A high BPR means less quantity of material for processing and small BPR means high material quantity introduced in milling chamber. One purpose of this study is to determine optimal condition of sand milling considering different BPR and milling times.

Powder Metallurgy and Advanced Materials – RoPM&AM 2017   Materials Research Forum LLC
Materials Research Proceedings **8** (2018) 95-104   doi: http://dx.doi.org/10.21741/9781945291999-11

The ball milling was found to be useful in sand purification and particle size reduction [10].

In the milled sand it is important to have an insight on the iron phase distribution in the particles [11]. Scanning electron microscopy (SEM) coupled with X-ray energy dispersive spectrometry (EDX) was found to be a suitable technique for sand characterization, as proved by several studies [12 - 16].

In the present study, the particle size, the iron distribution and iron quantity is studied for the quartz sand ball milled, in order to establish the optimum parameters for magnetic removal.

Experimental

The particle size analysis was performed on standard sieving method using sieves in the range 40 – 800 µm. The standard sieved sand quantity was 100 g. Multiple sieving were performed to obtain statistical data. Supplementary, particle size distribution analysis were recorded using an Analysette 22 Nano Tec particle analyzer. The size range was from 0.1µm to 135µm. From the particle size distribution, the parameters D10, D50 and D90 were determinate. D10, D 50 and D90 represents the mean powder diameter equal with a diameter more or equal with 10, 50 and 90 % from the total powder volume.

The resulting particle size ranges were morphological and compositional characterized by Scanning Electron Microscopy using a JEOL JSM 5600LV microscope equipped with an EDX spectrometer (Oxford Instruments, INCA 200 software). Crystallographic analysis was performed by X-ray diffraction on a INEL 3000 Equinox diffractometer, operating with CoKα radiation (λ = 1.79026 Å) in the angular range 2θ of 20 -110 °. Optical microscopy was performed on a VWR microscope. The milling experiments were performed with a planetary ball mill, Fritch Pulverisette 6. The main disc velocity was 350 rpm, for a hardened steel vial with 100 balls. The milling was conducted in air at several ball to powder weight ratio (BPR) for times up to 10 min.

Results and discussion

The milling experiments on quartz sand begun with the analysis of the as received sand. As the milling has its major influence on the particle size distribution, the distribution of the particle in the as-received sand is considered firstly, and is presented in figure 1.

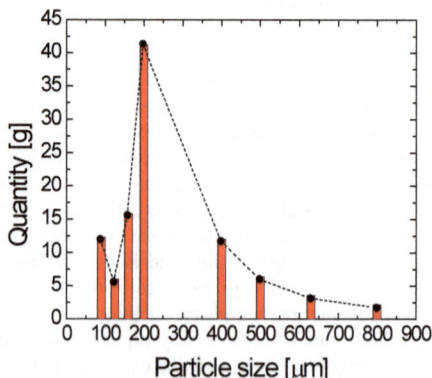

Fig. 1. Particle size distribution of the as-received sand. The distribution is obtained by sieving.

Powder Metallurgy and Advanced Materials – RoPM&AM 2017   Materials Research Forum LLC
Materials Research Proceedings **8** (2018) 95-104   doi: http://dx.doi.org/10.21741/9781945291999-11

Fig. 1 is obtained by sieving the sand with sieves having mesh between 125 μm and 630 μm. Analyzing the particle size distribution is found that the largest quantity has sizes in the range 200 – 400 μm. In addition, an important quantity has sizes below 125 μm. The small particles are important since if they contain iron phases can be more easily removed from the sand. Even the magnetization of the iron phase is smaller, if the weight of the particle is low enough the particle can be easily extracted at lower magnetic forces.

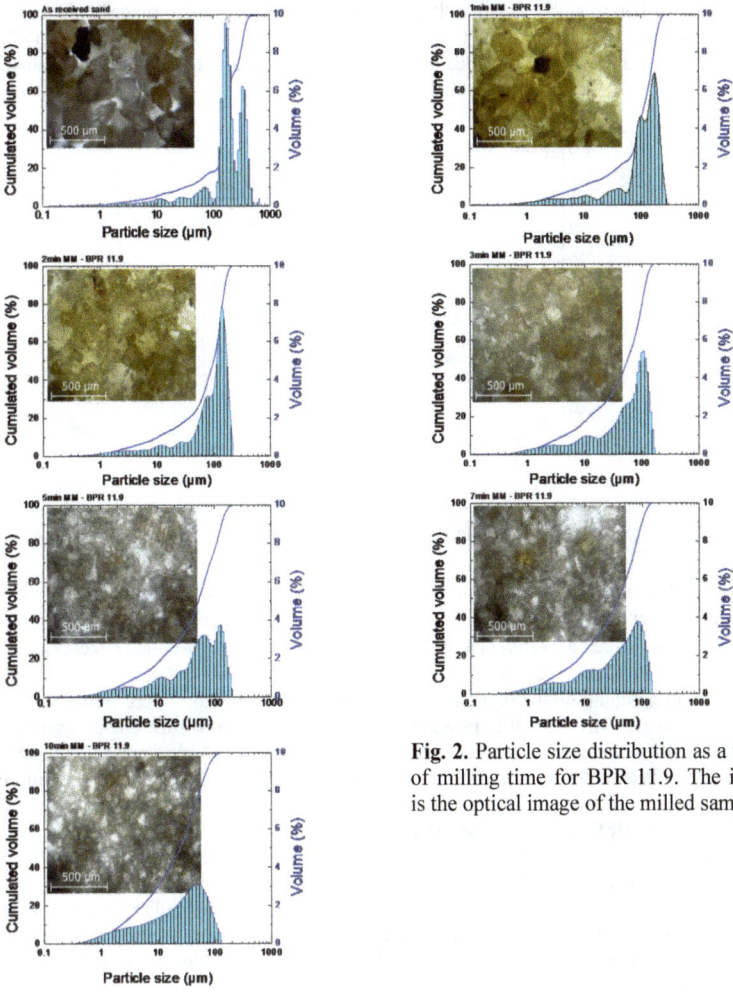

Fig. 2. Particle size distribution as a function of milling time for BPR 11.9. The insertion is the optical image of the milled sample.

Powder Metallurgy and Advanced Materials – RoPM&AM 2017   Materials Research Forum LLC
Materials Research Proceedings 8 (2018) 95-104   doi: http://dx.doi.org/10.21741/9781945291999-11

Milling of this as-received sand, leads to a shift in particle size, with influence on the sand color. In Fig. 2. The optical images and the particle size distribution is presented for several milling times. The milling conditions were the same for all samples.

As the particle modify their size, a color change occurs as the milling time increases. The color change is the effect of reflective index of light on the sand surface due to the size decrease. This color change is shown in insertions in fig.2. Milling of the sand in the 11.9 BPR, leads to a decrease of the particle size and the particle size distribution change from a 3 – 4 maxima distribution to a distribution with 2 maxima. It is expected that longer milling time should lead to a single peak distribution. A longer milling time is not performed to avoid iron contamination of the sand from the balls used in the milling process. In order to proper study the particle size change by milling, in Fig. 3, the D90 parameter is plotted versus milling time.

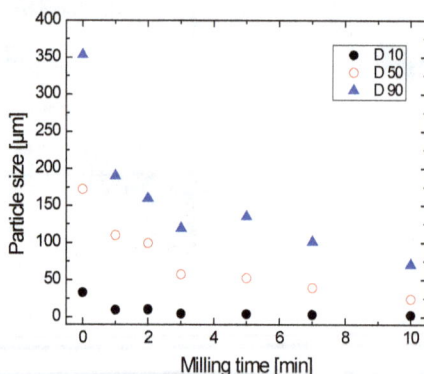

Fig. 3. D10, D50 and D90 evolution versus milling time.

It is observed that milling 10 min, the particle size decreases 5 times comparted with the unmilled sand (as-received) (from 353 to 70 μm). The medium particle size, 50 % volume powder decreases 7 times (from 172 to 24 μm). The low particle size fraction 10 % volume decreases even more of 14 times (from 33 μm to 2.3 μm).

To determine an optimum on the particle size, several milling conditions were considered. The main parameter changed was the ball to powder ratio (BPR). BPR controls the energy transferred to the powders, and in our case the particle shape of the sand. In fig. 4. Are presented the SEM images, recorded in backscattered mode and the particle size distributions for each chosen milling BPR.

It is seen that as the BPR value increases, the particles shapes tend to have a single distribution peak, as the particles are fragmented by collision impacts during the milling. It is assumed that for high BPR values, the energy transferred to the powders is higher. Analyzing Fig 2 and 3 it is concluded that for founding an optimum for the milling, can be considered either a change in BPR or in the milling time. Depending on the effect of iron contamination from the milling media, one of these situations can be prevail. The transferred energy is exemplified as well by the D90 parameter evolution versus BPR, Fig. 5

Powder Metallurgy and Advanced Materials – RoPM&AM 2017   Materials Research Forum LLC
Materials Research Proceedings **8** (2018) 95-104   doi: http://dx.doi.org/10.21741/9781945291999-11

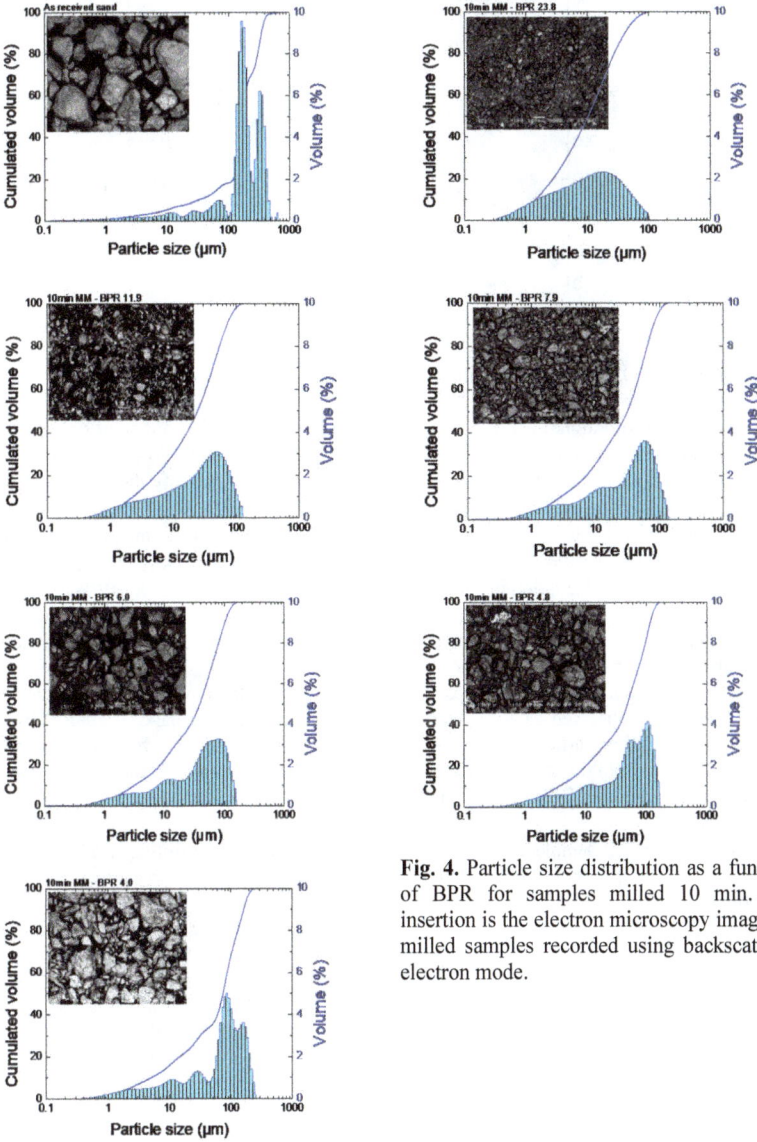

Fig. 4. Particle size distribution as a function of BPR for samples milled 10 min. The insertion is the electron microscopy image the milled samples recorded using backscattered electron mode.

Powder Metallurgy and Advanced Materials – RoPM&AM 2017   Materials Research Forum LLC
Materials Research Proceedings **8** (2018) 95-104   doi: http://dx.doi.org/10.21741/9781945291999-11

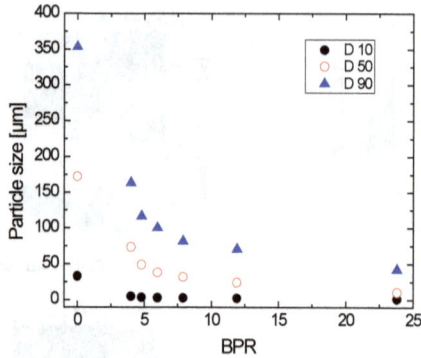

Fig. 5. D10, D50 and D90 evolution versus BPR for the samples milled 10 min.

In the case of BPR variation, the mean particle size decreases for 90 % volume of powder from 350 to 163 (at 4.0 BPR) down to 41 μm for 23.8 BPR. Meaning 2 times for lowest BPR (4.0) and 8 times for highest BPR (23.8). In the case of lower powder volume quantity the decrease is less pronounced. The D50 decrease from 172 μm (unmilled) to 73 μm (BPR 4.0) and 10 μm (BPR 23.8) – meaning 2 (BPR 4.0) and 17 (BPR 23.8) times compared with unmilled sample. The smallest particles (D10) decrease from 33 μm (unmilled) to 5 μm (4.0 BPR) to 1.5 μm (BPR 23.8). Computing the decrease ration is obtained a factor 6 for BPR 4.0 and 22 for BPR 28.9.

Since the milling of the sample is a step in the process of iron removal from the sand, distribution maps and semiquantitative analysis on the iron was performed. In Fig. 5, the iron and titanium distribution maps are presented for several BPR.

The iron and titanium distribution maps indicate that at high BPR, when the particles have small sizes, the iron is observed clearly in some small particles. On other hand at small BPR, the iron is embedded in larger particles. The occurrence of iron in small particles represents an easier way of removing him from the sand. However, computing the Fe amount from the milled samples, Fig. 6, it seems that al higher BPR the iron quantity increases, most probably due to contamination from milling balls.

Comparing to a sieved sample with particle size of 40 – 50 μm, the milling leads to a good homogenization of the iron and titanium content in the samples. This comparison suggest that Fe can be more effective removed in the samples containing small particles. If the particles are too small, and iron is not fully concentrated in individual particles the efficiency of removal process is decreased since almost all the particles will be deflected by the magnetic field. Supplementary the sieving of the sand at small particle size leads to high quantity of sand with large particle not suitable for the glass industry.

Powder Metallurgy and Advanced Materials – RoPM&AM 2017   Materials Research Forum LLC
Materials Research Proceedings **8** (2018) 95-104   doi: http://dx.doi.org/10.21741/9781945291999-11

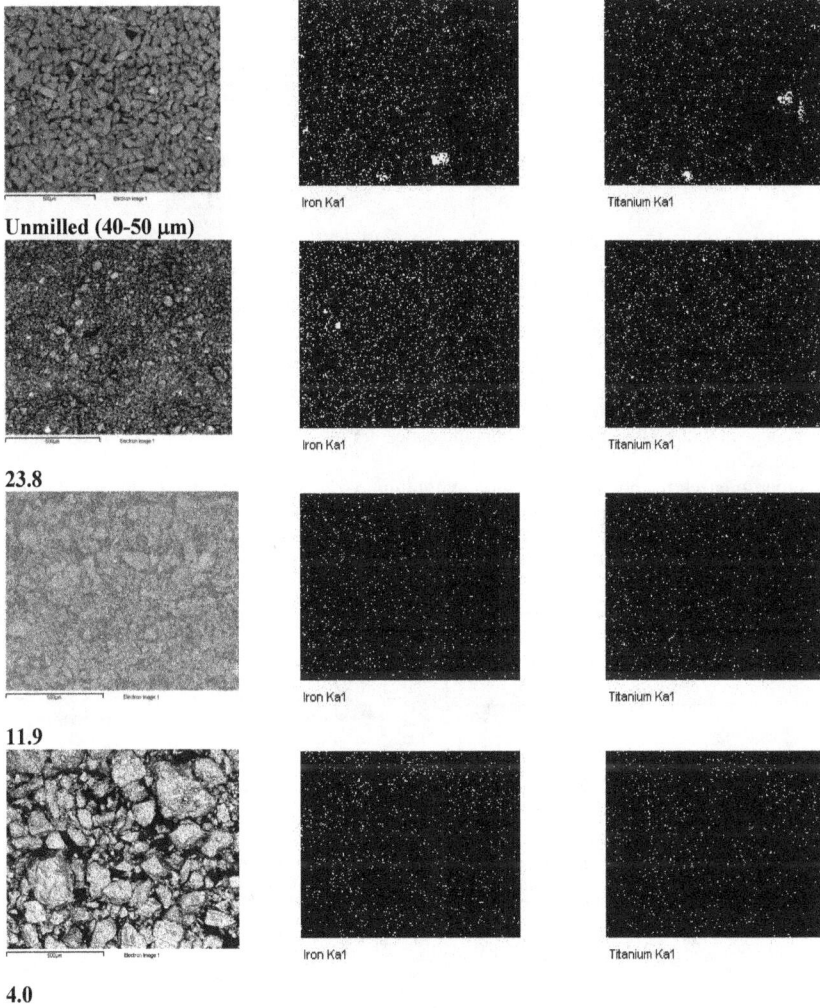

Unmilled (40-50 μm)

23.8

11.9

4.0

Fig.5. Elemental distribution maps for as received sample (sieved under 50 μm, a) and samples milled 10 min at BPR 23.8 (b), 11.9 (c) and 4.0 (d) respectively.

Powder Metallurgy and Advanced Materials – RoPM&AM 2017   Materials Research Forum LLC
Materials Research Proceedings **8** (2018) 95-104   doi: http://dx.doi.org/10.21741/9781945291999-11

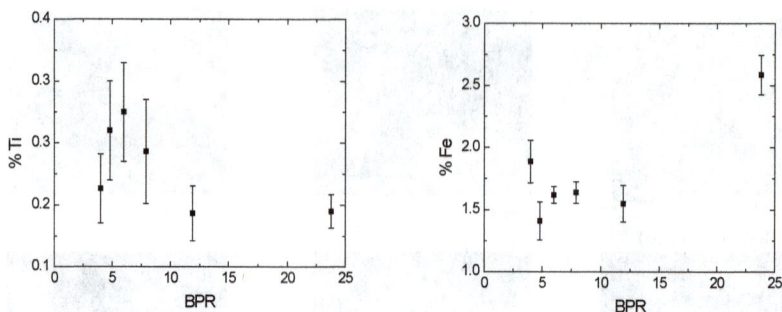

Fig. 6. Fe and Ti elemental weight evolution versus BPR.

Analyzing Fig. 6 is concluded that a BPR higher than 12 leads to iron contamination, and these milling conditions are not suitable for sand processing. Al lower BPR, the iron quantity stay at an almost constant value. Considering the titanium concentration, this element exhibits a maximum at BPR of 6, with a Gaussian distribution.

In order to check the phase modification by milling, X-ray diffraction were performed on each milling time and BPR. In Fig. 7 a are presented the X-ray pattern for the sample milled up to 10 min at a BPR of 11.9 and in Fig. 7.b are presented the X-ray patterns for the samples milled 10 min at different BPR.

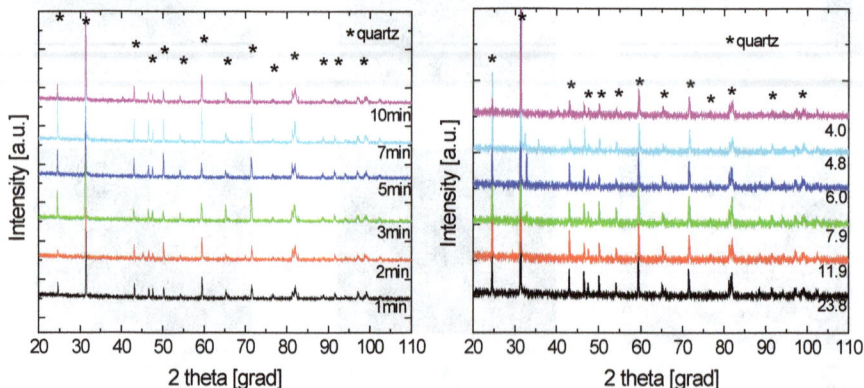

Fig. 7. X-ray diffraction patterns for samples milled up to 10 min (a) and for samples milled 10 min at different BPR.

From Fig. 7 it is concluded that the main phase in all the samples is quartz. All other minor phases from the as-received sand does not changes, neither the iron phases. The minor phases identified are TiO_2 – rutile, $CaCo_3$, anorthite $[Ca(Al_2Si_2O_8)]$, albite $[Na(AlSi_3O_8)]$ and FeSi, Fe_2O_3 (hematite) and Fe_3O_4 (magnetite) compounds.

Using a magnet, preliminary tests for iron removal were performed, and in Fig. 8 the removed iron quantity are plotted.

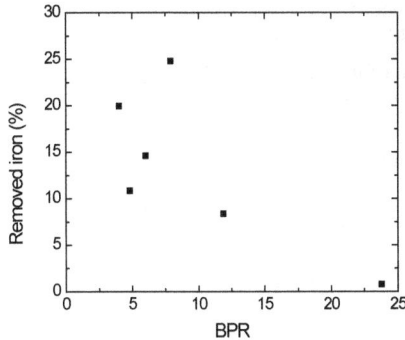

Fig. 8. Relative Fe quantity removed by magnetic separation as a BPR function.

At large BPR values, there are small differences between the iron contend before and after magnetic separation. At small BPR the differences are higher. The conclusion is that at high BPR the small particles can bond more easly with iron (from sand or from balls), making a more homogeneous sand. This observation is confirmed by particle size distribution with a single maximum. At low BPR the sand particles are fragmented, and the resulting pieces are ones with iron phases and others without iron phases, making more easily to be isolated by magnetic force.

Summary

The structural analysis of the as-received and milled samples indicates that the principal phase is quartz accompanied by small amounts of SiO2 phases and Fe2O3. The phase structure does not modify by milling. Particle size distribution decreases exponentially with milling time and with increasing the ball to powder weight ratio. The particle size distribution is changing by milling time and BPR. At higher BPR (23.8) smallest sand quantity processed a large maximum is recorded. The Occurrence of single peak distribution is connected with the highest energy transferred to the sand. Elemental distribution maps, shows that the smaller particles in the as-received sand can be formed almost integrally from iron phases. Processing of the sand at high BPR (low quantity sand) leads to iron homogeneous distribution due to smaller particles size resulted. Processing the sand at lower BPR (high quantity sand) non-homogenous iron phases are observed. The non-homogeneous iron distribution, but in size reduced particle sand is found to favors a better iron removal by magnetic forces. For low BPR values up to 20 % of initial iron is removed by magnetic separation.

Acknowledgement

This work was supported by a grant from the Romanian National Authority for Scientific Research, CNCS –UEFISCDI, project number PN-III-P2-2.1-BG-2016-0214.

Powder Metallurgy and Advanced Materials – RoPM&AM 2017   Materials Research Forum LLC
Materials Research Proceedings **8** (2018) 95-104   doi: http://dx.doi.org/10.21741/9781945291999-11

References

[1] J. Gotze, R. Mockel, Quartz: deposits, mineralogy and analytics, Springer Verlag Berlin Heidelberg, 2012. https://doi.org/10.1007/978-3-642-22161-3

[2] S. C. Rasmussen, How Glass Changed the World: The History and Chemistry of Glass from Antiquity to the 13th Century, Springer Heidelberg New York Dordrecht London, 2012. https://doi.org/10.1007/978-3-642-28183-9

[3] M. Ma, Froth Flotation of Iron Ores. Int. J. Mining Eng. Mineral Processing. 1 (2012) 56-61. https://doi.org/10.5923/j.mining.20120102.06

[4] D.W. Frommer, Iron ore flotation: practice, problems, and prospects. J. Am Oil Chem. Soc. 44 (1967) 270-274. https://doi.org/10.1007/BF02639274

[5] Z. Zhang, J. Li, H. Huang, L. Zhou, T. Xiong, High efficiency iron removal from quartz sand using phosphoric acid. Int. J. Miner. Process. 114-117 (2012) 30-34. https://doi.org/10.1016/j.minpro.2012.09.001

[6] J. Svoboda, Magnetic Techniques for the Treatment of Materials, Ed. Kluwer Academic; 2004.

[7] N. N. Konev, Magnetic enrichment of quartz sands. Analysis of separator operation, Glass and cermics 67 (2010) 132-137. https://doi.org/10.1007/s10717-010-9246-z

[8] C. Suryanarayana- Mechanical Alloying and Milling, Ed. Marcel Dekker, New York, 2004.

[9] E. Gaffet, G. Le Caer, Mechanical processing for nanomaterials, Encyclopedia Nanoscience Nanotechnology, vol X (2004) 1-39.\

[10] H. Nazratulhuda, M. Othman, Purification of Tronoh Silica Sand via preliminary process of mechanical milling, IOP Conf. Series: Mater. Sci. Eng. 114 (2016) 012114. https://doi.org/10.1088/1757-899X/114/1/012114

[11] M. Tashakor, B. Hochwimmer, S. Imanifard, Control of grain-size distribution of serpentinite soils on mineralogy and heavy metal concentration. Asian J Earth Sci. 8 (2015) 45-53. https://doi.org/10.3923/ajes.2015.45.53

[12] M. Sundararajan, S. Ramaswamy, P. Raghavan, Evaluation for the beneficiability of white silica sands from the Overburden of Lignite mine situated in Rajpardi district of Gujarat. J Miner. Mater. Charact. Eng. 8 (2009) 701-713. https://doi.org/10.4236/jmmce.2009.89061

[13] D. A. Fungaro, M.V. da Silva, M,V, Utilization of water treatment plant sludge and coal fly ash in brick manufacturing. Am. J Environ. Protection. 5 (2014) 83-88.

[14] H.A. Khwaja, O.S. Aburizaiza, D.L. Hershey, A. Siddique, D.A.P.E. Guerrieri, J. Zeb, M. Abbass, D.R. Blake, M.M. Hussain, A.J. Aburiziza, M.A. Kramer, I.J. Simpson, Study of black sand particles from sand dunes in Badr. Saudi Arabia using electron microscopy. Atmosphere. 6 (2015) 1175-1194. https://doi.org/10.3390/atmos6081175

[15] M. P. Mubiayi, Characterisation of Sandstones: mineralogy and physical properties. Proceedings of the World Congress on Engineering 2013 Vol III, 2013.

[16] M. Lucarz, The condition of silica sand grains surface subjected to reclamation treatment. Metalurgija. 45 (2006) 37-40.

Powder Metallurgy and Advanced Materials – RoPM&AM 2017   Materials Research Forum LLC
Materials Research Proceedings 8 (2018) 105-114   doi: http://dx.doi.org/10.21741/9781945291999-12

Fe_2O_3 hematite quantity increase in quartz sand by heat treatments

Ana COTAI, Traian Florin MARINCA, Florin POPA*

Materials Science and Engineering Department, Technical University of Cluj-Napoca, 103-105, Muncii Avenue, 400641 Cluj-Napoca, Romania

ana.cotai@yahoo.com, Florin.POPA@stm.utcluj.ro, Traian.MARINCA@stm.utcluj.ro

Keywords: Quartz sand, Milled sand, Heat treatments, Hematite (Fe_2O_3)

Abstract. Heat treatments were performed on the quartz sand to increase the quantity of Fe_2O_3 hematite phase. The heat treatments were performed on the as-received sand samples. The heating temperatures were chosen in the range of 120-600 °C and the time durations in the range of 1-24 h. The sand phases evolution on the temperature was followed by differential scanning calorimetry (DSC). Identification of the phases was realized by X-ray diffraction. The modifications of the iron quantity and distribution in the sand particles were identified by Energy Dispersive X-ray Spectroscopy (EDX) analyses. An optimum temperature/time for the annealing was identified, leading to highest Fe_2O_3 content. Testes for magnetic separation were performed to validate the method.

Introduction

At present, there is a steady increase in demand for high purity quartz worldwide [1]. Quartz is used frequently in glass, ceramic and even in nano-industries [2]. Quartz sand is the most common type of sand in the nature [3]. It is used all over the world in different applications because of distinct physical characteristics, like hardness, chemical and heat resistance, also low cost [4]. Depending on the training mode and where it is found, it appears in different shapes and colors [1].

The silicon dioxide that is used to manufacture glass is extracted almost all from quartz sand, which must have over 97 % SiO_2 [5]. Usually, the quartz is colorless or white, but the presence of the impurities can change the color. The iron oxide – hematite phase (Fe_2O_3) is one of the most frequent impurity and depending of the composition concentration, the quartz can alter the color up to yellow [3]. The quality of the sand is as better as the quantity of the iron oxide is smaller.

Despite the importance of the sand, the utilization is limited by the quality of the material which contains harmful mineral inclusions. The presence of the impurities, especially iron oxide, limit the sand utilization for high quality glass manufacturing [5]. A big part of the impurities released can be reduced or eliminated by physical operations, such as size separation, spiral concentration, magnetic separation, etc. [6]. The iron oxide from the sand can be reduced also by physicochemical method [4].

The most ecological method to improve the quality of the sand is the magnetic separation method. The magnetic separation is used to decrease and stabilize the iron content [7]. If the method is not effective enough, efficiency can be increased by a thermic treatment, mechanical milling or a specific granulometric class removal. The experiments presented in reference [5], shows that magnetic separation method removes about 80,49 % of iron oxide from sand and decrease the Fe_2O_3 content from 0,41 % down to 0,08%.

Powder Metallurgy and Advanced Materials – RoPM&AM 2017   Materials Research Forum LLC
Materials Research Proceedings 8 (2018) 105-114   doi: http://dx.doi.org/10.21741/9781945291999-12

A big part of the impurities presents in the quartz sand contain iron and they are finely dispersed and low magnetic. The special magnetic separators, characterized by high magnetic induction (> 0,6 T), are used to eliminate such impurities. Lately, there where fundamental changes on the separators, especially at the level of magnetic parts. The old magnetic separator systems where replaced by systems that are based on permanent magnets of iron-neodymium-boron type. Thus, this change improved the quantity of the sand and decreased the manufacture cost [8].

The present study is focused on the quartz sand evaluation regarding the iron content and its influence on the color and Fe_2O_3 phase content. Changes induced by annealing are considered and their effects on the Fe_2O_3 phase presence. Finally, a basic magnetic experiment is performed for evaluation of the proposed method efficiency.

Materials and Methods

The quartz sand (Cluj County, Romania) was used for all the experiments. Samples of quartz sands were heat treated at different annealing times from 1h up to 24h. For each test the required amount of samples were placed in a ceramic crucible heated in the furnace. Maximum temperatures of 120, 200, 300, 400, 500 and 600°C were considered. For the thermal treatments was used a programmable INDUSTRY furnace, in oxygen atmosphere.

The structural evolution of the samples was highlighted by X-ray diffraction (XRD) using the Cobalt Kα radiation (1.79026 Å) in an Inel Equinox 3000 powder diffractometer in the 2theta = 20-110° range. The occurring of transformation during the heating process were investigated by differential calorimetrical analysis (DSC) using Setaram Labsys equipment. The heating rate was 10 °C/min and the used atmosphere was argon gas. The morphology of the samples was investigated by the JEOL-JSM 5600 LV scanning electronic microscope (SEM) equipped with an EDX spectrometer (Oxford Instruments - INCA 200 software). The optical images were recorded using the optical microscope VisiScope TL384M (VWR) type at 40x magnification. The magnetic separation experiments were performed with a commercial NdFeB magnet.

Results and Discussions

The first effect of sand annealing was the sand colour change. The color of annealed samples for 24 h at various temperatures ranging from 120 °C up to 600 °C is presented in Fig. 1.

120⁰C/24h   200⁰C/24h   300⁰C/24h

400⁰C/24h   500⁰C/24h   600⁰C/24h

Fig.1. Color change observed in silica sand grains after heat treatment.

Powder Metallurgy and Advanced Materials – RoPM&AM 2017   Materials Research Forum LLC
Materials Research Proceedings **8** (2018) 105-114   doi: http://dx.doi.org/10.21741/9781945291999-12

Fig. 1 shows a color change of the sand with the increase of the temperature from yellow to pink-orange. Similar color change upon annealing was observed in [9].

Color change with temperature rise suggests that changes occur in the crystalline structure or impurity phase of sand beyond loss of pores moisture content and dehydration of iron deposits [9]. At temperatures above 250-300 °C, the color changes correspond to the dehydration of the iron compounds as indicated in [10]. The samples observed by eye, were analyzed by optical microscopy, and the recorded images at a magnification of 40X are presented in Fig. 2 for the samples annealed for 24h at different temperatures and in Fig. 3 for samples annealed at 600 °C at different durations.

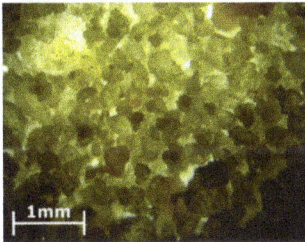

Un annealed

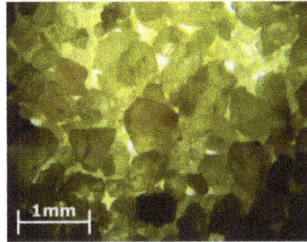

TT 120C°/24h

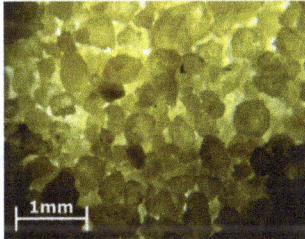

TT 200C°/24h

TT 300C°/24h

TT 400C°/24h

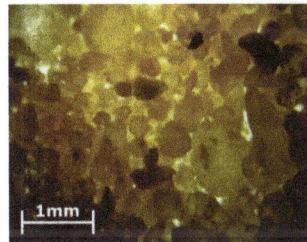

TT 600C°/24h

Fig. 2. Color change of grain sand with temperature rise during heat treatment (24h annealing time). Optical microscopy images, x40.

Powder Metallurgy and Advanced Materials – RoPM&AM 2017   Materials Research Forum LLC
Materials Research Proceedings 8 (2018) 105-114   doi: http://dx.doi.org/10.21741/9781945291999-12

In the images presented in Figs. 2 and 3, it is observed a change in the color of the sand grains upon increasing temperature and annealing time. For samples treated at 600 ˚C is observed the most intense pink-orange color. The change of sand color to red suggests the formation of Fe_2O_3 in sand from additional iron phases. X-ray diffraction (Fig.4) confirm the appearance of the Fe_2O_3 phase. The color change is observed for samples heated more than 300 °C. For temperatures smaller than 300 °C no color change is recorded. Later in our discussion the occurrence of this color change for high temperature relates to calorimetric measurements and some structural changes upon heating.

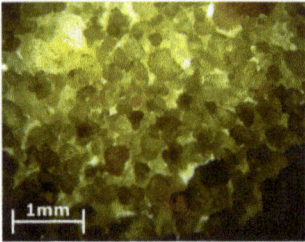

Un annealed

TT 600C%3h

TT 600C%6h

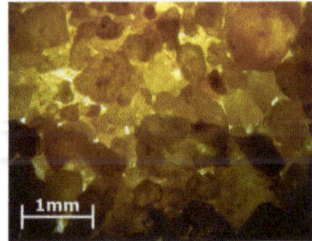

TT 600C%12h

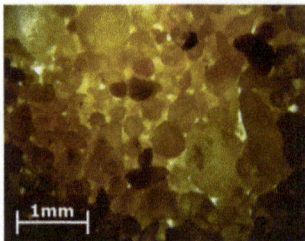

TT 600C%24h

Fig. 3. Change in the color of sand grains depending on the annealing time at TT of 600 ˚C.

Powder Metallurgy and Advanced Materials – RoPM&AM 2017   Materials Research Forum LLC
Materials Research Proceedings 8 (2018) 105-114   doi: http://dx.doi.org/10.21741/9781945291999-12

As concerning the annealing time, the color change is more visible after 12 h of heating as can be seen for the heat treatment at 600 °C (Fig. 3). Such behavior can be connected with the time needed for iron phase to transform under the temperature influence.

For confirming the possible structural changes in the quartz sand, X-ray diffraction studies were performed for samples annealed 24h at different temperatures, Fig. 4.

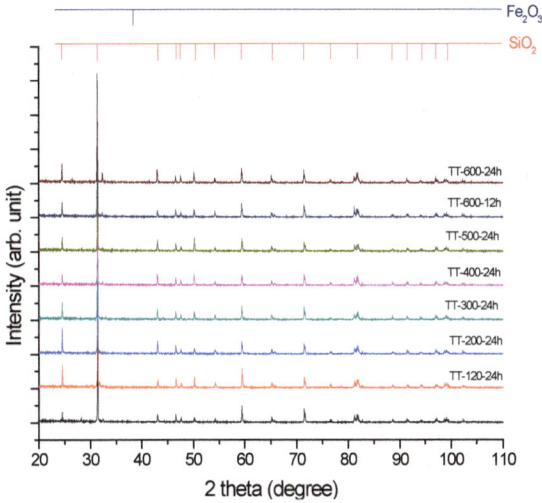

Fig.4. X-ray diffraction of sand after heat treatments at different annealing temperatures for 24h.

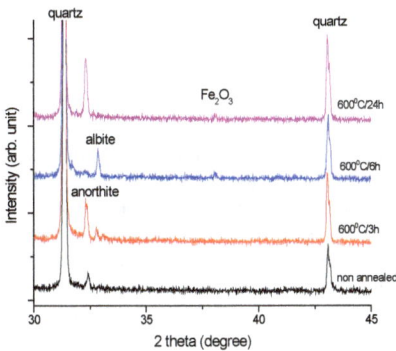

Fig.5a. Detailed view of the Fe_2O_3 peaks

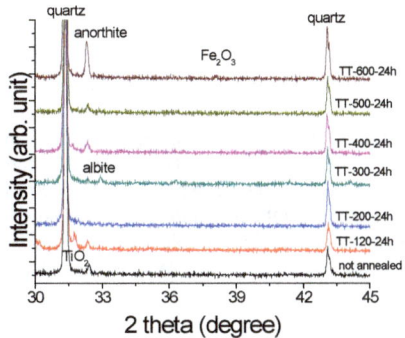

Fig.5b. Detailed view of the Fe_2O_3 peaks

Powder Metallurgy and Advanced Materials – RoPM&AM 2017   Materials Research Forum LLC
Materials Research Proceedings **8** (2018) 105-114   doi: http://dx.doi.org/10.21741/9781945291999-12

The Fig. 4 shows that the main phase present in all diffractograms is quartz. X-ray diffraction patterns on sand heat treated at different temperatures show that both iron oxides and other silicon minerals phases changes when temperature rise. This affirmation is sustained by the pattern study in the angular range 30 – 45 ° (Fig. 5 a and b). Due to the presence of a small amount of iron (about 0.6%), the iron oxide signal is weak, but visible for samples annealed more than 6 h at 600 °C. The occurrence of the Fe_2O_3 peak is an indication of this phase quantity increase upon annealing in air.

The formation of iron oxide (Fe_2O_3) is highlighted in the samples treated at 600°C (Fig. 5a) with a hold time longer than 6h at the 2theta diffraction angle of 38°. However, an analysis at smaller angles indicates the presence of other phases, as shown in Fig.5b. Concerning the annealing temperature, the Fe_2O_3 phase is visible only for the samples annealed at 600 °C. It is believed that at lower temperatures the iron phases receive less energy and is difficult to form Fe_2O_3 in large quantity. Other minor phases are, albite, and other complex phases containing Al, Si, K, Ca and O. To clarify the behavior in temperature, DSC experiments were carried out, and the results are shown in Fig. 6. In Fig. 6 are presented the DSC curves for the non-annealed sample and for a sample firstly annealed at 600 °C for 24h.

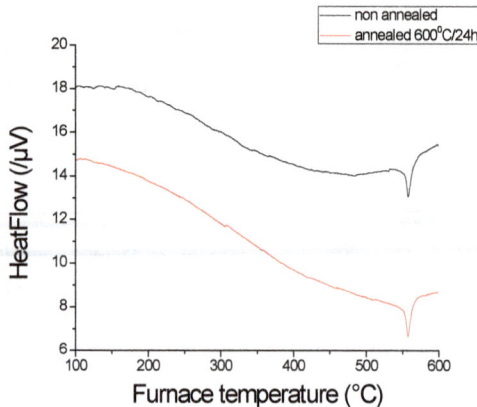

Fig. 6. DSC analysis of sand treated thermally for 24 hours. The heating was done at 600 °C.

In the DSC experiments an endothermic transformation peak is recorded at 575 °C. In order to explain this thermal event two hypothesis can be considered: (a) the presence of ferrimagnetic phase Fe_3O_4 magnetite which has the ferrimagnetic to paramagnetic transition at about this temperature; (b) the formation of FeO wüstite type phase by the reaction of the magnetite with the iron contained in the other phases in the sand or hematite Fe_2O_3 successive reduction. The reaction possibility is argued by the iron – oxygen phase diagram [11].

Powder Metallurgy and Advanced Materials – RoPM&AM 2017   Materials Research Forum LLC
Materials Research Proceedings 8 (2018) 105-114   doi: http://dx.doi.org/10.21741/9781945291999-12

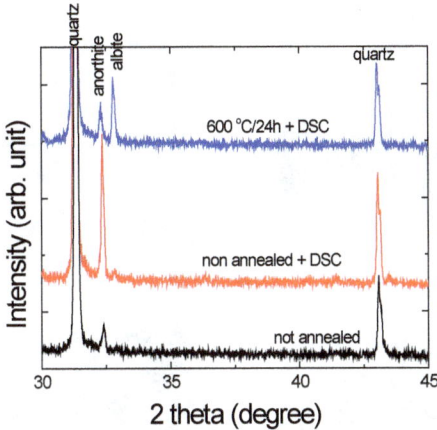

Fig. 7. X-ray diffraction patterns recorded after DSC experiments.

The formation of the FeO phase and the modification of the Fe-O phases ratio is sustained by the presence of different phases containing iron in the sand. However, the X-ray diffraction patterns recorded after DSC experiments shows no Fe_2O_3 peak or other iron-oxide phases (Fig. 7). This behavior is understandable if we consider the fact that the FeO phase formed at high temperature is not stable upon cooling in normal condition, but it can be encountered alongside other iron oxides. In the DSC experiment, the cooling take place in inert atmosphere and the FeO phase does not further oxidize, and the remains in the samples alongside of Fe_2O_3 and Fe_3O_4. Since the iron quantity is very small, the formation of the phases in DSC is smallest than the XRD experiment resolution. Another aspect than must be taken into account the non-stoichiometry of the FeO phase. This phase can occur in a large variety of $Fe_{1-x}O$. This hypothesis does not explain the occurrence of the Fe_2O_3 peak for the annealed samples. But a notable difference is between the two experiments the annealing is performed in air and the DSC is performed in argon gas. In air the FeO phase is further oxidized at cooling and forms in the presence of oxygen the Fe_2O_3 phase. Similar can happen with Fe_3O_4. In air cooling, the formed quantity of Fe_2O_3 exceeds the minimum quantity required for phase peak to be observable in the diffraction patterns.

To check the morphological changes induced by annealing, the SEM images were recorded for samples heated at 600 °C for different durations. The images are shown in Fig. 8.

Powder Metallurgy and Advanced Materials – RoPM&AM 2017   Materials Research Forum LLC
Materials Research Proceedings **8** (2018) 105-114   doi: http://dx.doi.org/10.21741/9781945291999-12

(a)   *(b)*

(c)   *(d)*

Fig. 8. Scanning electron images of the sand annealed at different temperatures up to 24 h. SEM images of sand treated at: (a)- 600 ^{0}C/ 3h, (b)- 600 ^{0}C / 12h, (c)- 600 ^{0}C / 24h. In (d) is presented an image of the un-annealed sample.

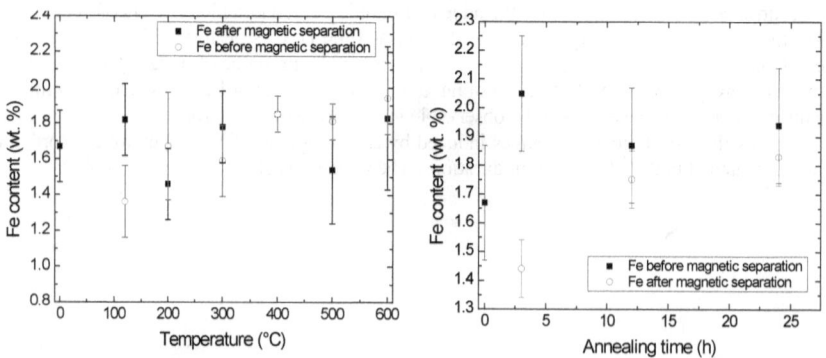

Fig. 9a. Modification of Fe concentration of sand upon temperature increase (annealing for

Fig. 9b. Modification of Fe concentration of

24h).   sand with annealing time (TT 600 °C).

As the annealing time increase, the surface of the particles became smoother. The surface change can be an indication of the phase modification inside the particles.

Performing an EDX analysis on multiple samples from the same annealing conditions, the quantity of iron was extracted and plotted in Fig. 9 versus annealing time (a) and versus annealing temperatures (b). In the same time, a magnetic separation was performed on the samples, and the iron quantity after separation is also shown for both experiments.

As expected during the annealing, the quantity of iron atoms remains almost constant in the samples. But a basic magnetic separation induces modifications. If we consider first the annealing temperature (Fig. 9a), the iron quantity removed variation is in connection with the color of the samples: up to 300 °C the removed quantity is either higher, either lower than before magnetic separation. Such variation can be the effect of sampling and low magnetic fraction of the iron oxide. Once the color of the sample is changed, for temperatures larger than 400 °C, the quantity of iron after magnetic separation is lower than before. Such modification is the effect of iron-oxygen phase change in this temperature range, as discussed for DSC measurements.

If we look at the annealing duration, at 600 °C, for all the annealing times, the iron quantity in the samples after magnetic separation is lower than before separation. The maximum difference occurs at low annealing times (3 h). Once again, the origin of this effect is connected with the Fe-O changes observed in the DSC measurements.

Summary
The quartz sand was studied from the point of view of phase and phase relation versus annealing at various temperatures and durations. The sand annealed at 600 ^{0}C is enriched in Fe_2O_3; annealed at lower temperatures the quantity of Fe_2O_3 is not visible by X-ray diffraction. The Fe_2O_3 phase is visible at annealing times larger than 6 h (at 600 ^{0}C). For lower temperatures, the Fe_2O_3 phase is not visible in the X-ray patterns. The DSC measurements indicates the occurring of an endothermic transition at 575 °C, with two possible cause: Curie temperature of Fe_3O_4 and FeO wustite phase formation. At cooling, in the DSC sample Fe_2O_3 phase is not visible in the X-ray diffraction patterns. The sample enrichment in Fe_2O_3 phase in air annealing is favored by oxygen presence and the reduction of high temperature Fe-O phases. This phenomenon does not occur in DSC experiments, performed in inert gas. The removal of iron is more efficient upon heating above 400 °C and for low annealing times.

References
[1] M. F. M. dos Santos, E. Fujiwara, E. A. Schenkel, J. Enzweile, C. K. Suzuki, Quartz resources in the Serra de Santa Helena formation, Brazil: A geochemical and technological study, J. South Am. Earth Sci. 56 (2014) 328-338. https://doi.org/10.1016/j.jsames.2014.09.017

[2] S. Beddiaf, S. Chihi, Y. Leghrieb, The determination of some crystallographic parameters of quartz, in the sand dunes of Ouargla, Algeria, J. Afr. Earth. Sci. 106 (2015) 129–133. https://doi.org/10.1016/j.jafrearsci.2015.03.014

[3] Information on https://owlcation.com/stem/Beaches-Unusual-Colour/

[4] T. R. Boulos, A. Yehia, M. B. Morsi, S. S. Ibrahim, High quality fused silica from egyptian silica sand, Int. J Sci. Eng. Investigation. 6 (2017) 160-166.

Powder Metallurgy and Advanced Materials – RoPM&AM 2017   Materials Research Forum LLC
Materials Research Proceedings 8 (2018) 105-114   doi: http://dx.doi.org/10.21741/9781945291999-12

[5] M. Bounouala, A. S. Chaib, Removal of iron from sandstone by magnetic separation and leaching: case of el-aouana deposit (Algeria), Mining Sci. 22 (2015) 33–44.

[6] F. Du, J. S. Li, X. X. Li, Z. Z. Zhang, Improvement of iron removal from silica sand using ultra-assisted oxalic acid, Ultrason. Sonochem, 18 (2011) 389–393. https://doi.org/10.1016/j.ultsonch.2010.07.006

[7] N. N. Konev, Magnetic Enrichment of quartz sands. Analysis of separator operation, Glass Ceram. 67 (2010) 132-137. https://doi.org/10.1007/s10717-010-9246-z

[8] N. N. Konev and I. P. Salo, Removal of iron-containing impurities by magnetic separation, Glass Ceram. 56 (1999) 32 – 33. https://doi.org/10.1007/BF02681401

[9] S.G.Zihms, C.Switzer, J. Irvine, M. Karstunen, Effects of high temperature processes on physical properties of silica sand, Eng. Geology 164 (2013) 139-145. https://doi.org/10.1016/j.enggeo.2013.06.004

[10] V. Brotons , R. Tomas, S. Ivorra, J.C. Alarcon, Temperature influence on the physical and mechanical properties of a porous rock: San Julian's calcarenite, Eng. Geology 167 (2013) 117–127. https://doi.org/10.1016/j.enggeo.2013.10.012

[11] ASM diagrams (CD).

Powder Metallurgy and Advanced Materials – RoPM&AM 2017   Materials Research Forum LLC
Materials Research Proceedings 8 (2018) 115-124   doi: http://dx.doi.org/10.21741/9781945291999-13

New weldability model based on the welding parameters and hardness profile

Marius BODEA[1, a]

[1]Faculty of Materials and Environmental Engineering, Technical University of Cluj, B-dul Muncii 103-105, 400641, Romania

[a]mbodea@stm.utcluj.ro

Keywords: Weldability, Maximum hardness, HAZ

Abstract. The weldability of the steels represents a problem of great interest in order to achieve welded structures that satisfy the high requirements on quality, imposed by the nowadays applications. In this paper has been proposed a more advanced model that has considered more factors of weldability influence, thus allowing a more detailed analysis based on the main welding process variables.

Introduction

Weldability is a general technological property commonly used in engineering, but very difficult to be defined and quantified in an exact manner. The American Welding Society has defined the weldability as being: "The capacity of a metal to be welded under the fabrication conditions imposed with a specific suitability designed structure and to perform satisfactorily in service" [1]. According to DIN 8528, Part 1 the weldability is seen as the output of the interaction of three main group factors, given in Table 1 [2].

Table 1. The weldability's factors of influence.

MATERIAL	MANUFACTURE	DESIGN
WELDING SUITABILITY	WELDING POSSIBILITY	WELDING SAFETY
• Chemical composition • Metallurgical properties • Physical properties	• Welding preparation • Welding execution • Heat treatments	• Design • Stress condition
• Tendency to hardening • Tendency to ageing • Tendency to hot cracking • Weld pool behaviour • Segregations • Inclusions • Grain size • Anisotropy • Expansion coefficient • Thermal conductivity • Melting point • Mechanical properties	• Welding technology • Groove shape • Preheating • Susceptibility to cracking • Heat input control • Welding position • Welding sequence • Weld penetration • Pool shape • Post weld heat treatment • Grinding • Pickling	• Material thickness • Notch effect • Stiffness differences • Joint geometry and displacement • Type and level of strains • Temperature • Corrosion • Loads and stress distribution • Weld bead shape

Powder Metallurgy and Advanced Materials – RoPM&AM 2017   Materials Research Forum LLC
Materials Research Proceedings 8 (2018) 115-124   doi: http://dx.doi.org/10.21741/9781945291999-13

From this definition it can observed that one should consider many factors in order to assess the weldability, like: filler and parent material properties, chemical compositions, metallurgical compatibility between them, metallic structure design, service conditions (load, temperature, environmental corrosion), welding parameters and manufacturing technology, but also the structure resistance to the service degradation in time. The welded structure performance is direct related to these factors that are controlling the weld mechanical properties, the nature and size of the imperfections produced in the weld and heat affected zone (HAZ), respectively the behaviour and life span of the welded structure in service.

In the literature can be found several methods for weldability assessment, based on the indirect methods, like: carbon equivalent formulae [3], peak hardness in HAZ [3, 4], or susceptibility cracking in HAZ due to hardening effect [4-6]. None of these methods have considered the simultaneous influence of different parameters in the proposed mathematical models, like it happens in the real phenomena. As a consequence, the validity of each method is limited priory, because there are many significant factors of influence, that have been ignored from the very beginning. In the paper it has been proposed a new weldability model, that takes into account the influence of the main factors, like: t_{8-5} cooling time, carbon equivalent, material thickness, parent material hardness and its microstructural features, respectively the mean grain diameter size. Thereby, the weldability estimated by this model can make a difference for the same materials, with the same chemical composition (C_{ech}), but having different microstructural features or welded with different technologies or welding parameters.

In this model, the weldability has been expressed by a single number between 0 and 1, the latter being assumed for an ideal weldability situation. Between the two extreme limits, there are unlimited intermediates values, corresponding for the real welding situations, the transitions between them, being made continuously. The weldability number calculated by the model proposed in the paper, can be used further in the estimating preheat temperature or for predicting mechanical properties of the welded joints.

Carbon Equivalent Formulae

In welding applications, carbon has a strong effect on the steel's weldability, reducing the ductility and material toughness. In order to assure high strength required in the many welded applications, new materials with very low carbon content has been developed lately, like the TMCP or HTUFF steels [7]. In the same time, beside the carbon equivalent index, the section size of the welded metallic structure plays a very important role, being related to the cooling rate, material properties and heat input. Hence, the carbon equivalent formula $Ceq(s)$ compensated from the material thickness perspective of view, is given in the eq (1), where the thickness material s is given in mm, [7]. For the carbon equivalent relation Ceq we have considered a Japanese formula that is evaluating the hardening of the welds, given in eq(2), [9].

$$Ceq(s) = Ceq + 0.00425 \cdot s \tag{1}$$

$$Ceq = C + \frac{Si}{24} + \frac{Mn}{6} + \frac{Ni}{40} + \frac{Cr}{5} + \frac{Mo}{4} + \frac{V}{14} \tag{2}$$

The compensated carbon equivalent formula is used in the weldability assessment as a direct parameter of influence, but also in the estimating hardness in HAZ, accordingly to the heat input, material properties and cooling rate t_{8-5}. The later are direct related to the strength and toughness in the HAZ and deposited metal.

Powder Metallurgy and Advanced Materials – RoPM&AM 2017   Materials Research Forum LLC
Materials Research Proceedings 8 (2018) 115-124   doi: http://dx.doi.org/10.21741/9781945291999-13

Cooling Time t_{8-5} and Hardness

The cooling time between 800 and 500°C has a very important significance over the HAZ maximum hardness and cracking susceptibility in this area [3 - 6]. For this reason, the weldability models based on the cooling time are frequently used in practice. There are two opposite effects related to the thermal history of the welding cycle. A low heat input lead to excessive hardness and increased risk of cold cracking in the HAZ or even in the weld, accordingly to the specific properties of the parent and filler materials, but also on the welding technology and stress distributions. On the other part, a heat input too high, lead to low mechanical properties due to grain coarsening effect in HAZ, but also toughness depreciation or other inconveniences, related to the welding quality and performance.

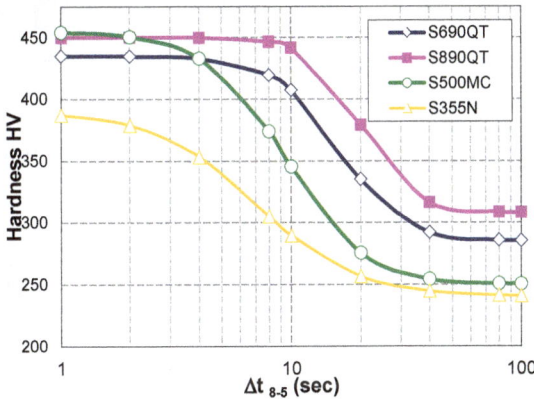

Fig. 1. HAZ hardening for different materials as a function of cooling time t_{8-5} after welding, calculated after [8].

From the Fig. 1, it can be observed the sigmoid evolution of the hardness in the HAZ based on the cooling time t_{8-5} for different steels used commonly in welded applications. The weldability can be related to this kind of evolution, but with considering other parameters of influence as well. The hardenability of material in the HAZ and also cracking susceptibility depends directly on the cooling time t_{8-5}. In another paper, these aspects have been analysed in the detail [8], the microstructural characteristics being essential for the most properties envisaged in welded applications. However, for the simplicity of the model, it has been considered only the cooling time t_{8-5} as essential variable from the thermal cycle related variables, the microstructure being considered as an output of all the factors involved from this point of view.

Based on the data extracted from the Fig. 1, it has been expressed the hardness HV vs. cooling time using the logistic functions with five parameters (5PL). The logistic function has the expression given in the eq(3), and the vector parameters p[a,b,c,d,e] has been given for each kind of material in the Table 2.

$$F(x; p) = a + \frac{d-a}{\left[1 + \left(\dfrac{x}{c}\right)^{b}\right]^{e}}$$

$$(3)$$

Table 2. Parameter vector for the 5PL functions used for HV hardness modelling.

No. crt.	Parent material	Vector parameters p[a,b,c,d,e]				
		a	b	c	d	e
		Minimum horizontal asymptote	Hill slope	Cooling speed for inflection point	Maximum horizontal asymptote	Asymmetry factor
1.	S500MC	250	2.4	10.5	455	1.2
2.	S355N	240	2	7	390	1
3.	S690QT	285	3	17.5	435	1.2
4.	S890QT	308	4	20	450	1

The maximum hardness in the HAZ can constitute a parameter for weldability assessment. As can be seen in the Fig. 2, the maximum HAZ hardness increases as the Ceq increases, between them existing a linear correlation, given in the eq(4), [9].

$$Hmax = (666Ceq + 40) \pm 40 \tag{4}$$

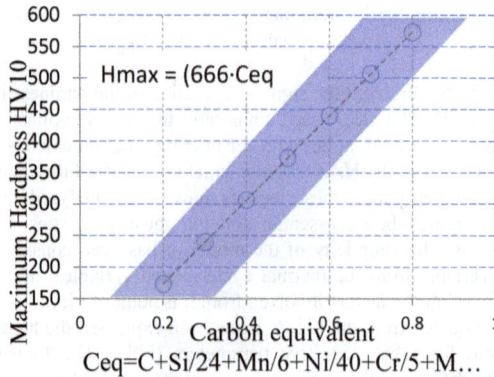

Fig. 2. Maximum HAZ hardness vs. Ceq of 20-mm thick mild steel, calculated after [9].

Powder Metallurgy and Advanced Materials – RoPM&AM 2017   Materials Research Forum LLC
Materials Research Proceedings **8** (2018) 115-124   doi: http://dx.doi.org/10.21741/9781945291999-13

Microstructure

The thermal welding cycle plays a key role in the microstructure characteristics in the HAZ and deposited metal. Different microstructural constituents can result, under different cooling conditions, accordingly to the chemical composition and former austenite grain size. Inclusions or inhomogeneity present in the parent material can also contribute significantly to the new product phases or constituents, resulted after austenite transformations during cooling.

Complex microstructures are formed, containing mixtures of upper bainite, lower bainite, granular bainite, ferrite laths, martensite, martensitic/austenitic (M/A) islands. For instance, in many rolled steels, bands reach in C and Mn can be found. Mn is a strong austenitizer element, decreasing the austenite decomposition temperature. If cooling conditions lead to an undercooling phenomenon, is very likely that in such cases the microstructure will be dominated by the bainite and martensite microstructural constituents. High cooling rates, very often meet in practice associated with low heat inputs, are reducing the time required by the diffusion-controlled processes. As result, the microstructure will be formed predominantly by martensitic and lower bainitic structures. It's very important to control the austenite grain size growing effect during heating cycle, because is in direct relation with the material hardenability.

Non-metallic inclusions play also a very important role in the nucleation process and resulting microstructure in the weld and HAZ, contributing to a grain refinement process. From these facts, it can be concluded that is essential for the weldability mathematic model to consider the influence of the parent initial microstructure, through the mean grain size of the ferritic matrix.

Weldability Mathematic Model

As has been stated before, the model is considering a continuously evolution of the weldability, that is expressed by a single number comprised between 0 and 1, existing a direct proportional variance between the weldability and its associated number.

As result, 0 means absolute no weldability, while 1 means an ideal weldability. The weldability variation must present saturation effect towards to the both limits. For this purpose, it has been considered a 5PL function that can be shaped very easily using a 5-parameter vector. Those parameters have been designed in order to account the influence of the main weldability factors, like: the t_{8-5} cooling time, carbon equivalent, material thickness, parent material hardness and the mean grain diameter size for the parent material. The function for the weldability proposed model, has the following equation:

$$WN = a + \frac{d-a}{\left[1+\left(\dfrac{Ceq}{c}\right)^{b}\right]^{e}}$$

(5)

where:
WN = weldability number (0-1).
a = minimum horizontal asymptote (a=0). Minimum theoretical weldability.
b = Hill slope, is expressing how fast the weldability changes around the inflection point.
c = weldability speed related to Ceq. Locates the inflection point.
d = maximum horizontal asymptote (d=1). Maximum theoretical weldability.
e = asymmetry factor for the weldability curve. (e=1, weldability curve is symmetrical in relation to the inflection point).

Powder Metallurgy and Advanced Materials – RoPM&AM 2017   Materials Research Forum LLC
Materials Research Proceedings **8** (2018) 115-124   doi: http://dx.doi.org/10.21741/9781945291999-13

Because of the very numerous factors that affect the weldability, instead a single weldability curve should be considered two curves, that are determining a weldability band, very likely to a hardenability band. Thus, for a specific material and welding conditions, the weldability falls within the weldability band illustrated in the Fig. 3.

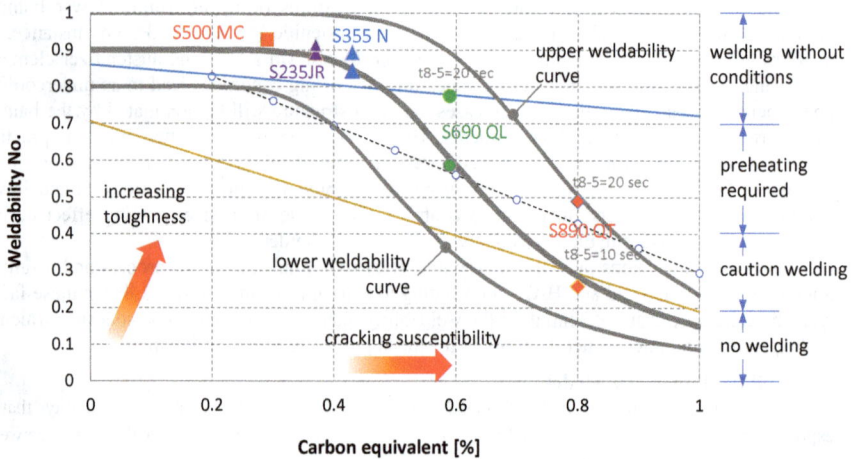

Fig. 3. Weldability band for steels fusion welding processes.

For the upper and lower weldability curves, the parameters vector p is as it follows:

Upper weldability curve	Lower weldability curve
a=0; b=8; c=0.7; d=1; e=0.5	a=0; b=7; c=0.45; d=0.8; e=0.4

The weldability can be improved if the grain size of the parent material is reduced. This effect has been considered by the parameter vector c, that determine the inflection point position. Lower values for d_F which stand for the ferritic mean grain size, is improving the weldability curve that is shifted to left or to right, accordingly to the eq.(6), for the c parameter, Fig. 4 and Fig. 5, where μ represent the optimum t_{8-5} cooling time and σ is the standard deviation.

$$c = c_{max} \cdot exp\left[-0.5 \cdot \left(\frac{t85-\mu}{\sigma}\right)^2\right] + 0.1 \cdot \frac{1}{\sqrt{dF}} \qquad (6)$$

Powder Metallurgy and Advanced Materials – RoPM&AM 2017   Materials Research Forum LLC
Materials Research Proceedings **8** (2018) 115-124   doi: http://dx.doi.org/10.21741/9781945291999-13

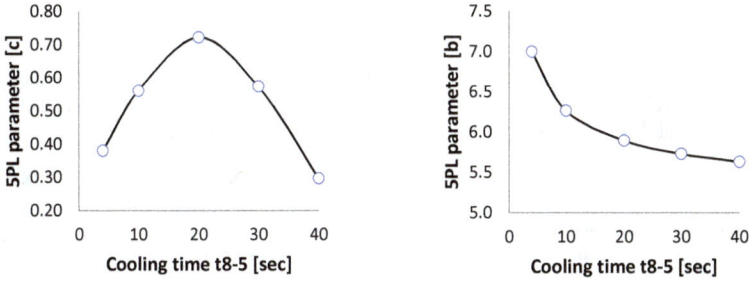

a) Parameter c vs. cooling time t_{8-5} and mean diameter grain size (dF)

b) Parameter b vs. cooling time t_{8-5}

Fig. 4. Parameter vector c and b for the 5PL weldability function.

The Hill slope given by the parameter b of the 5PL weldability curve is controlled by the cooling time t_{8-5}. As can be seen in the Fig. 1, for each material can be determined an optimum interval for the t_{8-5} cooling time, if we consider the maximum HV hardness in the HAZ. Shorter times will lead to hardening effect in the HAZ, while longer times will lead to insufficient weld toughness due to coarsening effect in the HAZ.

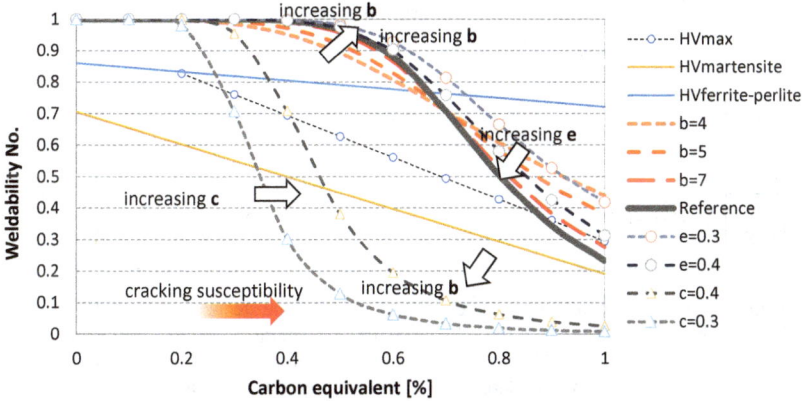

Fig. 5. The shape of the 5PL weldability curve vs. parameter vector p[a,b,c,d,e], the reference curve has p[a=0, b=5.6, c=0.5, d=1, e=0.5].

By modifying the Hill slope (b parameter), we can adjust the weldability curve accordingly to the t_{8-5} cooling time, Fig. 5. The parameter b is calculated using the eq. (7) and illustrated in the Fig. 4.

$$b = 5 + \frac{4}{\sqrt{t_{8-5}}} \qquad (7)$$

The asymmetry factor e has been determined based on the hardness HV of the parent material influence, that is related to the Ceq, carbon equivalent formulae. The asymmetry factor is stretching the weldability curve, accordingly to the e value, calculated with eq.(8), Fig. 5.

$$e = 0.4 + 0,2 \cdot Ceq \qquad (8)$$

Table 3. Weldability classification systems.

Observation	Conventional weldability classification based on the Ceq		New weldability classification based on the 5PL function	
Excellent weldability	$Ceq < 0,40\%$	No preheating required	$WN > 0.8$	Faster cooling is possible in order to achieve a good balance between toughness and strength.
Good weldability	$0,40\% < Ceq < 0,50\%$	Preheating of 200 to 250 °C for wall thickness > 25 mm	$0.7 < WN < 0.8$	No preheat required (WN>0.7), but very narrow welding range parameters available.
			$0.6 < WN < 0.7$	
Possible weldability	$0,50\% < Ceq < 0,60\%$	Preheating of 250 to 300 °C for wall thickness > 15 mm	$0.4 < WN < 0.6$	Preheat required, calculated accordingly to DIN EN 1011-2:2001
Poor weldability	$Ceq > 0,60\%$	Preheating of 300 to 400 °C for wall thickness > 15 mm	$0.2 < WN < 0.4$	The welding can be possible by selecting proper welding parameters and Tp(°C).
No weldability	$Ceq > 0,80\%$	No weldability	$WN < 0.2$	No weldability or very poor weldability

Beside the Ceq values, it can be made a new classification of the steels weldability based on the weldability number, calculated with the 5PL formula. Comparative to the conventional Ceq weldability classification, the weldability number is containing the influence of others main factors of influence for the steels weldability, allowing a more advanced analysis. With the new weldability model, for a given material with a particular Ceq, has been obtained a weldability range accordingly to the welding conditions, thus the weldability property being more precise evaluated. In the previous cases, when only one parameter was considered at a time, this was not being possible. For the very same material it could be used or obtained different t_{8-5} cooling times, grain size diameters, heat treatment conditions, thickness or welding parameters, that can make a difference in material weldability. Finally, for d parameter has been used the relation given in eq.(9), that determine a range of variation for this parameter between 1 and 0.8 for $0 < Ceq \leq 1\%$:

$$d = 1 - \frac{Ceq^2}{5} \tag{9}$$

In the Table 3, are presented comparatively the weldability classification in both systems, one based on the Ceq and the other one, on the 5PL weldability function. Also, for each case scenario are given the recommended measures for welding.

Summary
1. The steels weldability remains a problem of great interest, due to the higher requirements regarding the quality and performance of the welded structures.
2. The proposed model in this paper is evaluating more accurate the steel's weldability. In this model has been considered the influence of the main parameters: t_{8-5} cooling time, carbon equivalent, material thickness, parent material hardness and the mean grain diameter size of the parent material.
3. The weldability is confined between two limits, from the ideal weldability characterized by WN = 1, to the hypothetical case with zero weldability (WN = 0). The weldability is varying continuously between the two extremes limits, accordingly to the welding process parameters and materials properties.
4. One material can present different weldability values, according to welding parameters.
5. The shape of the 5PL weldability function can be easily adjusted accordingly to the parameter vector p[a,b,c,d,e] that is set based on the main process variables.
6. The weldability number can be used further for preheat temperature calculus.
7. The model can be improved further by considering the influence of other factors, related to the welding process or material properties.
8. The influence of parameter vector p[b,e] over the 5PL shape weldability function is more important for steels with higher Ceq (Ceq>0.4). The steels with Ceq<0.4% (or WN > 0.7) can be welded with more flexible welding parameters, without affecting the weldability steel in a significant manner.

References

[1] L. Jeffus, Welding and Metal Fabrication, ISBN-13: 978-1-4180-1374-5, Delmar Cengage Learning, (2012) 780-781.

[2] Springer Handbook of Mechanical Engineering, Volume 10, Karl-Heinrich Grote,Erik K. Antonsson, ISBN: 978-3-540-49131-6, (2009) 3-4.

[3] Carbon Equivalent to Assess Hardenability of Steel and Prediction of HAZ Hardness Distribution, T. Kasuya, Y. Hashiba, Nippon Steel Technical Report No. 95 (2007).

[4] Kasuya, Tadashi, Nobutaka Yurioka, and Makoto Okumura. "Methods for predicting maximum hardness of heat-affected zone and selecting necessary preheat temperature for steel welding." Nippon Steel Technical Report (1995) 7-14.

[5] H. Suzuki, Revised Cold Cracking Parameter P_{HA} and its Applications. Transactions of the Japan Welding Society, (1985) 40-49.

[6] N. Yurioka, Physical metallurgy of steel weldability. ISIJ international, (2001) 566-570. https://doi.org/10.2355/isijinternational.41.566

Powder Metallurgy and Advanced Materials – RoPM&AM 2017　　　　　Materials Research Forum LLC
Materials Research Proceedings 8 (2018) 115-124　　　　doi: http://dx.doi.org/10.21741/9781945291999-13

[7] G.T. Murray, Handbook of Materials Selection for Engineering Applications, ISBN: 0-8247-9910-0, (1997) 135-136.

[8] M. Bodea, R. Mureşan, Computation of HAZ hardness for low alloyed welded steels using fiveparameter logistic function, doi:10.4028/www.scientific.net/SSP.216.103, Solid State Phenomena Vol. 216, (2014) 103-109. https://doi.org/10.4028/www.scientific.net/SSP.216.103

[9] Kobelco, The ABC's of Arc Welding and Inspection, Kobe Steel Ltd. (2015) 81-83, http://www.kobelco-welding.jp/education-center/abc/ABC_2007-01.html.

Powder Metallurgy and Advanced Materials – RoPM&AM 2017   Materials Research Forum LLC
Materials Research Proceedings 8 (2018) 125-133   Doi: http://dx.doi.org/10.21741/9781945291999-14

Atomic force microscopy analyses on metallic thin films for optical MEMS

Violeta Valentina MERIE [1, a*], Marius Sorin PUSTAN [1, b], Gavril NEGREA [1, c], Corina BÎRLEANU [1, d],

[1]Technical University of Cluj-Napoca, 103-105 Muncii Avenue, 400641, Cluj-Napoca, Romania

[a]Violeta.Merie@stm.utcluj.ro, [b]pustan@mail.utcluj.ro, [c]Gavril.Negrea@ispm.utcluj.ro, [d]cbarleanu@mail.utcluj.ro

Keywords: Metallic films, Spectroscopy in point, Work of adhesion, Friction, Nanoindentation.

Abstract. This paper is a study on three metallic thin films usable for manufacturing optical MEMS. The films were deposited by thermal evaporation on glass substrates. They were characterized from the topographical, tribological and mechanical point of view at nanoscale. The results pointed out that the silver thin films present higher values of the tribological and mechanical properties than the other two films when testing at room temperature. Increasing the testing temperature from 20 to 100 °C caused a decreased of both hardness and Young's modulus with about 30 up to 55 %.

Introduction

The optical microelectromechanical systems (MEMS) are formed in general by multi-layers of metallic thin films characterized by good optical properties. Over the last decades, the attention of the researchers was focused on developing different devices known as microelectromechanical systems (MEMS) that are satisfying the demands of the customers. The properties of the materials employed for manufacturing such devices determine its properties and its performance [1]. The optical MEMS are a category of MEMS devices that are combining the optical, mechanical, and electronic properties in a single device. They are used in the manufacture of optical sensors, attenuators, micro-lenses, micro-mirrors, displays and so on [2-5]. Aluminum [6, 7], gold [8, 9] and silver [10, 11] are one of the most used materials for manufacturing optical MEMS due to their physical, chemical, mechanical, and optical properties. These materials can be obtained as thin films by different methods such as thermal evaporation [6], magnetron sputtering [7-9], electron beam deposition [12], and so on.

Arrazat and his colleagues reported their results concerning the evolution of gold thin films deposited by sputtering on silicon substrates. They investigated the deposited films by electron back scatter diffraction analyses that allowed them to study the reliability of micro-switches manufactured using gold thin films [13]. The growth of aluminum thin films and the interfacial precipitation between such films and the silicon substrates were studied by Dutta and his co-workers. They pointed out that at the interface between the aluminum thin films and the silicon substrate during the heat treatment, some silicon precipitates are formed. According to them, these precipitates are supplying the driving force necessary for the deposit of the aluminum thin films [14]. Hojabri and his team worked on determining the influence of substrate temperature on the morphological and structural characteristics of silver thin films deposited by direct current magnetron sputtering on silicon substrates. Their results showed that the substrate temperature

Powder Metallurgy and Advanced Materials – RoPM&AM 2017   Materials Research Forum LLC
Materials Research Proceedings 8 (2018) 125-133   doi: http://dx.doi.org/10.21741/9781945291999-14

strongly influence the growth of the silver thin films, their surface roughness, as well as their grain size [15].

This research is an experimental study regarding the deposition and characterization of aluminum, gold, and silver thin films deposited by thermal evaporation on glass substrates, these films being suitable for manufacturing optical MEMS.

Materials and Experimental Procedure

Aluminum, gold and silver targets with purity of 99.99 % were employed for the deposition of the three metallic films by thermal evaporation. The films were deposited on glass substrate. The substrates were cleaned in high purity alcohol (99.9 %), in an ultrasonic bath in order to remove any possible impurities. Further they were blown with compressed air. We used resistance heated tungsten sources ("boat" type) and a vacuum atmosphere ($5 \cdot 10^{-6}$ torr). A current of 60-80 A was applied. A distance of 50 mm was kept constant between the substrates and the resistors. The deposited thin films have a thickness of about 70 nm that was determined using a JEOL JSM 5600 LV scanning electron microscope from the Materials Science and Engineering Department, Technical University of Cluj-Napoca.

The so-obtained thin films were characterized from the topographical, tribological and mechanical point of view at nanoscale. The tests were performed on a XE70 atomic force microscope (AFM) from the Micro and Nano Systems Laboratory, Technical University of Cluj-Napoca, in a clean environment. An n-type silicon NSC35C cantilever was used for studying the topography and the tribological characteristics of the three metallic thin films. Its characteristics - as the manufacturer mentioned - are: length of 130 μm, thickness of 2 μm, width of 35 μm and force constant of 5.4 N/m. The set point used during tests was of 10 nN. The determination of the adhesion parameters was realized using a PPP-NCHR cantilever by spectroscopy in point. The relative humidity was 31 % and the testing temperature was varied between 20 and 100 °C, increasing it with 20 °C per testing. As the manufacturer indicated, the characteristics of this cantilever are: tip radius smaller than 10 nm, cantilever thickness of 4 μm, its width of 30 μm and length of 125 μm, tip height between 10 and 15 nm, force constant of 42 N·m-1. A TD21464 nanoindentor was employed for determining the mechanical characteristics (hardness and Young's modulus) of the deposited films. The tests were carried out at a relative humidity of 31 % and at different temperatures namely 20, 40, 60, 80 and 100 °C. The characteristics of this nanoindentor – as given by the manufacturer are: cantilever stiffness of 156 N/m; tip thickness of 19 μm; tip height of 103 μm; tip radius smaller than 25 nm and cantilever length of 581 μm. The tests were performed at a force limit of 50 nN. The obtained curves were interpreted using the XEI Image Processing Tool for SPM (scanning probe microscopy) data by both the Oliver and Pharr (for determining the values of the hardness) and the Hertzian (for determining the values of the Young's modulus) methods [16].

Theoretical formula

Based on the data obtained for the deflection of the tip when scanning a probe by contact mode, the friction force between the AFM tip and the deposited films was calculated using the equation (1) [17]:

$$F_f = \frac{d_z \cdot r \cdot G \cdot h^3 \cdot b}{l^2 \cdot s} \tag{1}$$

where F_f represents the friction force between the tip and the tested films, dz is the deflection of the tip, r is a constant (r=0.33), G is the shear modulus (for silicon: $G=50.92 \cdot 10^{-3}$ $N \cdot \mu m^{-2}$ [18]), s is tip height (s=15 μm) while h, b and l are the dimensions of the cantilever that are specified in the previous section.

The data obtained from the spectroscopy in point tests allowed us to determine the work of adhesion, W_a, using the following formula [17]:

$$W_a = F_{adh} \cdot (c \cdot \pi \cdot R)^{-1} \qquad (2)$$

where F_{adh} represents the adhesion force between the AFM tip and the investigated thin films, c is a constant and R is tip radius. There are two models that help us to calculate the work of adhesion namely the JKR (Johnson-Kendall-Roberts) and the DMT (Derjaguin-Müller-Toporov) models respectively. The difference between these two models is given by the value of the "c" constant. Thus, the JKR model considers that c=1.5 while the DMT model considers that c=2 [19]. The tip used for testing the deposited titanium nitride films has a radius of 10 nm.

Results and Discussions

Once the films were deposited, they were characterized from topographical, adhesion, tribological and mechanical point of view, these characteristics being very important in the operating of an optical MEMS device. The average roughness, friction force, adhesion force, work of adhesion, hardness and Young's modulus were determined, and we pointed out the influence of testing temperature on the adhesion and mechanical properties.

Topography and adhesion

The values of the topography parameter were determined to validate what we have found out from the 3D images. The values of the R_a (average roughness), R_q (room mean square roughness), R_{sk} (skewness roughness) and R_{ku} (kurtosis roughness) are given in Table 1. The average values determined for aluminum, gold and silver thin films were about 5.7 nm, 2.7 nm, and 4.8 nm respectively.

Table 1. Topographical parameters of the deposited thin films

Material	R_a (nm)	R_q (nm)	R_{sk} (-)	R_{ku} (-)
Aluminum	5.68	7.14	-0.381	2.948
Gold	2.68	3.47	-0.477	4.031
Silver	4.83	5.55	0.001	1.882

Adhesion is considered to be the major failure mechanism that affects the reliability of a MEMS device. In this regard, the adhesion force between each deposited thin film and the AFM tip was determined by spectroscopy in point. At room temperature, the smallest adhesion was marked out on the aluminum thin films (78.4 nN), while the highest is specific to the silver thin films (304.4 nN). We also studied the fluctuation of the adhesion force in terms of testing temperature. This fluctuation is graphically given in Fig. 1. In general, an exponential increase of the adhesion parameter with the increase of temperature from 20 to 100 °C can be observed. The most significant increase in adhesion force is specific to the gold thin films where the adhesion

Powder Metallurgy and Advanced Materials – RoPM&AM 2017   Materials Research Forum LLC
Materials Research Proceedings **8** (2018) 125-133   doi: http://dx.doi.org/10.21741/9781945291999-14

force increases from 110.2 nN (at 20 °C) to 991 nN (at 100 °C). The work of adhesion was calculated based on the average values that we've determined for the adhesion force. This characteristic was calculated using the two models: DMT and JKR (see section 3). The fluctuation of the work of adhesion is given in Fig. 2. In both cases, the silver thin films present the highest value of this property (9.69 $J \cdot m^{-2}$ for the DMT model and 12.93 $J \cdot m^{-2}$ for the JKR model). Instead the aluminum thin films are characterized by the smallest work of adhesion (2.5 for the DMT model and 3.33 $J \cdot m^{-2}$ for the JKR model). The values for the work of adhesion of gold thin films are slightly higher (about 0.7 times) than those specific to the aluminum thin films. These results are in accordance with the results that we've obtained for the friction force of these metallic thin film. Further research will aim at determining the experimental values for this parameter and to compare them to the theoretical values that are presented in this study.

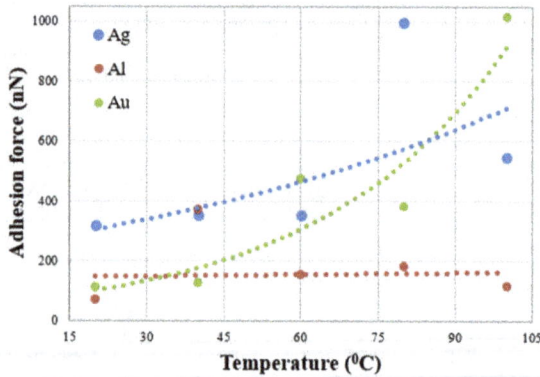

Fig. 1. Temperature influence on the adhesion force of the deposited aluminum, gold and silver thin films.

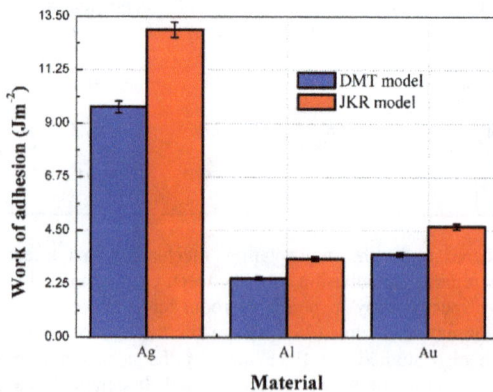

Fig. 2. Work of adhesion of the deposited aluminum, gold and silver thin film determined using the DMT and JKR models.

128

Powder Metallurgy and Advanced Materials – RoPM&AM 2017   Materials Research Forum LLC
Materials Research Proceedings **8** (2018) 125-133   doi: http://dx.doi.org/10.21741/9781945291999-14

Tribological characterization

The tribological characterization implied the determination of the friction force between the AFM tip and the surface of the deposited thin films.

The change in this friction force at room temperature is graphically given in Fig. 3. The average value if the friction parameter for the aluminum, gold and silver thin films is 0.87, 3.80 and 8.60 nN respectively. The fluctuation of friction force is, in general, of 5 % for each film. We must highlight that the friction force is strongly influence, at nanoscale, by the adhesion force. This claim confirms our previous results regarding the adhesion force of the three metallic films.

The friction force is strongly influenced at nanoscale by the adhesion force that occurs between the sample and the AFM tip. Besides the friction coefficient, this friction parameter is also directly proportional to the sum between the loading charge and the adhesion force.

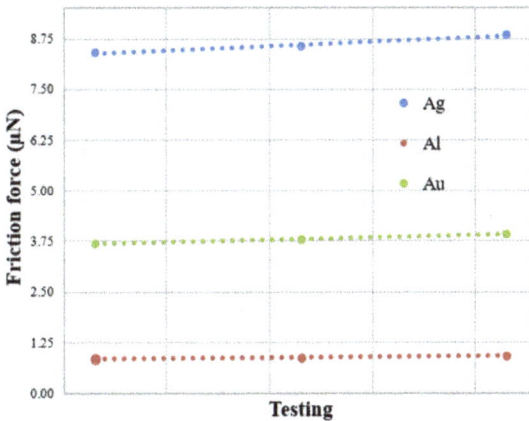

Fig. 3. The friction force between the AFM tip and the deposited aluminum, gold and silver thin films.

Mechanical characterization

The hardness and the Young's modulus were determined for each film to characterize them from the mechanical point of view. The determination of the hardness was realized using the Hertzian model while the force vs. Z scan curves were interpreted using the Oliver and Pharr model for determining the Young's modulus. The difference between these two models is that the Hertzian model assumes there is no plastic deformation after the nanoindentation of the sample. Instead, the other model assumes that the sample undergoes both plastic and elastic deformations.

This fluctuation of hardness according to the testing temperature is given in Fig. 4. As can be seen, the silver films are characterized by the highest hardness not only at room temperature (6.72 GPa) but also at 20, 40, 60, 80 and 100 °C. Although it must be pointed out that the variation in this mechanical parameter is more pronounced for the same silver films when the hardness varied from 6.72 GPa at 20 °C to 3.63 GPa at 100 °C. In particular, the values of the hardness at nanoscale for the aluminum and gold thin films are relatively closed especially when testing at 60 °C. The gold thin films are characterized by the best stability of this mechanical property, its values varying between 1.43 GPa at room temperature to 0.99 GPa at 100 °C.

Powder Metallurgy and Advanced Materials – RoPM&AM 2017   Materials Research Forum LLC
Materials Research Proceedings 8 (2018) 125-133   doi: http://dx.doi.org/10.21741/9781945291999-14

The fluctuation of Young's modulus is presented in Fig. 5. At room temperature, the aluminum, gold, and silver thin films were determined to have a Young's modulus of 21.4, 44.3, and 52.03 GPa respectively. Concerning the change in this property with the increase in temperature, the silver thin films presented a gradually decrease from 52.03 to 36.26 GPa. Conversely, the aluminum and gold thin films show between 40 and 60 °C a relative conservation of the elasticity followed by a drop in this mechanical property.

The obtained values for the mechanical characteristics of thin films determined at nano scale are smaller than the reported values for the bulk materials. This trend was also marked out by Chang-Wook Baek and his co-workers [20]. The determined values are in good agreement with the values reported in the scientific literature [12, 13, 21-23].

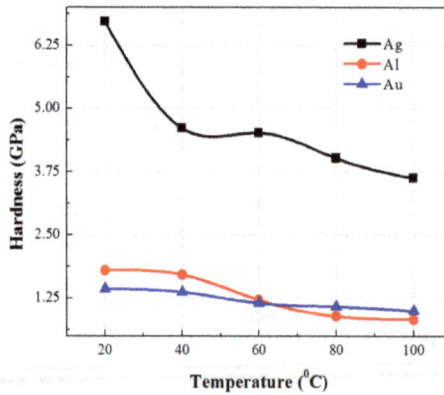

Fig. 4. Temperature influence on the hardness of the deposited aluminum, gold and silver thin films.

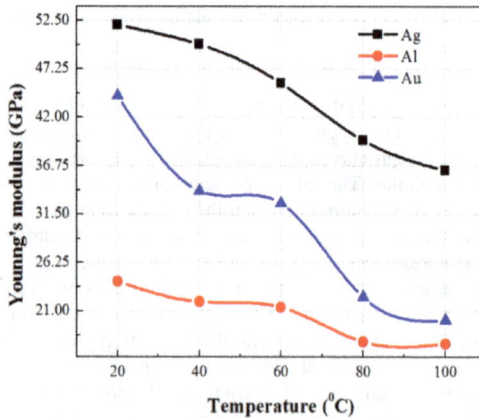

Fig. 5. Temperature influence on the Young's modulus of the deposited aluminum, gold and silver thin films.

Powder Metallurgy and Advanced Materials – RoPM&AM 2017   Materials Research Forum LLC
Materials Research Proceedings 8 (2018) 125-133   doi: http://dx.doi.org/10.21741/9781945291999-14

Conclusions

Aluminum, gold, and silver thin films were deposited by thermal evaporation on glass substrates. The deposited films were characterized by the topographical, tribological, adhesion and mechanical point of view at nanoscale. The gold thin films are characterized by the smallest value of average roughness (2.68 nm) while the highest value of this parameter (5.68 nm) was determined for the aluminum thin films. The increase of the testing temperature caused an increase of the adhesion force about 3.2 times for silver, 4.5 times for aluminum and 9 times for gold. The smallest and the highest work of adhesion was calculated for the aluminum thin films (2.5 to 3.33 J·m-2) and silver thin films (9.69 to 12.93 $J·m^{-2}$) respectively. The friction force between the deposited thin films and the AFM tip varied between 0.87 (aluminum) and 8.60 µN (silver). The silver thin films present the highest values for both the Young's modulus (52 GPa) and hardness (6.72 GPa) at room temperature. Increasing the testing temperature from 20 to 100 °C caused a decrease of the two mechanical characteristics but the decrease was more pronounced for the aluminum thin films (hardness) and gold thin films (Young's modulus).

Acknowledgement

This work was supported by a grant of the Romanian National Authority for Scientific Research and Innovation, CNCS-UEFISCDI, project number PN-II-RU-TE-2014-4-1271

References

[1] S.M. Han, R. Saha, W.D. Nix, Determining hardness of thin films in elastically mismatched film-on-substrate systems using nanoindentation, Acta Mater. 54 (2006) 1571-1581. https://doi.org/10.1016/j.actamat.2005.11.026

[2] C.L. Linslal, P.M. Syam Mohan, A. Halder, T.K. Gangopadhyay, Analysis and modeling of an optical fiber loop resonator and an evanescent field absorption sensor for the application for chemical detection, Sens. Actuators A 194 (2013) 160-168. https://doi.org/10.1016/j.sna.2013.01.021

[3] E.M. Bourim, H.-Y. Kim, J.-S. Yang, J.-W. Yang, K.-S. Woo, J.-H. Song, S.-K. Yun, Creep behavior of undoped and La–Nb codoped PZT based micro-piezoactuators for micro-optical modulator applications, Sens. Actuators A 155 (2009) 290-298. https://doi.org/10.1016/j.sna.2009.08.020

[4] W. Hortschitz, H. Steiner, M. Sachse, F. Kohl, J. Schalko, F. Keplinger, Hybrid optical MEMS vibration sensor, Procedia Eng. 5 (2010) 420–423. https://doi.org/10.1016/j.proeng.2010.09.136

[5] C. Winter, L. Fabre, F. Lo Conte, L. Kilcher, F. Kechana, N. Abelé, M. Kayal, Micro-beamer based on MEMS micro-mirrors and laser light source, Procedia Chem. 1 (2009) 1311–1314. https://doi.org/10.1016/j.proche.2009.07.327

[6] N. Barbosa, R.R. Keller, D.T. Read, R.H. Geiss, R.P. Vinci, Comparison of electrical and microtensile evaluations of mechanical properties of an aluminum film, Metall. Mater. Trans. A 38A (2007) 2160-2167. https://doi.org/10.1007/s11661-007-9112-y

[7] J. Martinez-Quijada, S. Caverhill-Godkewitsch, M. Reynolds, L. Gutierrez-Rivera, R.W. Johnstone, D.G. Elliott, D. Sameoto, C.J. Backhouse, Fabrication and characterization of aluminum thin film heaters and temperature sensors on a photopolymer for lab-on-chip systems, Sens. Actuators A 193 (2013) 170–181. https://doi.org/10.1016/j.sna.2013.01.035

Powder Metallurgy and Advanced Materials – RoPM&AM 2017   Materials Research Forum LLC
Materials Research Proceedings 8 (2018) 125-133   doi: http://dx.doi.org/10.21741/9781945291999-14

[8] O. Okman, J. W. Kysar, Fabrication of crack-free blanket nanoporous gold thin films by galvanostatic dealloying, J. Alloys Compd. 509 (2011) 6374–6381. https://doi.org/10.1016/j.jallcom.2011.02.115

[9] V. Raffa, B. Mazzolai, A. Mondini, V. Mattoli, A. Menciassi, P. Dario, Investigation on a sensitive Au thin film deposited on different substrates: Physical analysis via FIB and chemical analysis via evaluation of Au sensitivity to Hg0, Sens. Actuators B 122 (2007) 475–483. https://doi.org/10.1016/j.snb.2006.06.013

[10] Y. Nakanishi, K. Kato, H. Omoto, T. Tomioka, Improvement in salt-water durability of Ag thin films deposited by magnetron sputtering using argon and nitrogen mixed gas, Vac.87 (2013) 232-236. https://doi.org/10.1016/j.vacuum.2012.02.053

[11] I.C. Estrada-Raygoza, M. Sotelo-Lerma, R. Ramírez-Bon, Structural and morphological characterization of chemically deposited silver films, J. Phys. Chem. Solids 67 (2006) 782–788. https://doi.org/10.1016/j.jpcs.2005.10.183

[12] Y. Cao, S. Allameh, D. Nankivil, S. Sethiaraj, T. Otiti, W. Soboyejo, Nanoindentation measurements of the mechanical properties of polycrystalline Au and Ag thin films on silicon substrates: Effects of grain size and film thickness, Mater. Sci. Eng. A 427 (2006) 232–240. https://doi.org/10.1016/j.msea.2006.04.080

[13] B. Arrazat, V. Mandrillon, K. Inal, M. Vincent, C. Poulain, Microstructure evolution of gold thin films under spherical indentation for micro switch contact applications, J. Mater. Sci. 46 (2011) 6111–6117. https://doi.org/10.1007/s10853-011-5575-8

[14] I. Dutta, M. Burkhard, S. Kuwano, T. Fujita, M.W. Chen, Correlation between surface whisker growth and interfacial precipitation in aluminum thin films on silicon substrates, J. Mater. Sci. 45 (2010) 3367-3374. https://doi.org/10.1007/s10853-010-4359-x

[15] A. Hojabri, Z. Kavyani, M. Ghoranneviss, Effect of substrate temperatures on structural and morphological properties of nano-crystalline silver thin films grown on silicon substrates, J. Inorg. Organomet. Polym. Mater. 27 (2017) 53-60. https://doi.org/10.1007/s10904-016-0443-2

[16] W.C. Oliver, G.M. Pharr, An improved technique for determining hardness and elastic modulus using load and displacement sensing indentation experiments, J. Mater. Res.7 (1992) 1564-1583. https://doi.org/10.1557/JMR.1992.1564

[17] V. Merie, M. Pustan, G. Negrea, C. Birleanu, Research on titanium nitride thin films deposited by reactive magnetron sputtering for MEMS applications, Appl. Surf. Sci. 358 (2015) 525-532. https://doi.org/10.1016/j.apsusc.2015.07.063

[18] M.A. Hopcroft, W.D. Nix, T.K. Kenny, What is the Young's modulus of silicon?, J. Microelectromech. Syst. 19 (2010) 229-238. https://doi.org/10.1109/JMEMS.2009.2039697

[19] J. Drelich, Adhesion forces measured between particles and substrates with nano-roughness, Miner. Metall. Process. 23 (2006) 226-232.

[20] C.-W. Baek, Y.-K. Kim, Y. Ahn, Y.-H. Kim, Measurement of the mechanical properties of electroplated gold thin films using micromachined beams structures, Sens. Actuators A 117 (2005) 17-27. https://doi.org/10.1016/j.sna.2003.11.041

[21] H.D. Esponisa, B.C. Prorok, Size effects on the mechanical behavior of gold thin films, J. Mater. Sci. 38 (2003) 4125-4128. https://doi.org/10.1023/A:1026321404286

Powder Metallurgy and Advanced Materials – RoPM&AM 2017   Materials Research Forum LLC
Materials Research Proceedings 8 (2018) 125-133   doi: http://dx.doi.org/10.21741/9781945291999-14

[22] P. Peng, G. Liao, T. Shi, Z. Tang, Y. Gao, Molecular dynamic simulations of nanoindentation in aluminum thin film on silicon substrate, Appl. Surf. Sci. 256 (2010) 6284-6290. https://doi.org/10.1016/j.apsusc.2010.04.005

[23] S.W. Han, H.W. Lee, H.J. Lee, J.Y. Kim, J.H. Kim, C.S. Oh, S.H. Choa, Mechanical properties of Au thin film for application in MEMS/NENS using microtensile test. Curr. Appl. Phys. 6S1 (2006) e81-e85.

Powder Metallurgy and Advanced Materials – RoPM&AM 2017   Materials Research Forum LLC
Materials Research Proceedings 8 (2018) 134-142   doi: http://dx.doi.org/10.21741/9781945291999-15

Structural and optical characterization of titanium nitride thin films deposited by magnetron sputtering

Violeta Valentina MERIE [1, a *], Andreia MOLEA [1, b], Vlad Nicolae BURNETE [1, c],
Bogdan Viorel NEAMȚU [1, d] and Gavril NEGREA [1, e]

[1]Technical University of Cluj-Napoca, 103-105 Muncii Avenue, 400641, Cluj-Napoca, Romania

[a]Violeta.Merie@stm.utcluj.ro, [b] Andreia.Molea@chem.utcluj.ro, [c]
Nicolae.Vlad.Burnete@auto.utcluj.ro, [d]Bogdan.Neamtu@stm.utcluj.ro,
[e]Gavril.Negrea@ispm.utcluj.com

Keywords: Titanium nitride, Thin films, Magnetron sputtering, Reflectivity, Polarized, Semiconductor.

Abstract. Titanium nitride applicability covers different industries such as microelectronics, biomedicine and so on. This paper presents the analysis of the structural and optical properties of titanium nitride thin films for different deposition conditions. The samples were deposited by direct current magnetron sputtering on silicon substrates. The deposition was done at room temperature, on substrates preheated at 300 °C or on substrates that were polarized at -40 V and -90 V respectively. The results indicate a dependency of the structural orientation with respect to the deposition process when this takes place at room temperature. When the deposition was done on a preheated substrate there was no structural orientation. A negative polarization of the substrate leads to the formation of small sized crystallites. Regarding the optical properties, the films showed good semiconductor properties and a low reflectivity.

Introduction

Titanium nitride (TiN) thin films were studied by many researchers due to their excellent properties, especially mechanical and tribological properties, corrosion resistance, wear resistance and thermodynamic stability [1–3]. Due to these properties, titanium nitride thin films can be used in a wide range of applications like: diffusion barriers for micro-electric devices, optical coatings with antireflection and antistatic properties, electrodes, biomedicine and hard coatings for tools and so on [4–9].

The most often used methods to obtain titanium nitride films are: reactive magnetron sputtering, laser ablation, ion beam deposition or plasma assisted chemical vapor deposition and so on [10–14]. The physical-chemical and mechanical/tribological properties of titanium nitride films depend on the deposition parameters. In this regard, different researches present the influence of some deposition parameters such as the deposition rate, deposition time, substrate, the heating or the polarization of the substrate on the topographical, mechanical, tribological, adhesion properties for titanium nitride thin films deposited by DC (direct current) magnetron sputtering. All the results are pointing out a change in these properties with the change in deposition parameters. A possible explanation for this change can be the growth of the deposited films after different preferential orientations.

The present paper is a study concerning the deposition of titanium nitride thin films by DC magnetron sputtering on silicon substrates at different deposition parameters (substrate

Powder Metallurgy and Advanced Materials – RoPM&AM 2017   Materials Research Forum LLC
Materials Research Proceedings 8 (2018) 134-142   doi: http://dx.doi.org/10.21741/9781945291999-15

temperature, substrate bias voltage, deposition time) and the structural and optical characterization of the obtained thin films.

Materials and Methods
Deposition of titanium nitride thin films
The deposition of titanium nitride films was done by direct current reactive magnetron sputtering method, using 99.99 % purity titanium target and silicon Si (100) substrate. The experimental procedure details are presented in previously published work of the authors [4]. The parameters for titanium nitride films deposition were: (i) deposition time (10, 20 and 40 minutes) with the substrate (RT) at room temperature, (ii) deposition time (20 minutes) with the substrate at a temperature of 300 °C and (iii) deposition time (20 minutes) and polarization of the substrate at - 40 V and -90 V. These studies were conducted in order to determine the influence of the deposition conditions on the structural and optical properties of titanium nitride films destined for MEMS devices applications.

Characterization of titanium nitride thin films
X-ray diffraction analysis was carried out with an Inel Equinox 3000 diffractometer using a cobalt radiation (λ_1=1.7889 Å, λ_2=1.7928 Å). The patterns were evaluated using JPCDS 87-0633. The mean crystallite size and lattice strain were calculated using the Williamson-Hall method. According to this method, the broadening of the diffraction line, β, is given by the sum of the broadening introduced by the crystallite size, β_d, and the broadening introduced by the internal strain, β_ε (equation (1)). These two parameters can be calculated using the equations (2) and (3) as follows [15]:

$$\beta = \beta_d + \beta_\varepsilon \tag{1}$$

$$\beta_d = \frac{K \cdot \lambda}{D \cdot \cos\theta} \tag{2}$$

$$\beta_\varepsilon = 4 \cdot \varepsilon \cdot tg\theta \tag{3}$$

Where K is Scherrer's constant, λ is the wavelength of the X-ray radiation, D is the mean crystallite size, θ is the Bragg angle while ε represents the internal strain. Starting from the equations mentioned above and the graphical plotting of the dependency $\beta \cdot \cos\theta = f(4 \cdot \sin\theta)$, the mean crystallite size and the internal strain can be determined. Thus, the internal strain is given by the line's slope while the mean crystallite size is given by the intersection between the line and the Y axis ($\beta \cdot \cos\theta$).

UV-Vis spectroscopy studies were conducted in order to determine the optical properties of the films. Using an aluminum mirror as reference, the diffuse reflectance spectra were measured at 8° with a Lambda 35 Perkin-Elmer spectrometer equipped with an integrated sphere. The energy band gap of the titanium nitride films was determined based on the Kubelka-Munk function [16].

Results and discussions
XRD patterns of the titanium nitride films deposited at different deposition time are presented in Fig. 1. The crystalline phase of titanium nitride was identified based on JPCDS standard diffraction no. 87-0633. Based on the results obtained using the Williamson-Hall relation, it was observed that the increase in deposition time tends to decrease the structural parameters, i.e. crystallite size, lattice strain and cell volume (Table 1). It was also observed that for a deposition

Powder Metallurgy and Advanced Materials – RoPM&AM 2017   Materials Research Forum LLC
Materials Research Proceedings **8** (2018) 134-142   doi: http://dx.doi.org/10.21741/9781945291999-15

time of 10 minutes, the titanium nitride film has a textural orientation in the (111) crystalline plane, while when increasing the deposition time to 40 minutes, the preferred orientation of the deposited titanium nitride film changes to the (220) crystalline plane.

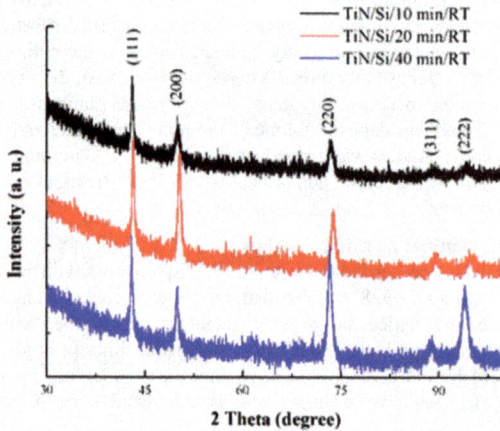

Fig. 1. XRD patterns of titanium nitride thin films deposited at different deposition time on a silicon substrate at room temperature.

The change in the preferred orientation, confirmed by other authors like [10, 14, 17], can be attributed to the increased energy of the bombarding particles from titanium target. Another cause for the change in the structural parameters with the increase in the deposition time can be the change in the preferred orientation, which can deform the cell parameters and to the increased thickness of the film [10]. According to the theoretical calculation, titanium nitride films thickness increased (0.27 μm, 0.55 μm and 1.10 μm) with the deposition time (10, 20 and 40 minutes). However, the change in lattice strain can be explain due to the presence of the $(TiN)_{.76}$ (JCPDS file no. 01-087-0626), $(TiN)_{.88}$ (JCPDS file no. 01-087-0630) and TiN (JCPDS file no. 01-087-0633) compounds in the structure of the deposited films. For instance, the presence of the $TiN_{0.76}$ leads to the change in 2θ angle from 50.68 ° (specific to the stoichiometric compound TiN) to 49.97 ° in the case of the (200) crystalline plane. The values of the lattice strain given in Table 1 are the average values obtained for each film and they are influenced by the quantity of non-stoichiometric compounds. The presence of non-stoichiometric compounds was also reported by other researchers [18, 19].

Fig. 2 presents the XRD patterns for the titanium nitride film deposited on the silicon substrate preheated at 300 °C and the titanium nitride film deposited on the silicon substrates negatively biased at -40 V and -90 V respectively, when the deposition time was kept constant at 20 minutes. It is obvious that the deposition parameters have a strong influence on the chemical composition and, implicitly, on the properties of the investigated thin films. As mentioned above, non-stoichiometric compounds exist in the deposited thin films. Their existence determines a shift to the left of the peaks. We assume that the shift is more significant in the case of titanium nitride

Powder Metallurgy and Advanced Materials – RoPM&AM 2017   Materials Research Forum LLC
Materials Research Proceedings 8 (2018) 134-142   doi: http://dx.doi.org/10.21741/9781945291999-15

thin films that are containing higher quantities of non-stoichiometric compounds. When the silicon substrate was polarized at -40 V and -90 V for a deposition time of 20 minutes, compared to the un-polarized film, it was observed a decrease in crystallite size titanium nitride films (Table 1) from 21 nm for the un-polarized titanium nitride film to 16-17 nm for titanium nitride film for the polarized substrate. According to Benegra et al. [20] a decrease of crystallite size with the increase of negative bias can be explained by the fact that the pulsed power may not only increase the impinging atom energy, but it may also increase the ion current densities.

Table 1. Structural parameters of titanium nitride thin films

Samples	Average crystallite size [nm]	Lattice strain [%]	Cell volume [Å³]
TiN/Si/10 min	37	0.379	12.73
TiN/Si/20 min	21	0.272	12.70
TiN/Si/40 min	17	0.160	12.69
TiN/Si/20 min/300 °C	17	0.210	12.70
TiN/Si/20 min/-40 V	17	0.585	12.73
TiN/Si/20 min/-90 V	16	0.930	12.73
TiN/JPCDS 87-0633	-	-	12.73

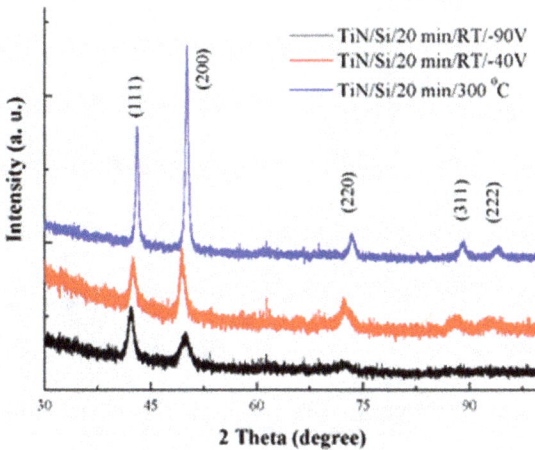

Fig. 2. XRD patterns of titanium nitride films deposited on substrates heated at a temperature 300 °C and polarized at -40 V and -90 V for 20 minutes.

Powder Metallurgy and Advanced Materials – RoPM&AM 2017   Materials Research Forum LLC
Materials Research Proceedings 8 (2018) 134-142   doi: http://dx.doi.org/10.21741/9781945291999-15

Fig. 3 shows the specular reflectance of titanium nitride films deposited on silicon substrates at room temperature at different deposition times. All the films exhibit a low reflectivity, between 300 nm and 500 nm, but as the wavelength increases to the infrared region, the reflectance also increases, especially for the titanium nitride film deposited for 20 minutes which has a of reflectance nearly 60 %.

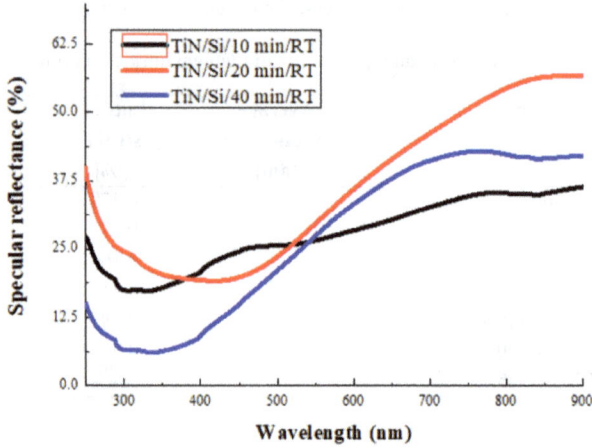

Fig. 3. Reflectance spectra of titanium nitride films deposited at different deposition time on a silicon substrate at room temperature.

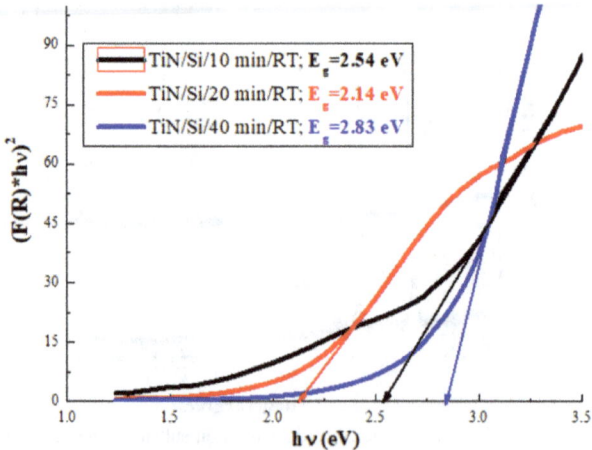

Fig. 4. Energy band gap of titanium nitride films deposited at different deposition time on silicon substrate at room temperature.

Powder Metallurgy and Advanced Materials – RoPM&AM 2017   Materials Research Forum LLC
Materials Research Proceedings 8 (2018) 134-142   doi: http://dx.doi.org/10.21741/9781945291999-15

The energy band gaps suggest that the titanium nitride films deposited at different deposition times are semiconducting materials (Fig. 4). The titanium nitride film deposited for 20 minutes exhibit a lower energy band gap (2.14 eV) than the films deposited for 10 and 40 minutes, respectively (2.54 eV and 2.83 eV).

A red-shifting in the optical energy of titanium nitride film deposited for 20 minutes, compared to the film deposited for 10 minutes can be attributed to the increase in the film thickness and with the change in the textural orientation from (111) crystalline plane to (220) crystalline plane, which also implied an increase in the roughness of the films [4]. According to Kiran et al. [21] the change in the titanium nitride films thickness causes a shift in the optical edge and therefore a change in the band structure of the films.

Fig. 5 presents the reflectance spectra of the titanium nitride films deposited for 20 minutes on a silicon substrate heated at 300 °C and on substrates, at room temperature, to which a negative bias-voltage of -40 V and -90 V respectively was applied. Compared to the titanium nitride film deposited on the silicon substrate heated at 300 °C, which have a reflectivity between 300 nm and 500 nm, the titanium nitride films deposited on a substrate with different voltage bias, the reflectivity increases to between 500 nm and 900 nm.

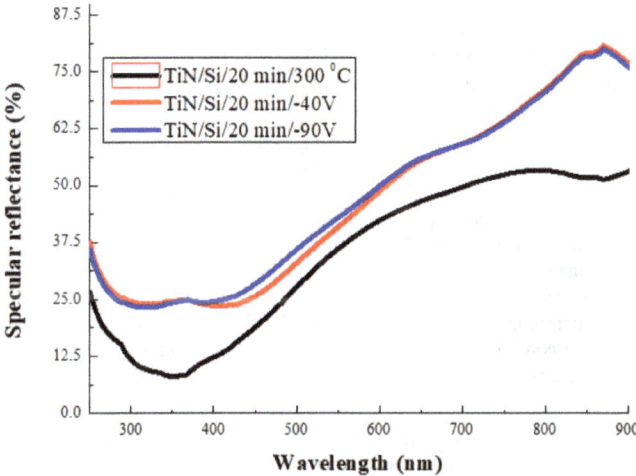

Fig. 5. Reflectance spectra of titanium nitride films deposited for 20 minutes on a silicon substrate preheated at 300 °C and deposited on a substrate at room temperature and polarized at -40 V and -90 V.

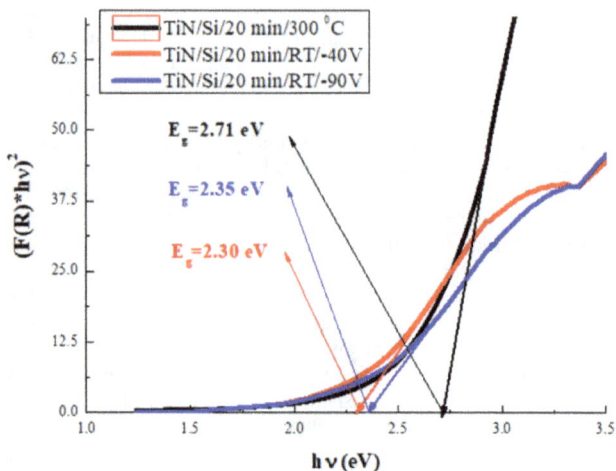

Fig. 6. Energy band gap of titanium nitride films deposited on a silicon substrate preheated at 300 °C and deposited on substrate at room temperature and a voltage bias of -40 V and -90 V.

The crystalline degree of the titanium nitride film deposited on a substrate heated at 300 °C increases compared to the titanium nitride film deposited on un-heated substrate. The energy band gap (Fig. 6) also increases due to the blue-shifting of the optical response. The latter is caused by a decrease in the crystallite size that implies band structure modifications due to cell size and by the fact that titanium nitride film deposited on a heated substrate do not present a textural orientation unlike the titanium nitride film deposited on an un-heated substrate [22]. The blue-shifting of the optical response was also observed for the films deposited at negative bias voltage. This can be also attributed to the decrease in the crystallite size.

It can be concluded that titanium nitride films exhibit semiconducting properties based on the structural and optical characterization of titanium nitride films deposited on silicon substrate at different deposition conditions. As a result, titanium nitride films may have possible applications in solar cells, optoelectronics or microelectronic-mechanical devices.

Summary
The deposition of titanium nitride films on a silicon Si (100) substrate at different deposition parameters influences the structural and optical properties of the material. Thus, when the deposition time increases from 10 to 40 minutes and the deposition is done on silicon substrates at room temperature, the textural orientation changes from (111) orientation to the (220) crystalline plane due to increase in both the energy of the bombarding particles of the titanium target and in the film thickness. Heating of the silicon substrate at 300 °C leads to an increase of crystallinity, but the titanium nitride film does not present a textural orientation. Applying a negative bias voltage to the substrate leads to the formation of titanium nitride films with small crystallite size. By optical means, titanium nitride films exhibit semiconducting properties, with a low reflectivity (between 300 nm and 500 nm), but they increase for a wavelength in the infrared region. The energy band gap of the films is also influenced by structural properties (crystallite size and textural orientation) and film thickness.

Powder Metallurgy and Advanced Materials – RoPM&AM 2017   Materials Research Forum LLC
Materials Research Proceedings 8 (2018) 134-142   doi: http://dx.doi.org/10.21741/9781945291999-15

Acknowledgement
This paper is written within the TUCN Internal Research Project Competition 2015 "Elaboration and structural, tribo-mechanical and optical characterization of some nitride thin films for MEMS applications (NitriMEMS)" (C.I.5/1.2/2015). The Internal Research Project Competition is funded by the Technical University of Cluj-Napoca in order to support and encourage young teams in addressing new topics in priority research areas of the university.

References

[1] P.J.J. Kelly, T. Vom Braucke, Z. Liu, R.D.D. Arnell, E.D.D. Doyle, Pulsed DC titanium nitride coatings for improved tribological performance and tool life (Report), Surf. Coat. Technol. 202 (2007) 774-780. https://doi.org/10.1016/j.surfcoat.2007.07.047

[2] H.Z. Durusoy, Ö. Duyar, A. Aydınlı, F. Ay, Influence of substrate temperature and bias voltage on the optical transmittance of TiN films, Vac. 70 (2003) 21–28. https://doi.org/10.1016/S0042-207X(02)00663-2

[3] K. Vasu, M.G. Krishna, K.A. Padmanabhan, Substrate-temperature dependent structure and composition variations in RF magnetron sputtered titanium nitride thin films, Appl. Surf. Sci. 257 (2011) 3069–3074. https://doi.org/10.1016/j.apsusc.2010.10.118

[4] V. Merie, M. Pustan, G. Negrea, C. Bîrleanu, Research on titanium nitride thin films deposited by reactive magnetron sputtering for MEMS applications, Appl. Surf. Sci. 358 (2015) 525–532. https://doi.org/10.1016/j.apsusc.2015.07.063

[5] X. Cao, T. Shao, S. Wen, Y. Yao, Micro/nanotribological and mechanical studies of TiN thin-film for MEMS applications, Tribol. Trans. 47 (2004) 227–232. https://doi.org/10.1080/05698190490439076

[6] N.Y. Kim, Y.B. Son, J.H. Oh, C.K. Hwangbo, M.C. Park, TiNx layer as an antireflection and antistatic coating for display, Surf. Coat. Technol. 128 (2000) 156–160. https://doi.org/10.1016/S0257-8972(00)00574-0

[7] D.-R. Deng, T.-H. An, Y.-J. Li, Q.-H. Wu, M.-S. Zheng, Q.-F. Dong, Hollow porous titanium nitride tubes as a cathode electrode for extremely stable Li-S batteries, J. Mater. Chem. A 4(2016) 16184–16190. https://doi.org/10.1039/C6TA07221F

[8] D. Starosvetsky, I. Gotman, Corrosion behavior of titanium nitride coated Ni–Ti shape memory surgical alloy, Biomater. 22 (2001) 1853–1859. https://doi.org/10.1016/S0142-9612(00)00368-9

[9] Z. Peng, H. Miao, L. Qi, S. Yang, C. Liu, Hard and wear-resistant titanium nitride coatings for cemented carbide cutting tools by pulsed high energy density plasma, Acta Mater. 51 (2003) 3085–3094. https://doi.org/10.1016/S1359-6454(03)00119-8

[10] S. Chinsakolthanakorn, A. Buranawong, S. Chiyakun, P. Limsuwan, Effects of titanium sputtering current on structure and morphology of TiZrN films prepared by reactive DC magnetron co-sputtering, Mater. Sci. Appl. 4 (2013) 689–694. https://doi.org/10.4236/msa.2013.411086

Powder Metallurgy and Advanced Materials – RoPM&AM 2017   Materials Research Forum LLC
Materials Research Proceedings 8 (2018) 134-142   doi: http://dx.doi.org/10.21741/9781945291999-15

[11] J. Bonse, H. Sturm, D. Schmidt, W. Kautek, Chemical, morphological and accumulation phenomena in ultrashort-pulse laser ablation of TiN in air, Appl. Phys. A 71 (2000) 657–665. https://doi.org/10.1007/s003390000585

[12] A.P. Serro, C. Completo, R. Colaço, F. dos Santos, C.L. da Silva, J.M.S. Cabral, B. Saramago, A comparative study of titanium nitrides, TiN, TiNbN and TiCN, as coatings for biomedical applications, Surf. Coat. Technol. 203 (2009) 3701–3707. https://doi.org/10.1016/j.surfcoat.2009.06.010

[13] G. Zhao, C. Zhao, L. Wu, G. Duan, J. Wang, G. Han, Study on the electrical and optical properties of vanadium doped TiN thin films prepared by atmospheric pressure chemical vapor, J. Alloys Compd. 569 (2013) 1–5. https://doi.org/10.1016/j.jallcom.2013.03.110

[14] S.H. Kim, H. Park, K.H. Lee, S.H. Jee, D.-J. Kim, Y.S. Yoon, H.B. Chae, Structure and mechanical properties of titanium nitride thin films grown by reactive pulsed laser deposition, J. Ceram. Process. Res. 10 (2009) 49–53.

[15] G.K. Williamson, W.H Hall, X-ray line broadening from filed aluminum and wolfram, Acta Metallurgica 1 (1953) 22-31. https://doi.org/10.1016/0001-6160(53)90006-6

[16] A.E. Morales, E.S. Mora, U. Pal, Use of diffuse reflectance spectroscopy for optical characterization of un-supported nanostructures, Rev. Mex. Fis. 53 (2007) 18-22.

[17] A. Yazdani, M. Soltanieh, H. Aghajani, S. Rastegari, A new method for deposition of nano sized titanium nitride on steels, Vacuum 86 (2011) 131-139. https://doi.org/10.1016/j.vacuum.2011.04.020

[18] V. Karagkiozaki, S. Logotheidis, N. Kalfagiannis, S. Lousinian, G. Giannoglou, Atomic force microscopy probing platelet activation behavior on titanium nitride nanocoatings for biomedical applications, Nanomedicine: Nanotechnology, Biology, and Medicine 5 (2009) 64-72. https://doi.org/10.1016/j.nano.2008.07.005

[19] C.-F. Hsieh, S. Jou, Titanium nitride electrodes for micro-gap discharge, Microelectronics Journal 37 (2006) 867-870. https://doi.org/10.1016/j.mejo.2006.03.003

[20] M. Benegra, D.G. Lamas, M.E. Fernández de Rapp, N. Mingolo, A.O. Kunrath, R.M. Souza, Residual stresses in titanium nitride thin films deposited by direct current and pulsed direct current unbalanced magnetron sputtering, Thin Solid Films 494 (2006)146–150. https://doi.org/10.1016/j.tsf.2005.08.214

[21] M.S.R.N. Kiran, M.G. Krishna, K.A. Padmanabhan, Growth, surface morphology, optical properties and electrical resistivity of ε-TiNx, Appl. Surf. Sci. 255 (2008) 1934–1941. https://doi.org/10.1016/j.apsusc.2008.06.122

[22] P. Patsalas, S. Logothetidis, Optical, electronic, and transport properties of nanocrystalline titanium nitride thin films, J. Appl. Phys. 90 (2001) 4725–4734. https://doi.org/10.1063/1.1403677

Powder Metallurgy and Advanced Materials – RoPM&AM 2017   Materials Research Forum LLC
Materials Research Proceedings 8 (2018) 143-151   doi: http://dx.doi.org/10.21741/9781945291999-16

Green synthesis of Reduced Graphene Oxide (RGNO) / Polyvinylchloride (PVC) composites and their structural characterization

Ferda MINDIVAN[1, a *], Meryem GOKTAS[2, b]

[1]Bilecik Seyh Edebali University, Biotechnology Research and Application Center, Bilecik, TURKEY

[2]Bilecik Seyh Edebali University, Vocational College, Department of Metallurgy, Bilecik, TURKEY

[a,*]ferda.mindivan@bilecik.edu.tr, [b]meryem.goktas@bilecik.edu.tr

Keywords: Reduced graphene oxide (RGNO), Vitamin C, Green synthesis, Polyvinylchloride (PVC), Composite.

Abstract. Graphene and graphene derivatives are widely used as fillers for polymer composite materials. The Reduced Graphene Oxide (RGNO) is usually considered as one kind of chemically derived graphene, just like Graphene Oxide (GO). However, very dangerous chemicals are used for synthesis of RGNO. Specially, hydrazine hydrate used to form RGNO is highly toxic and unstable. In this paper, a green strategy was reported for the synthesis of RGNO. To this aim, firstly GO was prepared from natural graphite by Hummers method and then obtained GO was reduced by vitamin C. Structural characterization results revealed that GO was successfully reduced to RGNO. RGNO filled polyvinylchloride (PVC) composites were prepared by colloidal blending method. The structural changes were observed in RGNO/PVC composites as a function of RGNO loading and confirmed by FTIR, XRD and SEM analyses. These analyses indicated that RGNO layers were fully exfoliated and well-dispersed in the PVC matrix.

Introduction

Graphene has recently attracted attention as inexpensive filler in composite materials that can be used in a wide range of potential applications due to its excellent structural, mechanical, thermal, and electrical properties [1, 2]. The most common method for producing graphene is chemical reduction of graphene oxide (GO) [3]. However, negative aspect of the reduction process is the high toxic nature of the reducing agents (hydrazine hydrate, di-methylhydrazine, hydroquinone, sodium borohydride, metal hydrides) [4]. These agents are harmful to both humans and the environment [5]. Specially, hydrazine hydrate is the widely used as reduction agent but it is highly toxic and dangerously unstable [6]. Therefore, it is very important to find harmless and efficient reducing agents to reduce GO [7]. Recently, nontoxic or natural products such as non-aromatic amino acids [3], leaf extracts of natural products [4], ethylene glycol [6], sodium carbonate [8], ascorbic acid [9], sugar [10], green tea [11] etc. that offer environmentally friendly approaches to reduce GO have been developed [12]. PVC is a common thermoplastic as a host polymer matrix for preparing polymeric composite materials because of its low cost and excellent chemical stability and biocompatibility [13, 14]. But PVC has disadvantageous properties such as poor processability, thermal stability and weatherability [15]. Recently, the improvement in thermal, electrical and mechanical properties of PVC matrix with graphene derivatives has been reported in many literatures [15 - 21]. However, there are no published data concerning the effect of the different amounts of RGNO that prepared with Vitamin C on the properties of RGNO/PVC composites. In the present study, we prepared RGNO filled PVC composites by a solution

Powder Metallurgy and Advanced Materials – RoPM&AM 2017   Materials Research Forum LLC
Materials Research Proceedings 8 (2018) 143-151   doi: http://dx.doi.org/10.21741/9781945291999-16

blending method and we are reporting a simple method of preparation of RGNO using Vitamin C. The effect of RGNO on structural properties has been investigated. This research opens a new route to fabricate composites with green synthesis of filler which can be a promising material for many applications.

Material and Method

Graphite powder (GF), concentrated sulfuric acid (98 % - H_2SO_4), potassium permanganate ($KMnO_4$), hydrogen peroxide (30 % - H_2O_2) solution, hydrochloric acid (HCl) and Vitamin C (L(+) - Ascorbic Acid) were of reagent grade and purchased from Merck. All the reagents were used without further purification. All solutions were prepared using deionized (DI) water. GO was prepared from natural graphite (45μm nominal particle size) by the Hummers method [22]. Graphite (1 g) was mixed with 69 mL of concentrated H_2SO_4 and the mixture was stirred in an ice bath for around 30 min. After homogeneous dispersion of the GF in the solution, $KMnO_4$ (8 g) was added slowly to the solution in an ice bath and the reaction mixture was stirred for 15 min. under a reaction temperature of 20°C. Then the ice bath was removed and the mixture was stirred at 35°C overnight to form thickened paste. Afterward 70 mL of de-ionized water was added slowly into the reaction solutions to avoid the reaction temperature rising to a limit of 98 °C. After 2 hours of vigorous stirring, 12 mL of 30 % H_2O_2 was added and the color turned golden yellow immediately. Finally, the mixture was then filtered and washed several times with 3 % HCl and DI water until pH 7 and dried at 65°C for 12 hours to obtain GO powder.

To prepare RGNO, 0.5 g of GO was dispersed in 100 mL of DI water. pH of the GO suspension was adjusted to ~10 by using ammonia solution. Then 0.75 g of Vitamin C was added to the mixture and heated at 95°C for 12 hours. After that the mixture was filtered and the RGNO was obtained as a black powder. This powder was washed with DI water several times.

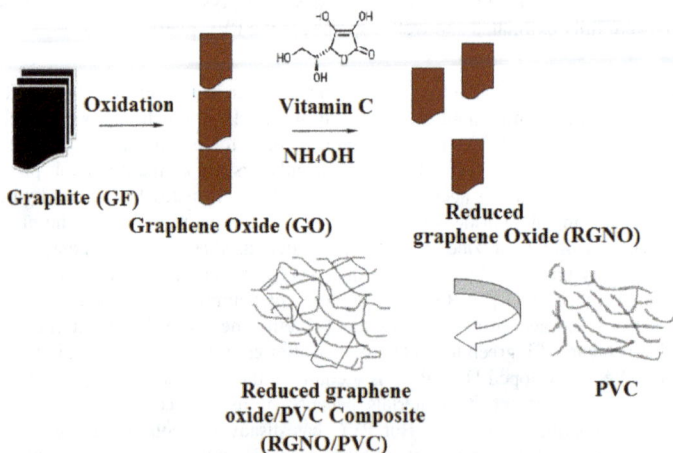

Fig. 1. Synthesis process of RGNO/PVC composites.

RGNO/PVC composites were prepared by a colloidal blending method. PVC (1 g) was first dissolved in Tetrahydrofuran (THF) at 70 °C and was cooled to room temperature. RGNO powder was separately dispersed in THF at 25°C. The two solutions were stirred for 2 hours at 60°C. The

Powder Metallurgy and Advanced Materials – RoPM&AM 2017   Materials Research Forum LLC
Materials Research Proceedings **8** (2018) 143-151   doi: http://dx.doi.org/10.21741/9781945291999-16

resulting homogeneous dispersion was poured into glass petri dish and kept in an oven at 60 °C for slow evaporation of the solvent to get RGNO/PVC composite. The synthesis process of RGNO/PVC composites is illustrated in Fig. 1. The RGNO content in the RGNO/PVC composite was varied from 0.1 – 1 wt. % (Table 1).

Table 1. Ratios and codes of RGNO in the composites.

Samples	RGNO Content (wt. %)
0.1-RGNO/PVC	0.1
0.3-RGNO/PVC	0.3
0.5-RGNO/PVC	0.5
1-RGNO/PVC	1

Structural analyses of the RGNO powders and composites were carried out by FTIR spectra (Spectrum 100, Perkin Elmer) in the range of 4000–400 cm^{-1} and X-Ray Diffraction (XRD, PAN analytical, Empyrean) in the range of 5 – 40°. The surface morphology was examined by a Scanning Electron Microscopy (SEM, Supra 40VP, Zeiss). EDS analyses were performed on the same instrument.

Results and Discussion
FTIR spectra for GO, RGNO, neat PVC and RGNO/PVC composites were presented in Fig. 2. Fig. 2 showed the stretching of hydroxyl group at 3214 cm^{-1}, the C=O carbonyl stretching at 1720 cm^{-1}, and the C–O epoxide group stretching at 1160 and 1040 cm^{-1} [6, 8]. These results suggested that the GO sample was oxidized and presented mainly oxygen-containing functional groups. After the reduction reaction, no obvious peak could be observed, which means that successful reduction of GO into RGNO [11].

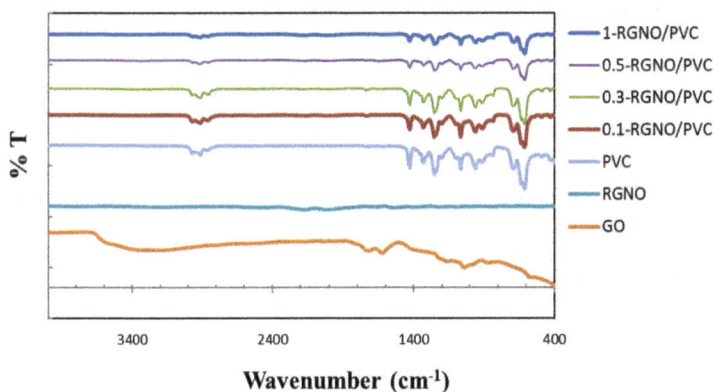

Fig. 2. FTIR spectra of GO, RGNO and RGNO/PVC composites.

Powder Metallurgy and Advanced Materials – RoPM&AM 2017   Materials Research Forum LLC
Materials Research Proceedings 8 (2018) 143-151   doi: http://dx.doi.org/10.21741/9781945291999-16

Characteristic peaks of carbon-oxygen functional groups of RGNO were just very weak. For neat PVC and the RGNO/PVC composites, the characteristic C–H in phase and out of phase stretching vibrations bands can be observed at 2911 cm^{-1} and 2859 cm^{-1}, respectively. The peaks at 1426, 1252, 956, 834, 611 cm^{-1} in the FTIR spectra of neat PVC and RGNO/PVC composites were attributed to the CH$_2$ deformation, CH-rocking, trans CH wagging, C-Cl stretching and cis CH wagging vibration, respectively [18, 23]. As seen from Fig. 2, FTIR spectra of RGNO/PVC composites showed a decrease in the intensities of the peaks with increase of RGNO loading content. This result indicated that RGNO prevented intermolecular vibrations with increase of RGNO loading content.

Fig. 3 showed XRD patterns of the GF and prepared GO and RGNO powders whereas Table 2 showed 2θ values and d-spacing data of the same materials.

Table 2. 2θ values and d spacing data obtained from XRD results.

Samples	2θ°	d (nm)
GF	26,4	0,337
GO	9,95	0,888

The crystalline GF had a strong characteristic peak at 2θ=26.4°. The corresponding interlayer distance was observed to be 0.337 nm. Fig. 3 exhibited a sharp peak at 9.95° corresponding to the (002) plane of GO and inter planner spacing of 0.888 nm, which confirms the successful preparation of GO from graphite powder by the modified Hummers method [6]. The increase in the interlayer distance from 0.337 nm to 0.888 nm was due to oxygen-containing functional groups intercalated within the layered structure (Table 1) [24]. For RGNO in Fig. 3, the diffraction peak at 9.95° has disappeared, indicating the complete reduction of GO to RGNO by Vitamin C and a broad peak appears at 24.04°. This indicated that the GO is completely exfoliated to a single layer of graphene [25, 26]. The XRD patterns of all RGNO/PVC composites were shown in Fig. 4 and no peak was seen for any of the RGNO-containing composites. XRD analysis result demonstrated that the amorphous structure of the PVC was maintained.

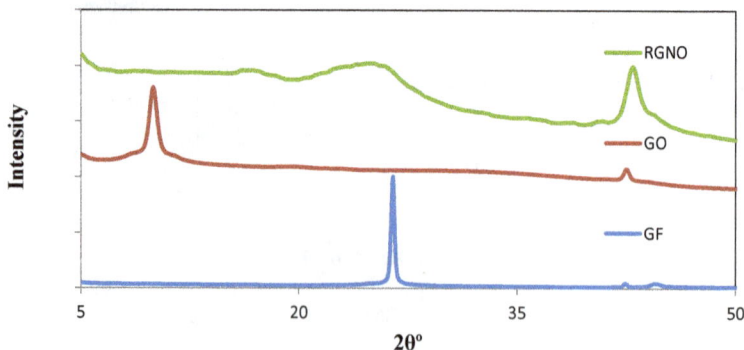

Fig. 3. X-ray diffraction patterns of GF, GO and RGNO powders.

Powder Metallurgy and Advanced Materials – RoPM&AM 2017   Materials Research Forum LLC
Materials Research Proceedings **8** (2018) 143-151   doi: http://dx.doi.org/10.21741/9781945291999-16

Fig. 5 presented the micrographs of GO and RGNO powders. It could be seen clearly from the SEM image of Fig. 5 a that the morphology of GO appeared as a randomly aggregated [27, 28]. As shown in Fig. 5b, the SEM image of the RGNO showed that wrinkle-like structure due to the rapid removal of oxygen containing functional groups in GO (hydroxyl, carbonyl and epoxy groups) [25]. According to EDS results of the GO and RGNO, oxygen content decreased from 46.78 atom % to 24.03 atom % (Table 3) which indicated successful reduction of GO.

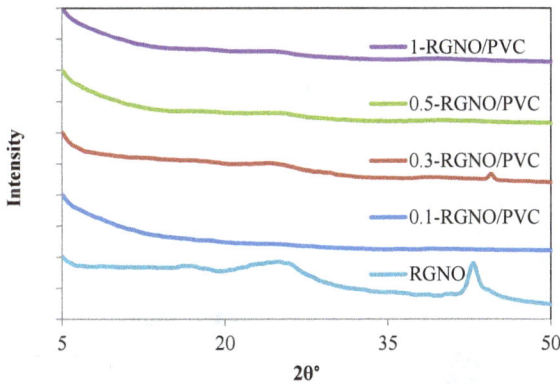

Fig. 4. X-ray diffraction patterns of RGNO and RGNO/PVC composites.

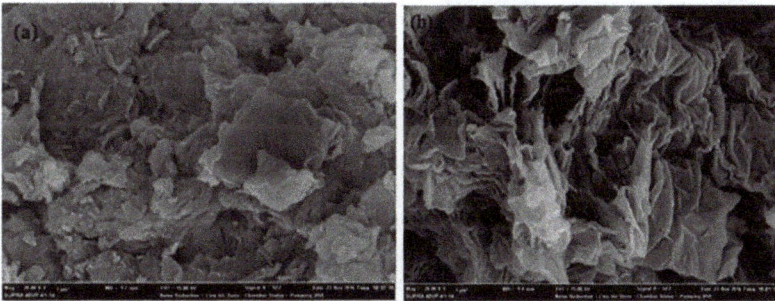

Fig. 5. SEM images of (a) GO, (b) RGNO (magnification 20.000 KX).

Table 3. EDS results of GO and RGNO

Samples	Oxygen Content (atom %)
GO	46.78
RGNO	24.03

Powder Metallurgy and Advanced Materials – RoPM&AM 2017   Materials Research Forum LLC
Materials Research Proceedings 8 (2018) 143-151   doi: http://dx.doi.org/10.21741/9781945291999-16

Fig. 6 showed the surface morphologies of the neat PVC and RGNO/ PVC composites. When compared to the straight surface of neat PVC shown in Fig. 6a, the PVC composite with 0.1 wt. % RGNO showed irregular and bumpy with a rough surface (Fig. 6b). The SEM image of the PVC composite with 0.3 wt. % RGNO showed that compact and highly porous (Fig. 6c). The PVC composite with 0.5 wt. % RGNO exhibited slightly lower porosity (Fig. 6d) owing to polymer growing in the pores and galleries of RGNO [17]. From the SEM image of the PVC composite with 1 wt. % RGNO (Fig. 6e), it could be seen that the porosity increased as the large-grained. It is evident that the RGNO led to porosity in the PVC matrix. The same structure was also observed previously study [29].

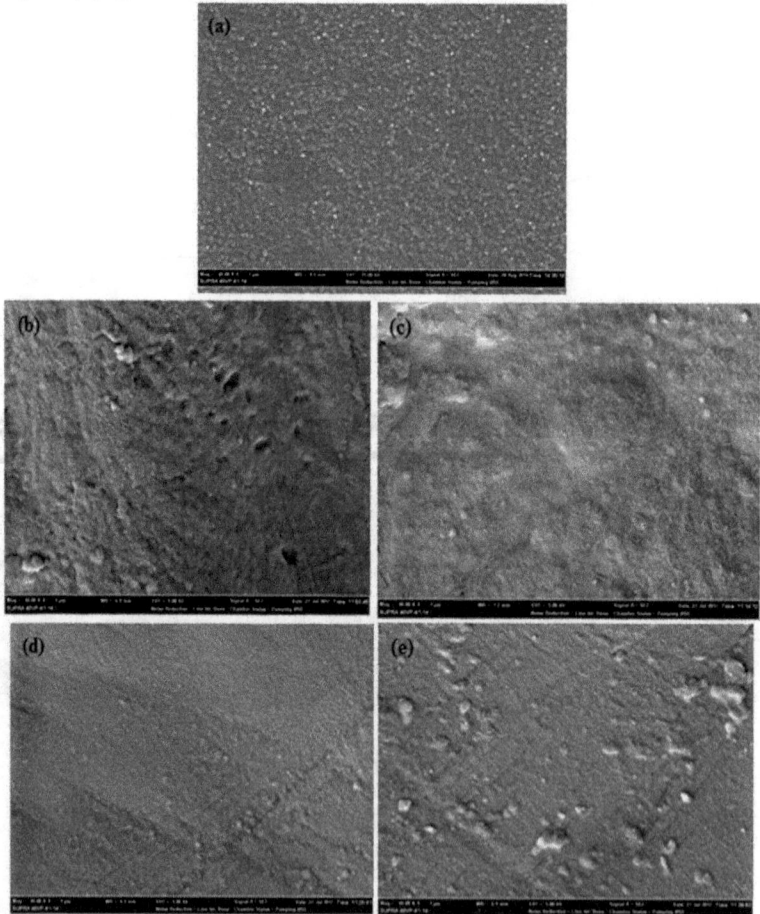

Fig. 6. SEM images of (a) Neat PVC, (b) 0.1-RGNO/PVC, (c) 0.3-RGNO/PVC, (d) 0.5-RGNO/PVC, (e) 1-RGNO/PVC (magnification 40.000 KX)

Powder Metallurgy and Advanced Materials – RoPM&AM 2017   Materials Research Forum LLC
Materials Research Proceedings 8 (2018) 143-151   doi: http://dx.doi.org/10.21741/9781945291999-16

Conclusion

In this study, the structural changes were observed in PVC composites with 0.1, 0.3, 0.5 and 1 wt. % RGNO. To this aim, firstly, GO was prepared from GF by Hummers method and then obtained GO reduced to RGNO with Vitamin C. FTIR, XRD and EDS results showed that GO and RGNO were successfully synthesized and SEM images proved their characteristic structures. RGNO/PVC composites with dispersion in THF had been synthesized by colloidal blending method. No changes were observed in XRD patterns, but FTIR spectra of RGNO/PVC composites showed a decrease in the intensities of the peaks with an increase of RGNO loading content. XRD and FTIR results of all composites indicated the RGNO layers well-dispersed in the PVC matrix. All composites had a different morphological structure when compared with the neat PVC. The SEM images confirmed the presence of PVC filling the pores and galleries of RGNO in the composites. All results of this study revealed that the structural changes of the RGNO/PVC composites must had influenced their thermal and mechanical properties. Therefore, these properties of these composites will be examined in a future work.

Acknowledgement

The authors thank the financial support of the research foundation (Project no.: 2015-02.BSEU.07-01) of Bilecik Seyh Edebali University.

References

[1] S. Stankovich, D. A. Dikin, R. D. Piner, K. A. Kohlhaas, A. Kleinhammes, Y. Jia, Y. Wu, S. T. Nguyen, R. S. Ruoff, Synthesis of graphene-based nanosheets via chemical reduction of exfoliated graphite oxide, Carbon. 45 (2007) 1558–1565. https://doi.org/10.1016/j.carbon.2007.02.034

[2] S. Park, J. An, J. R. Potts, A. Velamakanni, S. Murali, R. S. Ruoff, Hydrazine-reduction of graphite- and graphene oxide, Carbon. 49 (2011) 3019 –3023. https://doi.org/10.1016/j.carbon.2011.02.071

[3] D. N. H. Tran, S. Kabiri, D. Losic, A green approach for the reduction of graphene oxide nanosheets using non-aromatic amino acids, Carbon. 76 (2014) 193-202. https://doi.org/10.1016/j.carbon.2014.04.067

[4] S. Thakur, N. Karak, Green reduction of graphene oxide by aqueous phytoextracts, Carbon. 50 (2012) 5331-5339. https://doi.org/10.1016/j.carbon.2012.07.023

[5] Y.-K. Kim, M. -H. Kim, D.-H. Min, Biocompatible reduced graphene oxide prepared by using dextran as a multifunctional reducing agent, Chem. Commun. 47 (2011) 3195–3197. https://doi.org/10.1039/c0cc05005a

[6] Y. Liu, Y. Zhang, G. Ma, Z. Wang, K. Liu, H. Liu, Ethylene glycol reduced graphene oxide/polypyrrole composite for supercapacitor. Electrochim. Acta. 88 (2013) 519-525. https://doi.org/10.1016/j.electacta.2012.10.082

[7] S. Gurunathan, J. W. Han, A. A. Dayem, V. Eppakayala, M.-R. Park, D.-N. Kwon, J.-H. Kim, Antibacterial activity of dithiothreitol reduced graphene oxide, J. Ind. Eng. Chem. 19 (2013) 1280–1288. https://doi.org/10.1016/j.jiec.2012.12.029

[8] Y. Jin, S. Huang, M. Zhang, M. Jia, D. Hu, A green and efficient method to produce graphene for electrochemical capacitors from graphene oxide using sodium carbonate as a reducing agent, Appl. Surf. Sci. 268 (2013) 541– 546. https://doi.org/10.1016/j.apsusc.2013.01.004

[9] M.J. Fernandez-Merino, L. Guardia, J.I. Paredes, S. Villar-Rodil, P. Solis-Fernandez, A. Martinez-Alonso, J.M.D. Tascon, Vitamin C is an ideal substitute for hydrazine in the reduction of graphene oxide suspensions, J. Phys. Chem. C. 114 (2010) 6426–6432. https://doi.org/10.1021/jp100603h

[10] A.I. Kamisan, A.-S. Kamisan, R. Md. Ali, T.I. Tunku Kudin, O.H. Hassan, N. A. Halim, M.Z.A. Yahya, Synthesis of graphene via green reduction of graphene oxide with simple sugars, Adv. Mat. Res. 1107 (2015) 542-546.

[11] Y. Wang, Z. Shi, J. Yin, Facile Synthesis of soluble graphene via a green reduction of graphene oxide in tea solution and its biocomposites, ACS Appl. Mater. and Interfaces. 3 (2011) 1127–1133. https://doi.org/10.1021/am1012613

[12] Y. Guo, X. Sun, Y. Liu, W. Wang, H. Qiu, J. Gao, One pot preparation of reduced graphene oxide (RGO) or Au (Ag) nanoparticle-RGO hybrids using chitosan as a reducing and stabilizing agent and their use in methanol electrooxidation, Carbon. 50 (2012) 2513-2523. https://doi.org/10.1016/j.carbon.2012.01.074

[13] P. Li, X. Chen, J.-B. Zeng, L. Gan, M. Wang, Enhancement of the interfacial interaction between poly(vinyl chloride) and zinc oxide modified reduced graphene oxide, RSC Adv. 6 (2016) 5784–5791. https://doi.org/10.1039/C5RA20893A

[14] X. C. Ge, X. H. Li, Y. Z. Meng, Tensile Properties, Morphology and thermal behavior of PVC composites containing pine flour and bamboo flour, J. Appl. Polym. Sci. 93 (2004) 1804–1811. https://doi.org/10.1002/app.20644

[15] J. Hu, X. Jia, C. Li, Z. Ma, G. Zhang, W. Sheng, X. Zhang, Z. Wei, Effect of interfacial interaction between graphene oxide derivatives and poly(vinyl chloride) upon the mechanical properties of their nanocomposites, J. Mater. Sci. 49 (2014) 2943-51. https://doi.org/10.1007/s10853-013-8006-1

[16] H. J. Salavagione, G. Martínez, Importance of covalent linkages in the preparation of effective reduced graphene oxide_poly(vinyl chloride) nanocomposites, Macromolecules. 44, 2011, 2685–2692. https://doi.org/10.1021/ma102932c

[17] K. Deshmukh, G. M. Joshi, Thermo-mechanical properties of poly(vinyl chloride)/graphene oxide as high performance nanocomposites, Polym. Test. 34 (2014) 211–219. https://doi.org/10.1016/j.polymertesting.2014.01.015

[18] K. Deshmukh, S. M. Khatake, G. M. Joshi, Surface properties of graphene oxide reinforced polyvinylchloride nanocomposites, J. Polym. Res. 20 (2013) 286. https://doi.org/10.1007/s10965-013-0286-2

[19] M. Hasan, M. Lee, Enhancement of the thermo-mechanical properties and efficacy of mixing technique in the preparation of graphene/PVC nanocomposites compared to carbon nanotubes/PVC, Prog Nat Sci-Mater Int. 24 (2014) 579–587. https://doi.org/10.1016/j.pnsc.2014.10.004

[20] S. Vadukumpully, J. Paul, N. Mahanta, S. Valiyaveettil, Flexible conductive graphene/poly(vinyl chloride) composite thin films with high mechanical strength and thermal stability, Carbon. 49 (2011) 198-205. https://doi.org/10.1016/j.carbon.2010.09.004

Powder Metallurgy and Advanced Materials – RoPM&AM 2017   Materials Research Forum LLC
Materials Research Proceedings 8 (2018) 143-151   doi: http://dx.doi.org/10.21741/9781945291999-16

[21] H. Wang, G. Xie, M. Fang, Z. Ying, Y. Tong, Y. Zeng, Electrical and mechanical properties of antistatic PVC films containing multi-layer graphene, Compos. Part B-Eng. 79 (2014) 444–450. https://doi.org/10.1016/j.compositesb.2015.05.011

[22] W.S. Hummers, R. E. Offeman, Preparation of graphitic oxide, J. Am. Chem. Soc. 80 (1958) 1339. https://doi.org/10.1021/ja01539a017

[23] S. Ramesh, K. H. Leen, K. Kumutha, A.K. Arof, FTIR studies of PVC/PMMA blend based polymer electrolytes, Spectrochim. Acta A. 66 (2007) 1237–1242. https://doi.org/10.1016/j.saa.2006.06.012

[24] S. Gurunathan, J. W. Han, E. Kim, D.-N. Kwon, J.-K. Park, J.-H. Kim, Enhanced green fluorescent protein-mediated synthesis of biocompatible graphene, J. Nanobiotechnol. (2014) 12:41. https://doi.org/10.1186/s12951-014-0041-9

[25] C. Bora, P. Bharali, S. Baglari, S. K. Dolui, B. K. Konwar, Strong and conductive reduced graphene oxide/polyester resin composite films with improved mechanical strength, thermal stability and its antibacterial activity, Compos. Sci. Technol. 87 (2013) 1–7. https://doi.org/10.1016/j.compscitech.2013.07.025

[26] Y. Wu, , H. Luo, H. Wang, C. Wang, J. Zhang, Z. Zhang. Adsorption of hexavalent chromium from aqueous solutions by graphene modified with cetyltrimethylammonium bromide, J. Colloid. Interf. Sci. 394 (2013) 183–191. https://doi.org/10.1016/j.jcis.2012.11.049

[27] M. Safarpour, A. Khataee, V. Vatanpour, Thin film nanocomposite reverse osmosis membrane modified by reduced graphene oxide/TiO2 with improved desalination performance, J. Membrane Sci. 489 (2015) 43–54. https://doi.org/10.1016/j.memsci.2015.04.010

[28] D. Li, B. Zhang, F. Xuan, The sequestration of Sr(II) and Cs(I) from aqueous solutions by magnetic graphene oxides, J Mol. Liq. 209 (2015) 508–514. https://doi.org/10.1016/j.molliq.2015.06.022

[29] F. Mindivan, The Synthesis, Thermal and structural characterization of Polyvinylchloride/Graphene Oxide (PVC/GO) composites, Mater. Sci. Non - Equilib. Phase Transform. 3 (2015) 33-36.

Powder Metallurgy and Advanced Materials – RoPM&AM 2017   Materials Research Forum LLC
Materials Research Proceedings **8** (2018) 152-156   doi: http://dx.doi.org/10.21741/9781945291999-17

Research into the warm compaction of metal powders

Radu MURESAN

Technical University of Cluj-Napoca

Radu.Muresan@stm.utcluj.ro

Keywords: Metal powders, Warm compaction, Sintering.

Abstract. The technological method of warm powder compaction is similar to that of axial die compaction, except that both the powder and the die must be heated to a temperature between 100 and 160 °C. This temperature range was chosen for two reasons, i.e.: at temperatures below 100°C the effect is negligible, as the phenomena accompanying compaction are identical to those accompanying cold compaction. At temperatures above 160 °C the risk of particle oxidation increases, although there also is a positive aspect when exceeding such temperature, which is the facilitation of the process of lubricant elimination from the powder mass. The method ensures that the density of the compacts is over 99% of the theoretical density of the used powder mixtures. This paper intends to present a comparative study of the behaviour of the DensmixTM powder during cold and warm compaction.

Introduction

The warm compaction technique was proposed by the Höganäs Corporation in the U.S. at the beginning of the '90s, and since then it has received more and more attention. This technique can provide high density as compared to the classical compaction techniques. Generally, raw density in the case of warm compaction is influenced by a large number of plastic deformations, which are due to a decrease in the particles' compressive strength [2, 3]. Nevertheless, the main densification mechanism in the case of warm compaction is based on the plastic deformation of the particles. Since it was introduced, a decade ago, warm compaction has become increasingly popular, and it is now used extensively in the manufacture of high strength components. Besides the increase in density, it also results in high strength of raw materials and substantially increased lubrication of the die walls. The subsequent advantages were also used by combining warm compaction with other operations, such as mechanical processing of raw materials, high temperature sintering, sinterforging, surface densification, etc.

The main performance of this technology consists of the design of the powder mixture, the powder heating system and the used equipment. The good properties of the parts made by compaction at certain temperatures represent the success of mass production; this is also made possible by the use of adequate lubricants in the mixtures. Densmix™ type powders, which were especially designed for warm compaction. The subsequent advantages of the properties of such powders consist in the improvement of certain parameters, such as: dimensional variation, mass variation and specimen consistency.

Warm compaction is characterized by the heating to a temperature of approx. 150 °C. The mixtures designed for warm compaction may be based on any iron or steel powder, as well as alloyed steel powder. The lubricant is optimised to work at high temperatures at 0.6% content [1].

Warm compaction was introduced onto the market in 1994 and quickly gained ground as an effective method for the manufacture of high density parts used in different applications. Over 150 different components are currently developed and produced. The most commonly used parts that

Powder Metallurgy and Advanced Materials – RoPM&AM 2017   Materials Research Forum LLC
Materials Research Proceedings 8 (2018) 152-156   doi: http://dx.doi.org/10.21741/9781945291999-17

have been manufactured so far are different types of cogwheels for hand-held power tools and safety parts for automotive applications. The helical gears for different applications are a fast growing sector of warm compaction.

The weight of the parts ranges between 5 and 105 grams in Europe and Asia, while auto components weighing up to 1100 grams are produced in the U.S. The heating system used in Europe and Asia is the one developed by the Höganäs Company. Within this heating system, the powder is heated by heated oil.

Since it was introduced, warm compaction has gained much experience both in the laboratory, and in the part production [4-6]. The obtained advantages are not only the increase in density, raw strength and the static mechanical properties, but also a number of other favourable factors that will be important for the future growth in the competitiveness of powder metallurgy.

Fig. 1. The warm compaction system.

Experimental

We conducted experiments on the DensmixTM powder manufactured by the Höganäs AB Company, Sweden, powder which had the following characteristics:

- Iron content, 95.82% ;
- Copper content, min 2.97 %;
- Graphite content, 0.58 %;
- DM120 lubricant, 0.63% ;
- Apparent density during filling (at 120 °C), 3.30 g/cm^3;
- Fluidity (at 120 °C), 24 s/50 g;
- Density of compacts compacted at 600 MPa (at 120 °C), 7.30 g/cm^3;
- Theoretical density of the powder mixtures at 0% porosity (at 120 °C), 7.41 g/cm^3.

In order to determine the compactibility of the Densmix powder, we made tablets with an area of 1 cm^2 by cold compaction and tablets with an area of 0.5 cm^2 for warm compaction. Warm compaction was performed at the following temperatures: 80 °C, 100 °C, 120 °C, 140 °C, and 160 °C. For warm compaction purposes, the die and the powder were heated together using an electrical resistance heating device, Fig. 1. The powder was compacted in the die, and was kept for

Powder Metallurgy and Advanced Materials – RoPM&AM 2017   Materials Research Forum LLC
Materials Research Proceedings 8 (2018) 152-156   doi: http://dx.doi.org/10.21741/9781945291999-17

approximately 20 seconds at the chosen temperature at pressures of 100, 200 and 300 MPa, respectively. For the comparative study, the same powder type was cold compacted at the following pressures: 200, 300, 400, 500, and 600 MPa.

After sintered at a temperature of 1120 °C, in a belt oven in endogas atmosphere. The specimens were kept at the sintering temperature for 30 minutes. The density after sintering was determined by the hydrostatic weighing method. Resilience was determined by the Charpy method, and hardness was determined by the Brinell method.

Based on the obtained data, we plotted the variation curves of apparent density, density after sintering, resilience and hardness as functions of the temperature and the compaction pressure; the curves are shown in Fig. 2, 3, 4, 5 and 6 respectively.

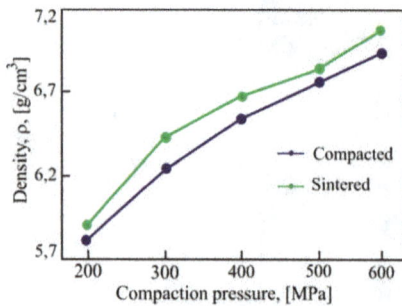

Fig. 2. Variation in the density of the Desmix™ powder after compaction and sintering.

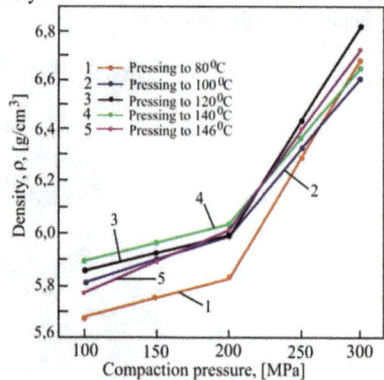

Fig. 3. Variation in the density of the Desmix™ powder as a function of the compaction temperature.

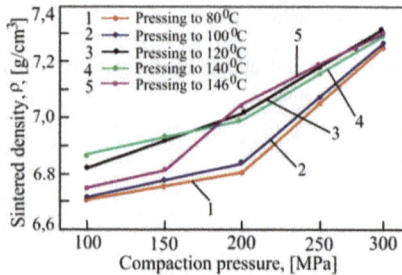

Fig. 4. Variation in the density of the warm compacted and sintered Desmix™ powder.

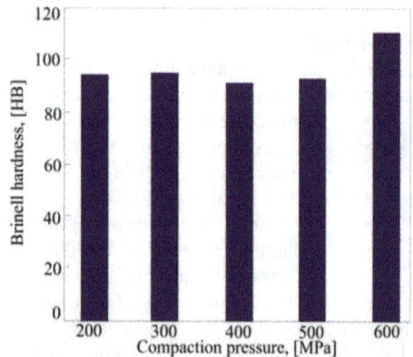

Fig. 5. Variation in the hardness of the cold compacted specimens as a function of the compaction pressure.

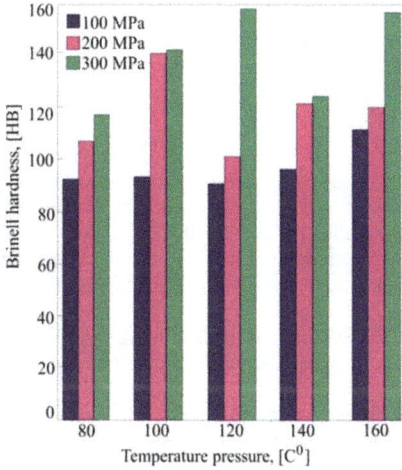

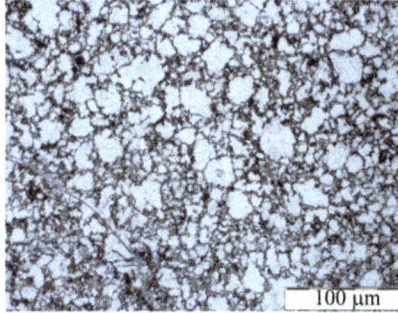

Fig. 6. Variation in specimen hardness after sintering as a function of the compaction pressure and temperature.

Fig. 7. Microstructure of the compacted and sintered Densmix™ specimen.

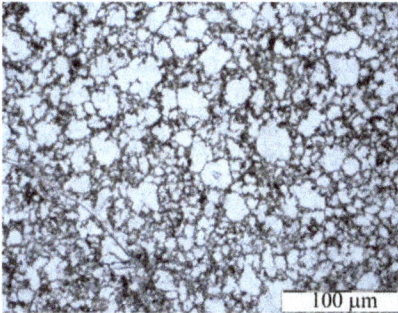

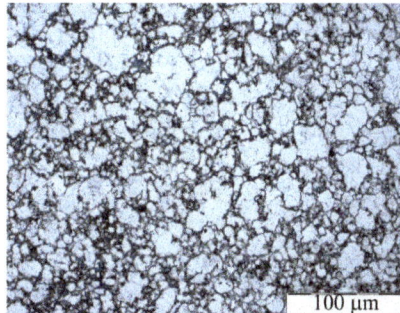

Fig. 7. Microstructure of the compacted and sintered Densmix™ specimen.

Fig. 8. Microstructure of the specimen compacted at a temperature of 80 °C and sintered.

Due to the complexity of the matter, we chose to study these properties, since they are the most commonly used in practical applications of warm compacted parts.

The metallographic study was conducted on all the specimens from which microstructures were selected for the specimens compacted at the temperatures of 80, 120, and 140 °C, and sintered, shown in Fig. 7, 8 and 9. In Fig. 7 we notice a finer structure, with larger agglomerations of solid

Powder Metallurgy and Advanced Materials – RoPM&AM 2017　　　　Materials Research Forum LLC
Materials Research Proceedings **8** (2018) 152-156　　　　doi: http://dx.doi.org/10.21741/9781945291999-17

solution and nonhomogeneous compactness. In Fig. 8 we notice that compactness increases gradually, the grains become smaller and smaller, and the agglomerates of solid solution become increasingly larger, while porosity decreases. Fig. 9 shows large agglomerates of solid solution surrounded by pearlitic structural constituents, isolated pores, but no major change from the specimens compacted at 120 °C.

Summary
Based on the experimental research conducted on the DensmixTM powder and the processing of the obtained experimental data, we concluded that the best mechanical properties are obtained by warm compaction at a temperature of 120 °C and a compaction pressure of 300 MPa.

The main densification mechanism in the case of warm compaction is based on the plastic deformation of particles. Consequently, the powder mixture intended for warm compaction must satisfy the following conditions:

• The shape of the particles must be spherical, when the friction is minimal and the rotation of particles is favourable, but, from the compactibility perspective, the surface of the powder has a defined rugosity.

• The grain distribution of the powder mixture must be similar in size.

• The basic powder for warm compaction must be a partially alloyed iron powder, which provides good compactibility to the mixture.

• In order to satisfy the sintering requirements, the lubricant must have a wide decay temperature range, and the decay must form volatile monomolecular chemicals, which have a favourable effect on the elimination of defects, such as void spaces, cracks and dimensional compact expansion during sintering, and last but not least prevent the formation of residual carbon, which precludes us from controlling the carbon content in the compact.

Particle rearrangement is one of the densification mechanisms in the case of warm compaction. However, plastic deformation of particles plays an important role in particle rearrangement.

References
[1] U. Engström, B. Johansson: Experiences with Warm Compaction of Densmix™ Powders in the Production of Complex Parts, Proceedings of Powder Metallurgy World Congress 2000, Kyoto, Japan, November 2000.

[2] J. Yi, T. Ye, Y. Peng, Effects of warm compaction on mechanical properties of sintered P/M steels, Journal of Central South University of Technology 14 (2001) 447-451. https://doi.org/10.1007/s11771-007-0087-z

[3] D.Y. Yoon, S.L.K.Y.Eun, Y.S. Kim, Densification Mechanism of warm Compaction for Iron-Based Powder Materials, Materials Science Forum, 534-536 (2007) 261-264. https://doi.org/10.4028/www.scientific.net/MSF.534-536.261

[4] G.F. Bocchini, The Warm Compaction Process. Basics, Advantages and Limitations, SAE Transactions, Society of Automotive Engineers, 107 (1998) 225-236.

[5] U. Engström, B. Johansson, P. Knutsson and H Vidarsson: Material Properties and Process Robustness obtained with Warm Compaction of Improved Densmix™ Powders, Proceedings of Powder Metallurgy World Congress 2002, Orlando, USA, June 2002.

[6] http://www.lindemetall.se/produkter/slot heater_e.htm.

Powder Metallurgy and Advanced Materials – RoPM&AM 2017   Materials Research Forum LLC
Materials Research Proceedings 8 (2018) 157-166   doi: http://dx.doi.org/10.21741/9781945291999-18

A comparative study of the Co-based amorphous alloy prepared by mechanical alloying and rapid quenching

Bogdan Viorel NEAMŢU [1,a*], Traian Florin MARINCA [1,b], Horea Florin CHICINAŞ[1,c], Florin POPA [1,d], Ionel CHICINAŞ [1,e], Olivier ISNARD [2,f], Gabriel ABABEI [3,g] and Mihai GABOR [1,h]

[1]Materials Science and Engineering Department, Technical University of Cluj-Napoca, 103-105 Muncii Avenue, 400641 Cluj-Napoca, Romania

[2]Université Grenoble Alpes, Institut NEEL, CNRS, 25 rue des martyrs, BP166, F-38042, Grenoble, France

[3]National Institute of Research and Development for Technical Physics, 47 Mangeron Boulevard, 700050 Iasi, Romania

[4]Center for Superconductivity, Spintronics and Surface Science, Technical University of Cluj-Napoca, Memorandumului Street, No 28, 400114 Cluj-Napoca, Romania

[a*]Bogdan.Neamtu@stm.utcluj.ro, [b]Traian.Marinca@stm.utcluj.ro, [c]Horea.Chicinas@stm.utcluj.ro, [d]Popa.Florin@stm.utcluj.ro, [e]Ionel.Chicnas@stm.utcluj.ro, [f]olivier.isnard@neel.cnrs.fr, [g]gababei@phys-iasi.ro, [h]mihai.gabor@phys.utcluj.ro

Keywords: Amorphous materials, Mechanical alloying, Rapid quenching, Kinetic of crystallization, Magnetic properties

Abstract. Co-Fe-Ni-M-Si-B (M = Zr, Ti) amorphous alloys were prepared via mechanical alloying and rapid quenching. The influence of the preparation conditions over the alloy's characteristics was investigated in the light of several techniques, such as X-ray diffraction, scanning electron microscopy, differential scanning calorimetry, magnetization measurements and in-situ high-temperature X-ray diffraction. It was found that the preparation technique has a significant influence on the thermal stability, magnetic characteristics and the crystallisation kinetic of the amorphous phase. Substitution of transition metals (Ti or Zr) for Si is investigated for the amorphous powders and ribbons. The influence of substitution type over the alloy characteristics is presented and discussed in this paper.

Introduction

The amorphous materials are widely studied in the last decades due to their potential application as novel materials in various fields. The amorphous soft magnetic materials are characterised by low or even zero magnetocrystalline anisotropy and high electrical resistivity. This characteristic derives from the lack of the long-range atomic order, and imposes them as viable candidates for a series of modern applications.

Among the techniques that were employed to produce amorphous alloys, mechanical alloying and rapid quenching are by far the most used ones [1, 2]. The amorphisation via rapid quenching it is possible if the cooling rates are in the range of $10^5 – 10^6$ K/s. The amorphisation via mechanical milling, on the other hand, is possible due to the large amount of crystallographic defects (dislocations, grain boundaries, stacking faults, anti-site defects etc.) induced in the powder which destabilizes the crystalline phase [1]. Both techniques have advantages and drawbacks. For example: mechanical alloying can produce amorphous alloys in powder form which is, from the

Powder Metallurgy and Advanced Materials – RoPM&AM 2017   Materials Research Forum LLC
Materials Research Proceedings 8 (2018) 157-166   doi: http://dx.doi.org/10.21741/9781945291999-18

applicative viewpoint, very important considering the versatility of the powder technology. On the other hand, the milling time required for powder amorphisation (the milling experiment) is far longer as compared to rapid quenching process. The rapid quenching technique produces amorphous ribbons that are very difficult to be used in order to create magnetic cores with complicated 3D geometry [3]. The alloy contamination risk is reduced in the case of using rapid quenching technique while the contamination with debris from milling bodies must be taken into account when mechanical alloying is used to induce the amorphisation [4]. Moreover, a suitable chemical composition, generally close to deep eutectic composition is needed when the alloy is processed via rapid quenching while the mechanical alloying was proved to induce amorphisation in alloys with chemical composition farther from eutectic compositions [1, 2].

This paper aims to highlight the influence of the processing technique on the crystallisation kinetics and magnetic properties of several Co-based alloys.

Experimental procedure

In order to prepare the reference alloy $Co_{70}Fe_4Ni_2Si_{15}B_9$ (at. %) as well as the $Co_{70}Fe_4Ni_2Si_{10}B_9Zr_5$ and $Co_{70}Fe_4Ni_2Si_{10}B_9Ti_5$ (at.%) alloys, a mixture of elemental powders was prepared. The elemental powders used in this study were: Co (99.8% purity, particle size range 45-150 μm), Fe type NC 100.24 (98.5% purity, particle size < 150 μm), Ni type 123-carbonyl (99.8% purity, particle size < 7 μm), Si (99.9% purity, particle size < 150 μm), amorphous B (amorphous powder, 99.8% purity, particle size < 45 μm), Ti (99.5% purity, particle size < 150 μm) and Zr (99.2% purity, particle size < 45 μm). The mixtures of elemental powders were subjected to wet mechanical alloying (MA). Benzene was used as process control agent (PCA). 1 ml of C_6H_6 was added every time sampling was performed in order to avoid the excessive cold welding phenomena and to counterbalance the PCA evaporation. The samples were milled up to 40 h in identical conditions using tempered steel balls and vials. The ball to powder ratio (BPR) was chosen to be 16:1 and the rotational speed of the disk was set at 350 rpm. Argon atmosphere was used in order to avoid powder oxidation during the milling process.

The amorphous ribbons were prepared by rapid quenching (RQ) from master alloys that were subjected to arc melting in an argon atmosphere. The following parameters were used for the RQ experiments: high purity argon atmosphere, the wheel was manufactured by copper, the crucible was made from quartz, tangential wheel speed was 32 m/s, the heating and melting of the sample was made by induction, the diameter of the nozzle was 600 μm and the disc – crucible distance was 200 μm. This parameters lead to ribbons of 10-15 mm thickness and 0.9-1 mm width.

The structural characterisation of the powders and ribbons was performed by X-ray diffractions (XRD) using the Co Kα radiation (λ=1.7903 Å). An Inel Equinox 3000 diffractometer was used and the scanned interval was 2Θ = 20 – 110 °. The high-temperature X-ray diffraction (HT-XRD) was performed using the same diffractometer and an Anton-Paar high-temperature chamber (HTK 1200 N advanced). The XRD patterns were recorded up to the temperature of 900 °C with a heating rate of 10 °C/min. One diffraction per minute was the acquisition frequency of the XRD patterns. Preliminary vacuum (10^{-2} Torr) was used as the atmosphere inside the high-temperature chamber.

The morphological analysis of the powders and ribbons was performed by scanning electron microscopy (SEM), using a Jeol-JSM 5600 LV electronic microscope. The chemical composition and homogeneity of the samples were analysed using an EDX spectrometer

The differential scanning calorimetry (DSC) was performed with a Setaram Labsys equipment in order to investigate the thermal stability of the amorphous powders and ribbons. The

Powder Metallurgy and Advanced Materials – RoPM&AM 2017   Materials Research Forum LLC
Materials Research Proceedings **8** (2018) 157-166   doi: http://dx.doi.org/10.21741/9781945291999-18

investigations were performed in the 20-900 °C temperature range under Ar with a heating/cooling rate of 20 °C/min.

The magnetic characteristics of the samples in the amorphous state were investigated by a vibrating sample magnetometer (VSM) by recording their hysteresis loops. A VSM equipment, produced by Lake Shore Cryotronics Inc, was used and the maximum applied field was 800 kA/m.

Results and discussions

The XRD pattern of the $Co_{70}Fe_4Ni_2Si_{15}B_9$ (at. %) starting sample (the mixture of elemental powders) and the XRD patters corresponding to the samples obtained by wet milling up to 40 hours are presented in Fig. 1. In the starting sample, the XRD peaks corresponding to elemental Co, Si, Fe and Ni can be noticed. It worth to be mentioned that in the case of Co, two types of crystallographic structures can be observed: hexagonal Co- hcp (P63/mmc (#194) space group) and cubic Co - fcc (Fm-3m (#225) space group). The presence of small quantities of impurities and the crystallographic stresses lead to partial transformation of elemental hcp Co to fcc Co [5, 6]. The Bragg peaks corresponding to B cannot be noticed in the XRD pattern of the starting sample due to the amorphous state of the used B powder. After 2 hours of milling, the XRD peaks corresponding to Fe and Ni are no longer visible and the peaks of Si start to vanish indicating the progressive dissolution of Fe, Ni and Si into Co. After 10 hours of milling, the only Bragg reflexions that are clearly visible correspond to a fcc Co based solid solution. In the experimentally limits of the XRD technique, we can assume that all the elements are dissolved in Co after 10 hours of milling.

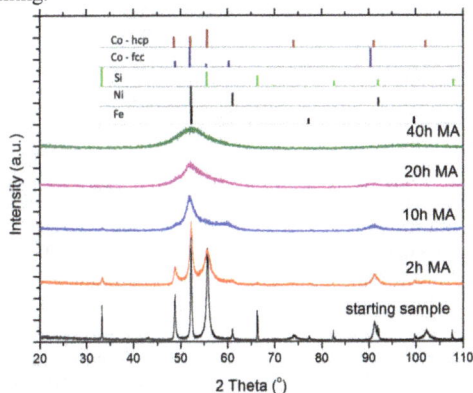

Fig. 1. XRD pattern of the $Co_{70}Fe_4Ni_2Si_{15}B_9$ (at. %) starting sample and the XRD patters corresponding to the samples obtained by wet milling up to 40 hours.

The XRD pattern of the sample milled up to 25 hours contains a single broad peak situated at the same position in 2 theta as the most intense peak of the fcc Co based solid solution. Even if the peak is very broad and with low intensity, its asymmetric shape (towards higher angles) indicate that the sample contains a mixture of phases. Most probably, after 25 hours of milling, the sample is formed by an amorphous phase that embeds a nanocrystalline fcc Co based solid solution. The XRD pattern of the sample milled up 40 hours contains a single peak that is very broad, with low intensity and with a symmetrical shape. Such an XRD pattern suggests the lack of the long-range

Powder Metallurgy and Advanced Materials – RoPM&AM 2017   Materials Research Forum LLC
Materials Research Proceedings 8 (2018) 157-166   doi: http://dx.doi.org/10.21741/9781945291999-18

atomic order indicating the amorphous nature of our sample. In conclusion, we can affirm that 40 hours of wet milling are necessary in order to form the alloy from elemental powders and to induce its amorphisation.

In the case of samples in which 5 at.% of Si was substituted with the same amount of Ti or Zr, the evolution of the structure of the alloy versus milling time is similar to the one above mentioned in the case of $Co_{70}Fe_4Ni_2Si_{15}B_9$ (at. %) alloy. Due to this reason, it will not be further presented in details. The XRD patterns of the samples wet milled up to 25 hours are presented in Fig. 2. For comparison reasons, the XRD pattern of the $Co_{70}Fe_4Ni_2Si_{15}B_9$ sample wet milled up to 40 hours is given in Fig. 2.

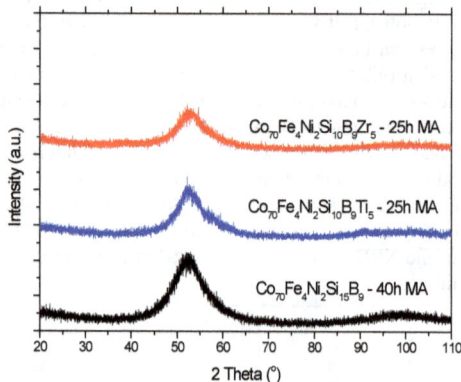

Fig. 2. XRD patterns of the $Co_{70}Fe_4Ni_2Si_{15}B_9$ (at. %), $Co_{70}Fe_4Ni_2Si_{10}B_9Zr_5$ and $Co_{70}Fe_4Ni_2Si_{10}B_9Ti_5$ (at. %) alloys obtained after wet milling up to 25 hours

It was noticed that replacing 5 at.% of Si with 5 at.% of Ti or Zr reduce the milling time needed for the alloy amorphisation. Since the milling parameters were identical with the ones used for the preparation of $Co_{70}Fe_4Ni_2Si_{15}B_9$, we assume that the introduction of the new atomic species (large atoms like Zr or Ti) contribute to the faster structural disordering of the alloy and thus the alloy amorphisation. It is widely accepted that increasing the dissimilarity between atomic sizes of the constituents increase the chances to induce the alloy amorphisation [7].

For comparison reasons, we have prepared amorphous ribbons of the same compositions as the alloys prepared via wet mechanical alloying. The XRD patterns of the ribbons obtained by rapid quenching are presented in Fig. 3. The shape of the XRD patterns of the ribbons demonstrates the amorphous structure obtained after rapid quenching. At the angle of 45.2 degrees in two-theta, a sharp peak can be noticed in all XRD patterns of the amorphous ribbons. This corresponds to the sample holder (incomplete covering of the sample holder with the ribbon) which, in our case, is made of aluminium.

The XRD investigations showed that after 2 hours of milling the powder is not homogeneous. This was also confirmed by the EDX investigations. In the Fig. 4a are presented the distribution maps for the powder wet milled up to 2 hours. It can be observed that Fe, Ni and Si are not uniformly distributed on the analysed surface, their distribution maps presenting areas with higher concentrations.

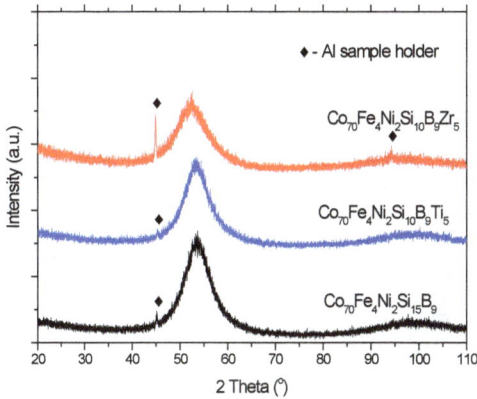

Fig. 3. XRD patterns of the $Co_{70}Fe_4Ni_2Si_{15}B_9$ (at. %), $Co_{70}Fe_4Ni_2Si_{10}B_9Zr_5$ and $Co_{70}Fe_4Ni_2Si_{10}B_9Ti_5$ (at. %) amorphous ribbons.

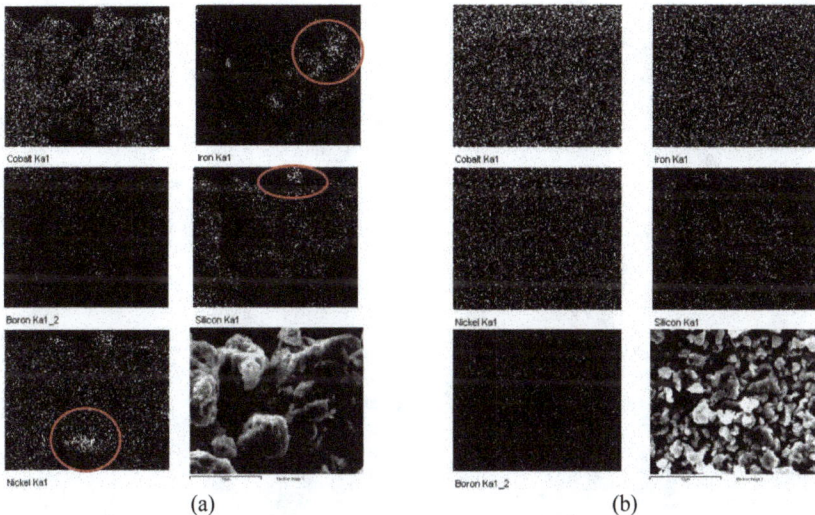

(a)   (b)

Fig. 4. EDX analysis of the $Co_{70}Fe_4Ni_2Si_{15}B_9$ sample wet milled up to 2 h (a) and 40 h (b).

Increasing the milling time leads to the homogenisation of powder composition and after 40 hours of milling the EDX distribution maps shows a uniform distribution of the Co, Fe, Ni, Si and B in the sample. From the SEM images presented also in Fig. 4, it can be noticed that the particle size decreases upon increasing milling time from 2 hours to 40 hours. According to the literature, wet milling process favours the fracturing phenomena in detriment of cold welding [1]. Also, increasing the milling time leads to cold worked particles with a decreased plasticity favouring thus the fragmentation processes. All the above mentioned leads to the decrease of the particles size.

AMORPHOUS POWDERS   **AMORPHOUS RIBBONS**

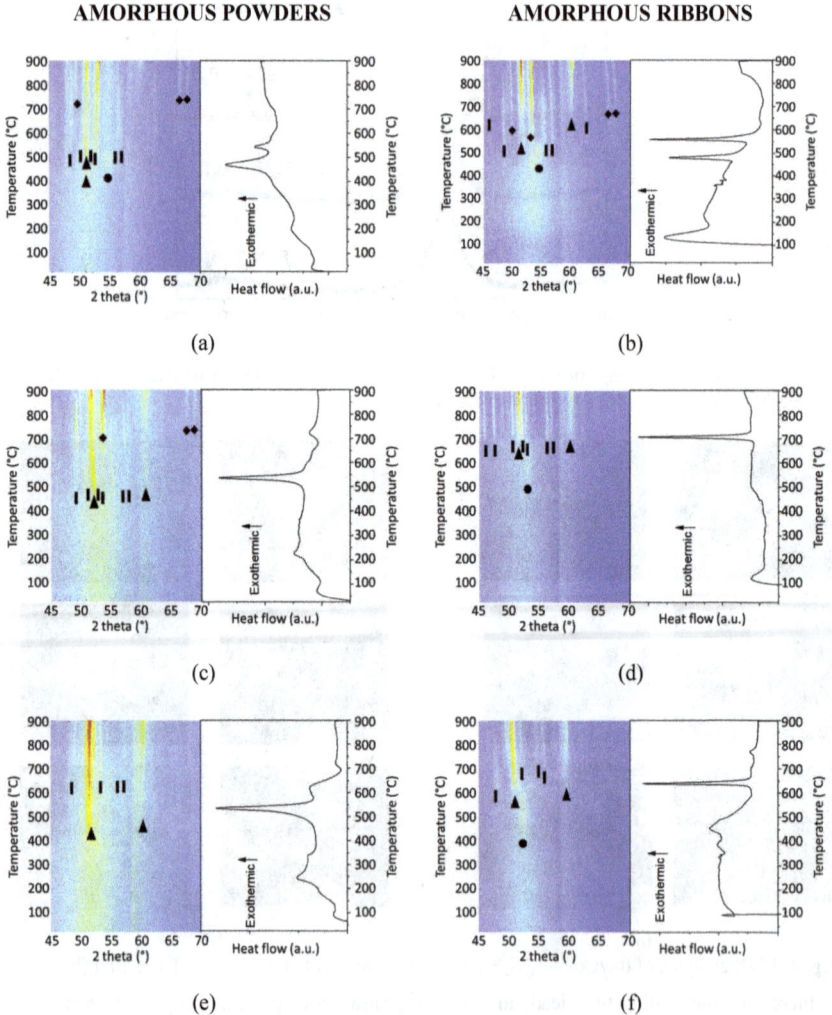

(a)   (b)

(c)   (d)

(e)   (f)

Fig. 5. Thermal stability of amorphous powders (a, c, e) and ribbons (b, d, f) investigated by HT-XRD and DSC. The symbols indicate the positions of the following phases: ● – hcp Co based solid solution, ▲ – fcc Co based solid solution, ◆ - Co_2B and ▮ - Co_2Si.

Powder Metallurgy and Advanced Materials – RoPM&AM 2017   Materials Research Forum LLC
Materials Research Proceedings 8 (2018) 157-166   doi: http://dx.doi.org/10.21741/9781945291999-18

The thermal stability and the crystallisation kinetics of the amorphous powders and ribbons were investigated by DSC and HT-XRD. A comparison between amorphous powders and ribbons for each composition is presented in Fig. 5.

It is noticeable that in the case of $Co_{70}Fe_4Ni_2Si_{15}B_9$ powders (Fig. 5a), the amorphous phase is stable up to the temperature of 410 °C. At this temperature, in the HT-XRD plot is visible that the broad peak characteristic of amorphous phase turns into a series of narrower peaks. At the same temperature, in the DSC curve, the onset of a large exothermic reaction can be noticed indicating the primary crystallisation of the amorphous powder. According to the XRD patterns, at this temperature, a mixture of Co – hcp, Co – fcc based solid solutions and Co_2Si is formed. At higher temperatures, around 530 °C, a new exothermic reaction can be noticed in the DSC curve. According to the HT-XRD analysis, at this temperature, the peaks corresponding to the Co-hcp solid solution disappear. The second exothermic reaction observed on DSC curve was attributed to the allotropic transformation of Co-hcp solid solution. It is known that the pure cobalt presents an allotropic transformation (hcp to fcc) at the temperature of 422 °C. However, in our case, the allotropic transformation occurs at higher temperatures due to the presence of Si which stabilise the hcp phase according to Co-Si phase diagram [8]. At this temperature (530 °C), the sample consists in a mixture of Co fcc solid solution and Co_2Si embedded into an amorphous matrix. At the temperature of 675 °C, the onset of another exothermic reaction is observed in the DSC curve of the $Co_{70}Fe_4Ni_2Si_{15}B_9$ powder. In the HT-XRD plot presented in Fig. 5a, a series of new Bragg reflexions appear. The new Bragg reflexions correspond to the Co_2B phase.

The thermal stability of the $Co_{70}Fe_4Ni_2Si_{15}B_9$ amorphous ribbons investigated by DSC and HT-XRD is presented in Fig. 5b. It can be observed that the DSC curve of the amorphous ribbon and the DSC curve of the amorphous powder are quite similar. The onset temperature of the two main exothermic reactions noticed on the DSC curve are 465 °C and 546 °C. The different onset temperatures of the two exothermic reactions as compared to the ones corresponding to the powder can be related to the preparation technique. In our opinion, the amount of stresses induced by mechanical milling is larger as compared to the amount of stresses induced by rapid quenching. In such a case, a lower temperature for the transformations (i.e. crystallisation) should be expected as a result of higher energy stored in the powders as compared to the ribbons. Similar behaviour was reported in the case of amorphous Fe-based alloys prepared by rapid quenching and mechanical alloying. During the first exothermic reaction, the crystallisation of an hcp Co based solid solution occurs according to HT-XRD analysis presented in Fig. 5b. The XRD peaks corresponding to any Co based solid solution cannot be noticed as was observed in the case of amorphous powders. It was previously shown that the formation of the fcc-Co based solid solution is favoured by the presence of large amount of stresses. It is assumed that larger amount of stresses are induced by milling process as compared to rapid quenching. The second exothermic peak corresponds to the allotropic transformation of Co and the crystallisation of the Co_2Si phase. At the temperature of 630 °C the crystallisation of the Co_2B phase can be noticed. In conclusion, it can be mentioned that the crystallisation kinetic of the amorphous powder is slightly different as compared to the crystallisation kinetic of the amorphous ribbons.

The investigations concerning the crystallisation of the amorphous $Co_{70}Fe_4Ni_2Si_{10}B_9Zr_5r$ are presented in Fig. 5c. It can be noticed that the DSC curve present a single large exothermic reaction heaving the onset at the temperature of 480 °C. According to HT-XRD plot, at this temperature, the crystallisation of an fcc-Co based solid solution and Co_2Si occurs. The higher crystallisation temperature of these amorphous powders as compared to the Zr free amorphous powders can be explained by the so-called geometrical effects. The random packing of an amorphous material can be increased by the addition of atoms with different atomic radii [9].

Powder Metallurgy and Advanced Materials – RoPM&AM 2017   Materials Research Forum LLC
Materials Research Proceedings 8 (2018) 157-166   doi: http://dx.doi.org/10.21741/9781945291999-18

Naturally, the atomic diffusion in such a structure is slower as compared to the one that occurs in a less dense random packed structure. In this case, the stability of the amorphous structure is enhanced. The occurrence of the hcp Co based solid solution cannot be noticed in the case of $Co_{70}Fe_4Ni_2Si_{10}B_9Zr_5$ powders. This can be explained taking into account at least two factors that promote the formation of fcc Co based solid solution in detriment of hcp-Co based solid solution:

(i)   Diminution of the Si content (5 at.% of Si was replaced with 5 at.% of Zr);

(ii) Stresses induced by milling.

At the temperature of 685 °C, the onset of a small exothermic reaction can be noticed. The HT-XRD plot revealed that the crystallisation of Co_2B phase takes place at this temperature.

The HT-XRD analysis reveals that, in the case of amorphous ribbon of $Co_{70}Fe_4Ni_2Si_{10}B_9Zr_5$, the primary crystallisation leads to the formation of the hcp-Co based solid solution embedded into an amorphous phase. The separation of the hcp phase is difficult to be noticed on the DSC curve but is better visible in the HT-XRD plot as a broad peak on the right side (higher angle) of the amorphous halo. This hcp structure is stable up to the temperature of 692 °C. At this temperature, the transition hcp-Co to fcc-Co occurs and it is accompanied by the formation of the Co_2Si phase. At the temperature of 692 °C, in the DSC curve, a large exothermic reaction can be noticed.

In the case of $Co_{70}Fe_4Ni_2Si_{10}B_9Ti_5$ amorphous powders, the HT-XRD analysis evidenced that the fcc-Co based solid solution is the main phase that is formed when the powder is heated up to 900 °C. Is worth mentioning that, the hcp-Co based solid solution is not formed in the case of $Co_{70}Fe_4Ni_2Si_{10}B_9Ti_5$ amorphous powders. The reason for this can be the Ti addition and the high level of stresses induced by milling, both favouring the formation of the fcc-Co based structure. According to the Co-Ti phase diagram, Ti stabilises the fcc phase, reducing the temperature range (with about 100 °C) in which the hcp structure is stable [8]. A small amount of Co_2Si was detected by the XRD analysis. However, it seems that the Ti addition hinder the crystallisation of Co_2B phase when the powder is heated up to 900 °C.

Regarding the $Co_{70}Fe_4Ni_2Si_{10}B_9Ti_5$ amorphous ribbon, it can be noticed that its crystallisation is different from that of amorphous powders of the same composition suggesting once again the influence of the preparation technique over the sample characteristics. At the temperature of 435 °C, a small exothermic reaction began as can be seen on the DSC curve of this sample. At the same temperature in HT-XRD plot, the crystallisation of the hcp-Co based solid solution is noticed. The hcp structure is stable up to the temperature of 625 °C. At this temperature, takes place the hcp – fcc transition as can be noticed in the HT-XRD plot. Also, at this temperature, a large exothermic reaction can be observed in the DSC plot. At 756 °C, a new set of XRD peaks can be noticed in the HT-XRD plot, indicating the crystallisation of Co_2Si phase. As in the case of powders, the crystallisation of the Co_2B phase cannot be detected.

The influence of the preparation technique and the substitutions on the magnetic characteristics of the amorphous powders and ribbons is presented in Fig. 6. It can be generally noticed that the saturation magnetisation of the amorphous powder is lower than the saturation magnetisation of the amorphous ribbons. Also, the coercive field of the amorphous ribbons is about one-third of the coercive field measured on amorphous powders. The lower saturation magnetisation of the amorphous powders as compared to saturation magnetisation of the ribbons can be explained taking into account several aspects as follow:

(i)   The large number of structural defects and high level of stresses induced by milling [1];

(ii) The presence of a certain amount of superparamagnetic particles. Wet milling route promotes the fracturing phenomena leading to the decrease of the particles size. For long milling duration, the occurrence of particles with superparamagnetic behaviour is plausible [10].

Powder Metallurgy and Advanced Materials – RoPM&AM 2017   Materials Research Forum LLC
Materials Research Proceedings 8 (2018) 157-166   doi: http://dx.doi.org/10.21741/9781945291999-18

(iii) Powder contamination. It was previously proved that during wet milling the powder contamination takes place in two ways as follows: (1) surface contamination – the PCA is adsorbed on the particles surface; (2) chemical composition alteration – atoms resulted from the decomposition of the PCA are incorporated into alloy during milling [4, 11]. The above mentioned will lead to a decreased magnetic moment per unit mass and thus a decreased saturation magnetisation

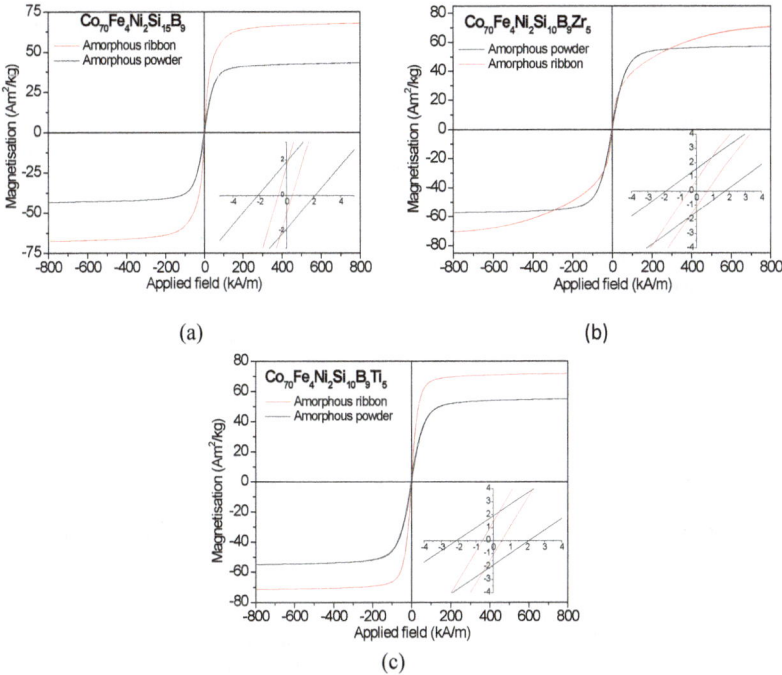

(a)

(b)

(c)

Fig. 6. Hysteresis loops corresponding $Co_{70}Fe_4Ni_2Si_{15}B_9$ (at. %), $Co_{70}Fe_4Ni_2Si_{10}B_9Zr_5$ and $Co_{70}Fe_4Ni_2Si_{10}B_9Ti_5$ (at. %) amorphous powders (black curves) and ribbons (red curves).The $0x$ axis in the insets of each graph (6a, 6b and 6c) is in kA/m.

According to the literature, larger coercive field is expected for powders as compared to amorphous ribbons due to the increased number of domain wall pining centres. Mechanical alloying route induces a larger amount of free volume content as compared to rapid quenching [12]. Also, it was above mentioned that mechanical alloying leads to samples with a large number of structural defects and high level of stresses as compared to rapid quenching that naturally increases the coercivity.

Summary
Co-Fe-Ni-M-Si-B (M = Zr, Ti) amorphous alloys were successfully prepared via wet mechanical alloying and rapid quenching. The alloy $Co_{70}Fe_4Ni_2Si_{15}B_9$ is obtained in an amorphous state after 40 hours of wet mechanical alloying. Substitution of Si with Zr or Ti leads to a significant

reduction of the milling duration for the alloy amorphisation. It was shown that the alloys prepared in powder form via mechanical alloying crystallise at a lower temperature as compared to their counterpart in form of amorphous ribbons. Substitution of Si with Zr or Ti leads to increased thermal stability of the amorphous phase. Also, it was evidenced by HT-XRD and DSC measurements that the preparation technique and the substitution made (Si for Zr or Si for Ti) modifies the crystallisation kinetic of the alloys. Magnetic measurements revealed that the saturation magnetisation of the amorphous powder is lower than the saturation magnetisation of the amorphous ribbons. Also, it was noticed that the coercive field of the amorphous ribbons is about one third of the coercive field measured on amorphous powders and was explained based on the particularities of the wet mechanical alloying process.

Acknowledgements
This work was supported by a Grant of the Romanian National Authority for Scientific Research CNCS - UEFISCDI, Project number PN II-RU-TE-2012-3-0367.

References
[1] C. Suryanarayana, Mechanical alloying and milling, Prog. Mater. Sci., 46 (2001), 1-184. https://doi.org/10.1016/S0079-6425(99)00010-9

[2] C. Suryanarayana, A. Inoue, Bulk Metallic Glasses, CRC Press, Boca Raton, London, New York, 2010.

[3] C. Suryanarayana, A. Inoue, Iron-based bulk metallic glasses, Int. Mater. Rev., 58 (2013) 131–166. https://doi.org/10.1179/1743280412Y.0000000007

[4] B.V. Neamţu, H.F. Chicinaş, T.F. Marinca, O. Isnard, O. Pană, I. Chicinaş, Amorphisation of Fe-based alloy via wet mechanical alloying assisted by PCA decomposition, Mater. Chem. Phys., 183 (2016) 83-92. https://doi.org/10.1016/j.matchemphys.2016.08.005

[5] J. Y. Huang, Y. K. Wu and H. Q. Ye, Allotropic transformation of cobalt induced by ball milling, Acta Mater. 44 (1996) 1201 – 1209. https://doi.org/10.1016/1359-6454(95)00234-0

[6] O. S. Edwards, H. Lipson, Imperfections in the structure of cobalt. I. Experimental work and proposed structure, Proc. Royal Soc. A, 180 (1942) 268 – 277. https://doi.org/10.1098/rspa.1942.0039

[7] A. Inoue, Stabilization of metallic supercooled liquid and bulk amorphous alloys, Acta Mater. 48 (2000) 279–306. https://doi.org/10.1016/S1359-6454(99)00300-6

[8] Hiroaki Okamoto, Mark E. Schlesinger, ASM Handbook Vol. 3, Alloy Phase Diagrams, Ohio, 2016.

[9] B.V. Neamţu, H.F. Chicinaş, T.F. Marinca, O. Isnard, I. Chicinaş, Preparation and characterisation of Co–Fe–Ni–M–Si–B (M = Zr, Ti) amorphous powders by wet mechanical alloying, J. Alloy Compd., 673 (2016) 80-85. https://doi.org/10.1016/j.jallcom.2016.02.233

[10] J.Y. Yang, J.S. Wu, T.J. Zhang, K. Cui, Multicomponent mechanical alloying of Fe–Cu–Nb–Si–B, J. Alloy. Compd. 265 (1998) 269–272. https://doi.org/10.1016/S0925-8388(97)00308-3

[11] B.V. Neamţu, O. Isnard, I. Chicinaş, C. Vagner, N. Jumate, P. Plaindoux, Influence of benzene on the Ni$_3$Fe nanocrystalline compound formation by wet mechanical alloying: An investigation combining DSC, X-ray diffraction, mass and IR spectrometries, Mater. Chem. Phys., 125 (2011) 364–369. https://doi.org/10.1016/j.matchemphys.2010.10.056

[12] A.H. Taghvaei, M. Stoica, K.G. Prashanth, J. Eckert, Fabrication and characterization of bulk glassy Co40Fe22Ta8B30 alloy with high thermal stability and excellent soft magnetic properties, Acta Mater. 61 (2013) 6609-6621. https://doi.org/10.1016/j.actamat.2013.07.045

Powder Metallurgy and Advanced Materials – RoPM&AM 2017   Materials Research Forum LLC
Materials Research Proceedings 8 (2018) 167-172   doi: http://dx.doi.org/10.21741/9781945291999-19

Comparative study of structural and optical properties of zirconium-doped indium oxide synthesized by solid state reactions and sol-gel technique

Liliana BIZO[1,a], Klara MAGYARI[2,b], Julieta Daniela CHELARU[1,c], Ovidiu NEMEŞ[3,d]

[1]Universitatea Babeş-Bolyai, Faculty of Chemistry and Chemical Engineering, Department of Chemical Engineering, 11 Arany Janos Street, RO-400028, Cluj-Napoca, Romania

[2] Universitatea Babeş-Bolyai, Institute for Interdisciplinary Research on Bio-Nano-Sciences, 42 Treboniu Laurian Street, RO–400271, Cluj-Napoca, Romania

[3]Technical University of Cluj-Napoca, 28 Memorandumului Street, RO-400114, Cluj-Napoca, Romania

[a]lbizo@chem.ubbcluj.ro, [b]klara.magyari@ubbcluj.ro, [c]jdchelaru@chem.ubbcluj.ro, [d]ovidiu.nemes@sim.utcluj.ro

Keywords: Advanced Electronic Materials, Transparent Conducting Oxides (TCOs), Zr-Doped Indium Oxide, Structural Properties, Optical Properties

Abstract. In the present work the effect of zirconium doping (1.5 at.%) on the structural and optical properties of indium zirconium oxide was studied. Zr-doped indium oxides were prepared by using two different methods, solid state reactions and sol-gel technique. The compositions with bixbyite structure have been synthesized by two different methods, solid state reactions in air and sol-gel process. X-ray powder diffraction (XRPD) used for phases analysis confirm the validity of the cubic bixbyite-type structure of In_2O_3. The optical properties of the prepared composition were considered in terms of their diffuse reflectance spectra (DRS). The morphology of crystals was evidenced by SEM analyses and reveal agglomeration of particles with their size ranges in the micrometer domain.

Introduction

Transparent conducting oxides (TCOs) are an important class of materials which have attracted much attention in the last years due to their main properties, low resistivity and high optical transparency. The dominant TCOs are mainly zinc oxide (ZnO), indium oxide (In_2O_3) and tin oxide (SnO_2), as well as subsequent mixtures of these, such as well-known indium tin oxide (ITO) [1]. In_2O_3, a wide-bandgap n-type semiconductor, is widely used in various applications due to their properties including solar applications, the transparent electrodes in various optoelectronic devices, the flat panel liquid crystals displays, the barrier layers in tunnel junctions, the active layers of gas sensors or the material for ultraviolet lasers [2]. Therefore, the physical properties of the In_2O_3, like high electrical conductivity and optical transparency in the visible range, strongly depend on the preparation method. For this reason there are many investigations which are studying their properties in dependence on the synthesis methods. These methods include physical methods such as sputtering, evaporation, pulsed laser deposition, spray pyrolysis, as well as chemical methods like chemical vapor deposition, sol-gel, bath deposition and electroplating [3–8]. On the other hand electrical properties of In_2O_3 can be improved by doping with metallic donor impurity. For this reason the system ZrO_2–In_2O_3 was mostly investigated [8-10].

Powder Metallurgy and Advanced Materials – RoPM&AM 2017   Materials Research Forum LLC
Materials Research Proceedings 8 (2018) 167-172   doi: http://dx.doi.org/10.21741/9781945291999-19

In the present paper we comparatively investigated the structural evolution, optical and morphological properties of indium zirconium oxide prepared by two methods, solid state synthesis and sol-gel route.

Materials and Methods
Synthesis

Two different preparation methods were employed in the synthesis of zirconium-doped indium oxide: solid state synthesis and sol-gel method.

Compositions belonging to $In_{2-x}Zr_xO_3$ ($0.025 \leq x \leq 0.15$) system with the bixbyite structure were prepared by solid state reactions from mixtures of pure In_2O_3 (Alfa Aesar 99.995%) and ZrO_2 (Alfa Aesar 99%) at 1400 °C, in air.

In the sol-gel synthesis indium (III) nitrate hydrate ($In(NO_3)_2 \cdot xH_2O$) was used as a precursor and zirconium (IV) chloride ($ZrCl_4$) as a doping source at the molar ratios required to obtain In_2O_3 doped with 1.5 at % ZrO_2. Aqueous solutions of the two salts were stirred on a warming plate at 80 °C in order to mix the solution uniformly. The sol was prepared using isopropyl alcohol as solvent and ethylene glycol or sucrose as polymerization agent. The sucrose solution was added gradually into the aqueous solutions under stirring until colloidal suspensions were obtained. The sol continuously transformed into wet gel and then dried gel under heating at 100 °C during 24h. The resulted dried materials were annealed at 600 °C in air for one hour to form the required Zr-doped indium oxide. Fig. 1 presents schematic flow chart of indium zirconium oxide obtained by solid state synthesis.

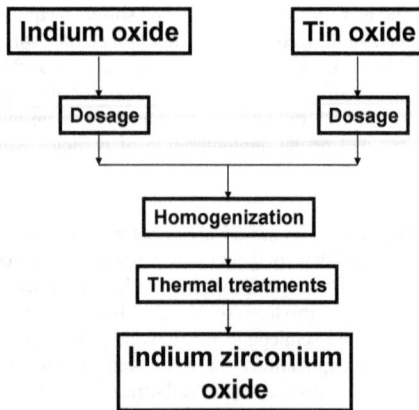

Fig. 1. Schematic flow chart of indium zirconium oxide obtained by solid state synthesis.

Characterization Methods

The X-ray powder diffraction (XRPD) data were collected in the $2\theta = 10\text{-}80°$ angular domain with a Bruker D8 Advance diffractometer, using CuKα1 radiation (40 kV; 40 mA).

Optical properties were investigated using a Jasco V-650 spectrophotometer (Japan) equipped with an ISV-722 Integrating Sphere. The diffusse reflectance spectra (DRS) were recorded in the range 200-800 nm with a scan rate of 400 nm/min.

The samples morphology was investigated using a Hitachi SU8230 Scanning Electron Microscope (SEM).

Results and discussion
X-ray powder diffraction (XRPD) analysis
The X-ray diffraction patterns of zirconium doped In_2O_3 prepared by solid state reaction with different concentrations are shown in Fig. 2. All samples showed similar XRPD patterns, indicating the high crystallinity of our bulk materials. No extra peaks are observed due to the addition of zirconium in indium oxide films which indicates the absence of an impurity phase in the prepared samples.

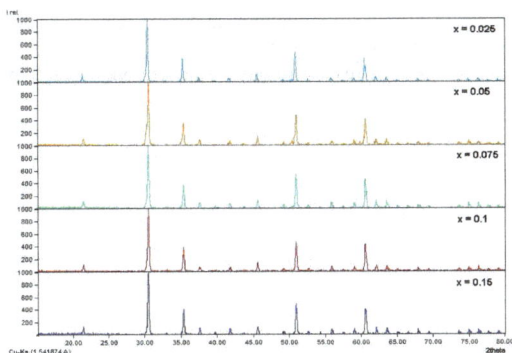

Fig. 2. XRPD patterns of $In_{2-x}Zr_xO_3$ ($0.025 \leq x \leq 0.15$) system prepared by solid state reaction.

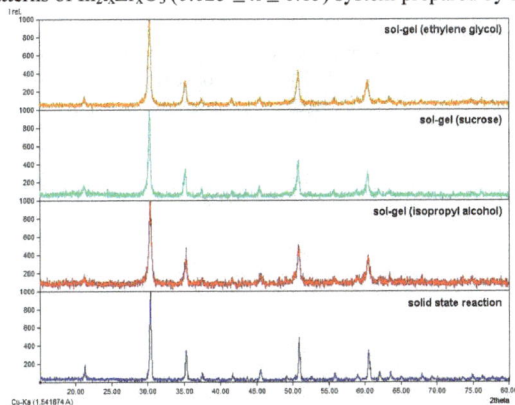

Fig. 3. XRPD patterns of x = 0.075 composition prepared by solid state reaction (a) and sol-gel method using (b) isopropyl alcohol, (c) sucrose or (d) ethylene glycol.

Powder Metallurgy and Advanced Materials – RoPM&AM 2017   Materials Research Forum LLC
Materials Research Proceedings 8 (2018) 167-172   doi: http://dx.doi.org/10.21741/9781945291999-19

There are four prominent peaks as shown in Fig. 2, which can be indexed to the (222), (400), (440) and (622) crystal planes of the In_2O_3 cubic structure. XRPD of x = 0.075 sample prepared via sol-gel route and solid state reaction are presented comparatively in Fig. 3. From the figure, it was clear that the intensity of peaks is increased in the x = 0.075 composition prepared by solid state reaction. The increase in intensity of the peaks is due to the improvement of sample crystallinity.

UV-VIS spectroscopy analysis

Fig.4 show the difusse reflectance spectra (DRS) of zirconium doped In_2O_3 prepared by sol-gel method using isopropyl alcohol as solvent and ethylene glycol or sucrose as polymerization agent. The maximum percent reflectance around 450 nm, observed for sample prepared using sucrose, is increasing for samples prepared using isopropyl alcohol as solvent and ethylene glycol as polymerization agent. Simultaneously, a shift of the optical bandgap to higher wavelengths occurs. The reflectance spectra of x = 0.075 composition prepared by solid state reaction and sol-gel method using sucrose are displayed comparatively in Fig. 5. One indeed observes that sol-gel preparation of zirconium doped indium oxide decreases the maximum percent reflectance around 450 nm, by approximately 20%, with respect to same composition prepared by solid state reaction.

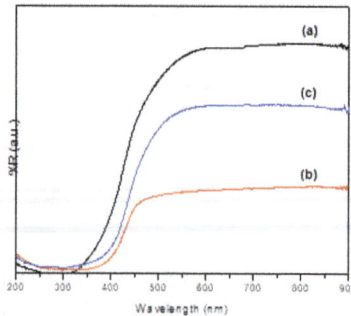

Fig. 4. Measured DRS in the compositions x = 0.075 prepared by sol-gel method using (a) isopropyl alcohol, (b) sucrose or (c) ethylene glycol.

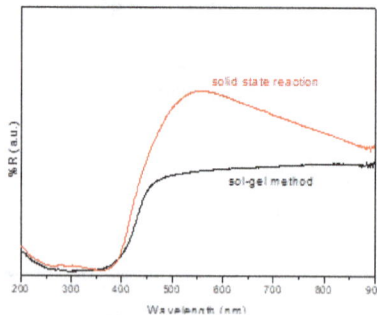

Fig. 5. Measured DRS in the composition x = 0.075 prepared by solid state reactions (a) and sol-gel method using sucrose (b).

SEM microscopy analysis

Scanning electron microscopy (SEM) was used to study the particle morphology of the studied samples. SEM images of the samples recorded for x = 0.075 composition prepared by solid state reaction and sol-gel route are shown in Fig. 6. Smooth surface and uniform distribution of grain are observed for composition synthesized by solid state reaction (Fig. 6 a).

Fig. 6. SEM images at x100 (up) and x10.0k (down) magnification of x = 0.075 composition prepared by solid state reaction (a) and sol-gel method using (b) isopropyl alcohol, (c) sucrose or (d) ethylene glycol.

Powder Metallurgy and Advanced Materials – RoPM&AM 2017   Materials Research Forum LLC
Materials Research Proceedings 8 (2018) 167-172   doi: http://dx.doi.org/10.21741/9781945291999-19

Particles with an angular shape are observed for the sol-gel synthesized samples using sucrose or ethylene glycol, as seen in Fig. 6c and 6d. The micrographs of zirconium indium oxide prepared by sol-gel with isopropyl alcohol as solvent (Fig. 6b) showed agglomeration of the particles. In all the cases the size of grains is situated in the micrometer range.

Conclusions

Structural evolution of the bulk materials in the candidate TCOs zirconium-doped indium oxide prepared by solid state reaction at high temperature and the sol-gel method were investigated. XRPD analysis shows that cubic bixbyite structure was confirmed with pure phases obtained in all the prepared samples. Measured diffuse reflectance spectra (DRS) show a decrease of max. %R for compositions prepared by the sol-gel method; moreover the sucrose addition as polymerization agent improve the optical properties. The surface morphology of the all prepared samples appears homogeneous and consists of small grains with size ranges in the micrometer domain.

References

[1] D. S. Ginley, H. Hosono, D. C. Paine, "Handbook of Transparent Conductors", Springer, Berlin, 2010.

[2] J. Xu, Y. Chen, Y, J. Shen, Ethanol sensor based on hexagonal indium oxide nanorods prepared by solvothermal methods, Mater. Lett. (2008), 1363-1365. https://doi.org/10.1016/j.matlet.2007.08.054

[3] S.M. Rozati, S. Mirzapour, M. G. Takwale, B. R. Marathe and V. G. Bhide, Influence of annealing on evaporated indium oxide thin films, Mater. Chem. Phys. 36 (1994) 252-255. https://doi.org/10.1016/0254-0584(94)90038-8

[4] Z. Yuan, X. Zhu, X. Wang, X. Cai, B. Zhang, D. Qiu, H. Wu, Annealing effects of In_2O_3 thin films on electrical properties and application in thin film transistors, Thin Solid Films, 519 (2011) 3254-3258. https://doi.org/10.1016/j.tsf.2010.12.022

[5] R.K. Gupta, N. Mamidi, K. Ghosh, S.R. Mishra, P.K. Kahol, Growth and characterization of In_2O_3 thin films prepared by pulsed laser deposition, J. Optoelectron. Adv. Mater., 7 (2007) 2211-2216.

[6] M. Girtan , G. Folcher, Structural and optical properties of indium oxide thin films prepared by an ultrasonic spray CVD process, Surf. Coat. Technol. 172 (2003) 242-250. https://doi.org/10.1016/S0257-8972(03)00334-7

[7] J. Joseph Prince, S. Ramamurthy, B. Subramanian, C. Sanjeeviraja, M. Jayachandran, Spray pyrolysis growth and material properties of In2O3 films, J. Cryst. Growth, 240 (2002) 142-151. https://doi.org/10.1016/S0022-0248(01)02161-3

[8] S. Jana, P. K. Biswas, Effect of Zr (IV) doping on the optical properties of sol–gel based nanostructured indium oxide films on glass, Mater. Chem. Phys., 117 (2009) 511-516. https://doi.org/10.1016/j.matchemphys.2009.06.038

[9] S. B. Qadri, H. Kim, H. R. Khan, A. Pique, J. S. Horwitz, D. Chrisey, W. J. Kim, E. F. Skelton, Transparent conducting films of In_2O_3–ZrO_2, SnO_2–ZrO_2 and ZnO–ZrO_2, Thin Solid Films 377-378 (2000) 750-754. https://doi.org/10.1016/S0040-6090(00)01328-6

[10] Y.C. Liang, Surface morphology and conductivity of zirconium-doped nanostructured indium oxide films with various crystallographic features, Ceram. Int. 36 (2010) 1743–1747. https://doi.org/10.1016/j.ceramint.2010.03.004

Powder Metallurgy and Advanced Materials – RoPM&AM 2017   Materials Research Forum LLC
Materials Research Proceedings 8 (2018) 173-181   doi: http://dx.doi.org/10.21741/9781945291999-20

Obtaining of W/Cu nanocomposite powders by high energy ball milling process

Claudiu NICOLICESCU[1, a *], Victor Horia NICOARĂ[2,b], Florin POPA[3,c], Traian Florin MARINCA[4,d]

[1,2]University of Craiova, Faculty of Mechanics, Department of Engineering and Management of the Technological Systems, Drobeta-Turnu Severin, Romania

[3,4]Technical University of Cluj-Napoca, Faculty of Materials and Environmental Engineering, Department of Materials Science and Engineering, Technical University of Cluj-Napoca, Cluj-Napoca, Romania

[a]nicolicescu_claudiu@yahoo.com, [b]victorczh@gmail.com, [c]florin.popa@stm.utcluj.ro, [d]traian.marinca@stm.utcluj.ro

Keywords: Powder metallurgy, Tungsten nanopowders, W/Cu nanocomposite powders, Mechanical milling

Abstract. The morphology of the particles is important in the process of obtaining alloys based on W/Cu, thus this investigation is focused on the influence of the copper content on the properties of W/Cu nanocomposites powders obtained after 20 hours of high energy ball milling. The experimental results regarding the obtaining of W_{100-x}/Cu_x nanocomposites (x between 20 and 45 wt. %) are presented. Composition of the mixtures influenced the particle size distribution namely, the higher is Cu content the larger dimensions of the particles will be attained. After 20 hours of high energy ball milling the crystallites size was about 30 nm for copper respectively 12 nm for tungsten and Cu atoms entered in the W structure.

Introduction

One of the specific classes of materials which are suitable for elaboration by Powder Metallurgy (PM) consists in pseudo-alloys based on W-Cu due to their mutual insolubility. These materials are very important due to their wide field of applicability such as: welding electrodes, nozzle liners for rockets and missiles, heat sink materials, high power electrical contacts, fission reactors and so on, applications that require high mechanical properties conferred, in this case by tungsten, combined with high electrical and thermal conductivity which are conferred by copper [1-6]. The properties of these materials are in correlation with their composition and morphology and because of that is very important to choose the right composition function of the application [7]. In the field of high power electrical contacts, W-Cu materials must have high arc erosion resistance high temperature strength and high tribological properties to ensure an as long as possible lifetime [8, 9]. Particle size of the component elements plays an important role in the final properties of W-Cu alloys [10]. One of the techniques used for fabrication of the W-Cu materials is the infiltration one, which consists in the formation of a porous skeleton by tungsten which will be filled with molten copper. Vacuum pulse carburisation was reported [11] to be an infiltration method that leads to the formation of W-30wt.%Cu material with core-shell structure which presents high electrical conductivity (46.55%IACS) compared to the national standard (GB/T8320-2003 – 42%IACS) and a friction coefficient $\mu=0.64$. To improve the sinter ability of W-Cu materials it can be introduced some activators such as Ni, Fe or Co which can be grain growth inhibitors [12]. Using of this activators can lead to a decreasing of electrical and thermal conductivity of W-Cu

Powder Metallurgy and Advanced Materials – RoPM&AM 2017   Materials Research Forum LLC
Materials Research Proceedings **8** (2018) 173-181   doi: http://dx.doi.org/10.21741/9781945291999-20

materials [13, 14].

A technique used for mass production which is suitable to produce W-Cu complex shape parts is Metal Injection Moulding (MIM) [15].

Another method to produce W-Cu materials is Mechanical Alloying (MA) which ensures obtaining of the nanocomposite powders and structural homogeneity which leads to an improvement of the sintering process by reducing the sintering activation energy [16-19]. Compared with micron powders, by using submicron powders (400nm) in the case of infiltrated W-25wt%Cu alloys can be obtained better properties such as: microstructural homogeneity, relative density (98.9%), hardness (230 HB) [10]. By MA, in the case of W-Cu system, Cu phase can dissolute in the W phase [20]. In most of cases the MA process is carried out at room temperature [21, 22].

In the present work nanocomposite powders W_{100-x}/Cu_x (x=20-45 wt. %) were prepared by mechanical milling (MM) process having as starting powders W nanometric and Cu micrometric powders. The influence of the MM times and composition of the mixture on the morphology, phase transformations and particle size distribution were investigated.

Experimental work

As raw materials were used tungsten nanopowders prepared by MM (35 hours of milling) and copper micrometric powders (type SE Pometon). The morphologies of the initial powders at the same magnification (1500x) are presented in (Fig. 1).

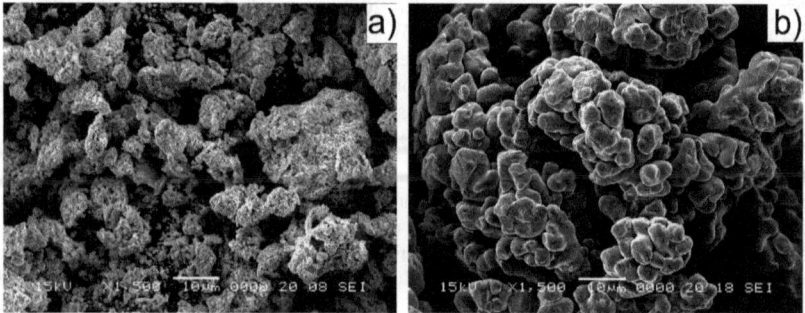

Fig. 1. SEM images of initial powders: a) W nanopowders; b) Cu powders.

For MM process of the six mixtures W_{100-x}/Cu_x (x=20, 25, 30, 35, 40, 40 and 45 wt. %) a Pulverissete 4 Vario planetary ball mill made by Fritsch was used. The parameters used for MM were: bowls volume - 250 ml; material of the bowls - stainless steel; balls diameter - 10 mm; balls material - stainless steel; milling type - dry; milling medium – argon (type 5.0, purity 99.999%); material/ball weight ratio - 1/2; speed: 400 rpm for the main disk and -800 rpm for the planets; milling time - 20 hours; from 5 to 5 hours were taken samples to be analysed.

In order to establish the cycles for MM, which means to not have higher temperature and pressure inside the grinding bowls, a GTM system made by Fritsch was used. In (Fig. 2) is presented the evolution of temperature and pressure for a MM of 10 minute with a break of 2 minutes.

Powder Metallurgy and Advanced Materials – RoPM&AM 2017   Materials Research Forum LLC
Materials Research Proceedings 8 (2018) 173-181   doi: http://dx.doi.org/10.21741/9781945291999-20

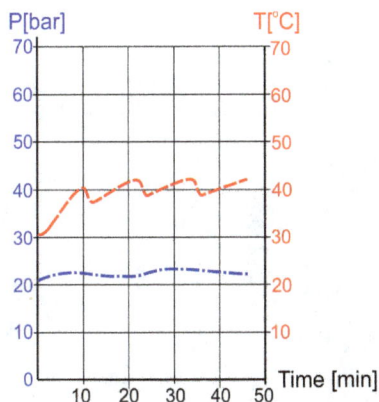

Fig. 2. Evolution of temperature and pressure inside the grinding bowl

As it can be seen in (Fig. 2), the temperature didn't exceed 50 °C and the pressure is almost constant due to the dry milling type. These parameters must be controlled because if the temperature is higher, it can be damage the bowl and the equipment too.

For morphological aspects of the powder mixtures was used a JEOL microscope JSM-5600 LV.

Evolution of particle size distributions and polydispersity were studied by dynamic laser scattering (DLS) using a 90Plus particle size analyser, Brookhaven Instruments Corporation, USA, equipped with 35 mW solid state laser, having 660 nm wavelength. The temperature was 25 °C and the scattering angle was 90°. The dilutions of the powders mixtures were made in water and the solutions of each sample were subjected to ultrasonic treatment for 5 minutes to avoid flocculation of the particles.

The investigation by X-ray diffraction has been performed using an Inel diffractometer, model Equinox 3000 working in reflection and using Co radiation. The 2theta investigated interval was 20-110 degree.

Results and discussion

In (Fig. 3) are presented morphologies of the as initial homogenous mixture and for the samples milled for 20 hours, respectively. From (Fig. 3 a, c, e, g, i, k) it is obvious that the initial homogenised mixtures have tungsten particles with lower particle size (nanoscale) compared with those of copper. Also, in (Fig. 3 k) it is observed higher covering degree of tungsten nanopowders on the copper particles, which is in accordance with composition of the mixture (80W/Cu). After 20 hours of MM (Fig. 3 b, d, f, h, j, l) all the samples elemental individual morphology is changed to a homogenous state. Most probably a mixing and welding of different particles being realized. Also the powders tend to agglomerate.

Fig. 4 shows the XRD pattern of the mixtures in different stages. In the figure are presented the X-ray diffraction patterns of the W-Cu mixtures not milled samples and samples milled for 5, 10, 15 and 20 h. Alongside of these patterns, in the same figure are presented the evolution of the mean crystallite size of the W and Cu upon increasing the milling time. In the X-ray diffraction patterns of the not milled samples on can observe the peaks characteristic for the W bcc structure

Powder Metallurgy and Advanced Materials – RoPM&AM 2017   Materials Research Forum LLC
Materials Research Proceedings 8 (2018) 173-181   doi: http://dx.doi.org/10.21741/9781945291999-20

from Im-3m space group and Cu fcc structure from Fm-3m space group according to JCPDS files 04-0806 and 04-0836 respectively.

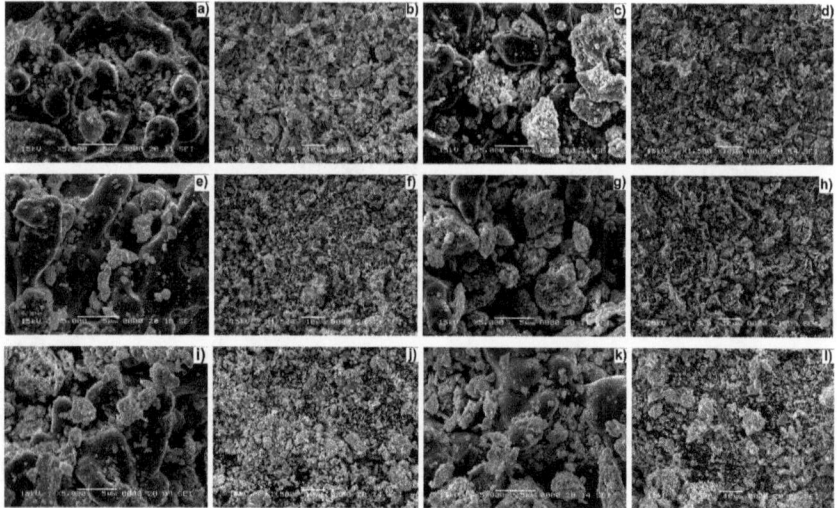

Fig. 3. SEM images of the: a) 55W/Cu homogenised mixture; b) 55W/Cu after 20 hours of MM; c) 60W/Cu homogenised mixture; d) 60W/Cu after 20 hours of MM; e) 65W/Cu homogenised mixture; f) 65W/Cu after 20 hours of MM; g) 70W/Cu homogenised mixture; h) 70W/Cu after 20 hours of MM; i) 75W/Cu homogenised mixture; j) 75W/Cu after 20 hours of MM; k) 80W/Cu homogenised mixture; l) 80W/Cu after 20 hours of MM;

The ratio of the W and Cu peaks intensities differ upon varying the amount of W and Cu in the mixtures, as expected. One can also observe is the broadening of the tungsten diffraction peaks due to the nanocrystalline state of the tungsten powder used in the mixture. The mean crystallite size of the tungsten powder (computed with the Scherrer method) is at 12 ± 2 nm. After 5 h of milling it can be noticed that the diffraction peaks of the copper are also broadened indicating the copper crystallite decreases. Further increase of the milling time lead also to the enlargement of the diffraction peaks of the copper and up to the final milling time no other peaks are identified in the diffraction patterns for all the ratios between W and Cu. It can be observed that independent on the amount of Cu in the material its peaks are observed in the diffraction pattern. The fcc Cu-based structure is present in the material after 20 h of milling. Known that in equilibrium condition there is no solubility between W and Cu and known also that by mechanically milling solid solution between immiscible elements can be obtained, it can be assumed that after 20 of milling some Cu atoms entered in the W structure. At the end of milling time independent on W to Cu ratio according to X-ray investigation the material is a nanocomposite one consisting in a W-Cu solid solution and Cu nanocrystallites. It can be remarked that the tungsten crystallites does not have significant variation upon milling together with copper. It remains at about 12 ± 2 nm. The mean

Powder Metallurgy and Advanced Materials – RoPM&AM 2017   Materials Research Forum LLC
Materials Research Proceedings **8** (2018) 173-181   doi: http://dx.doi.org/10.21741/9781945291999-20

crystallite size of the copper present a significant variation upon increasing the milling time. It can be noticed that during milling from 0 to 5 h a decrease from micrometer up to 65 ± 2 nm occurs. Further increase of the milling time lead to a further refinement of the copper crystallites size up to 56 ± 2 nm after 10 h of milling and reach 32 ± 2 nm after 15 h of milling. At 20 h of milling the copper crystallite size is at about 30 ± 2 nm.

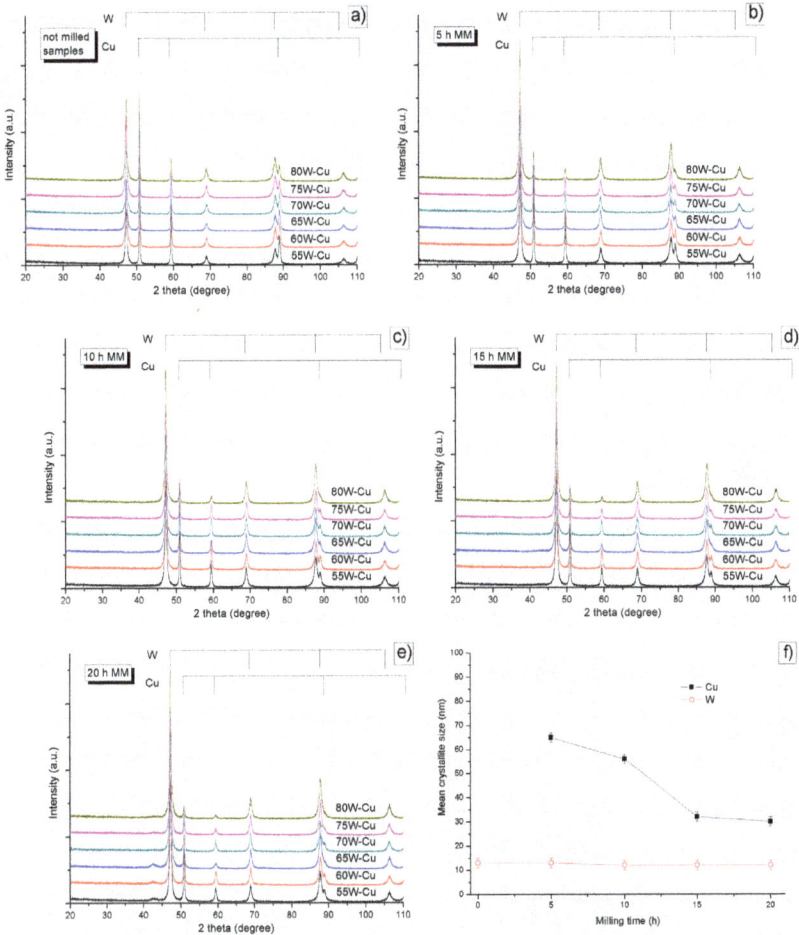

Fig. 4. X-ray diffraction patterns of the W-Cu mixtures: a) not milled - homogenised samples, b) samples milled for 5 h, c) samples milled for 10 h, d) samples milled for 15 h, e) samples milled for 20 h and f) evolution of the mean crystallite size of W and Cu upon milling

Powder Metallurgy and Advanced Materials – RoPM&AM 2017   Materials Research Forum LLC
Materials Research Proceedings **8** (2018) 173-181   doi: http://dx.doi.org/10.21741/9781945291999-20

In (Fig. 5) are presented the histograms of the number distribution versus particle size for all the mixtures after 20 hours of MM.

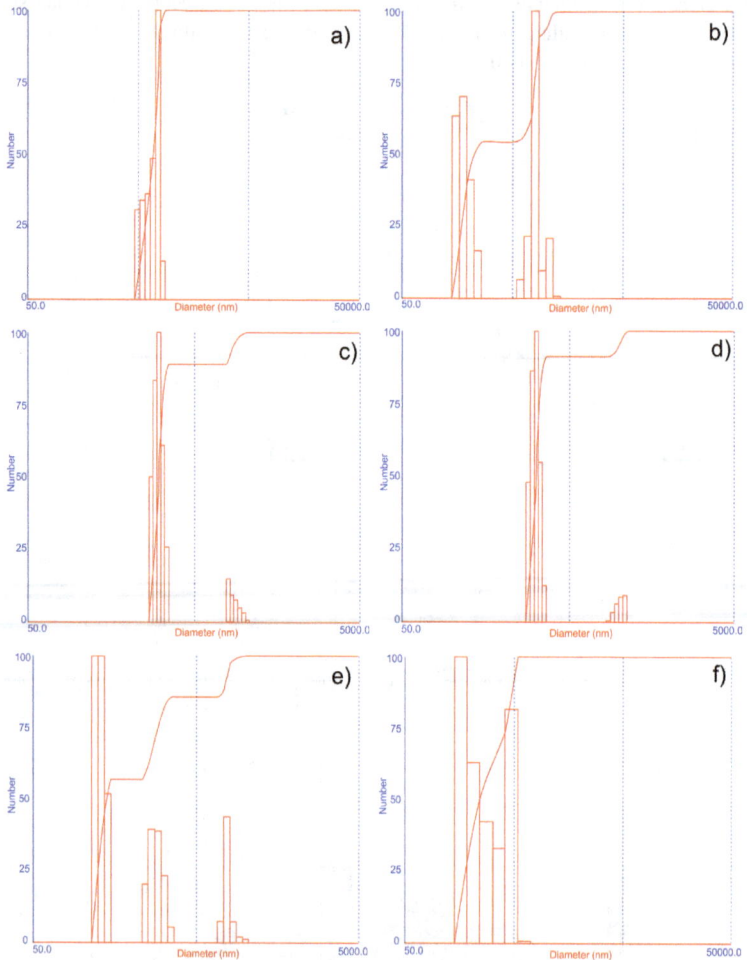

Fig. 5. Histograms of the particle size distributions of the mixtures after 20 h of MM: a) 55W/Cu; b) 60W/Cu; c) 65W/Cu; d) 70W/Cu; e) 75W/Cu; f) 80W/Cu;

In the case of 80W/Cu, (Fig. 5 f), according to the DLS measurement the particles after 20 h of MM are in the range of (161-613) nm. The distribution presents the following structure: highest percent from the total number of the particles (31.15%) is reported at 161 nm; 16.63% - 210 nm; 13.08% - 275; 10.28% - 359 nm; 25.55% - 469 nm; 0.31% - 613 nm. The presence of highest values of the distribution can be explained by the flocculation of these and to resolve this is better to find the proper time for ultrasonic treatment respectively to find anti flocculation agents in order to introduce them in the solution. For the mixture 75W/Cu, according to (Fig. 5 e) the particle size distribution presents three space-bars as following: one having small particles (119-143) nm, second between (243-346) nm and the third between (704-1004) nm. According to (Fig. 5 d) the particle size distribution of the 70W/Cu is a bimodal one composed from small particles (262-353) nm and large particles (862-1077) nm. The percent of the number for large particles is under 3%. The mixture 65W/Cu, (Fig. 5 c), presents a bimodal particle size distribution with small particles (275-340) nm respectively large prticles (800-991) nm. The large particles are under 5%. In the case of 60W/Cu, (Fig. 5 b) the DLS shows a bimodal particle size distribution with small particles (152-238) nm respectively large prticles (583-1232) nm. The DLS analysis of the mixture 55W/Cu, (Fig. 5 a) shows that the particles size is in the range of (488-821) nm. Compared to the diameter of the particles from SEM analysis, the hydrodinamic diameter measured by DLS can be a little larger due to the fact that by DLS it is measured the diameter of a sphere which has the same average diffusion coeficient as the particle which is measured. According to Stokes-Einstein equation the hydrodynamic diameter is inversely proportional to the diffusion coefficient and it depends on the size of the particle core, its surface structure, concentration and type of ions in the medium [23, 24]. The polydispersity index decreases with particle size decreasing.

Summary

Based on the results it can be underline the following conclusions:
- Nanocomposite powders based on W/Cu were obtained after 20 hours of MM process having homogenous compositions, consist in W/Cu solid solution and nanocrystaline Cu;
- Copper crystalite size decrease from micronic size in the initial homogenous mixture up to about 30 ± 2 nm after 20 hours of MM. This process was very fast in the first 10 hours;
- The mixtures with higher concentration of Cu present larger particle size at each time of MM;

By using these nanocomposite powders, according to [10] the sintering treatment can be improved leading to better properties of the final products.

Acknowledgement

This work was supported by the strategic grant POSDRU/159/1.5/S/133255, Project ID 133255 (2014), co-financed by the European Social Fund within the Sectorial Operational Program Human Resources Development 2007 – 2013.

References

[1] D.G. Kim, G.S. Kim, M.J. Suk, et al., Effect of heating rate on microstructural homogeneity of sintered W-15 wt% Cu nanocomposite fabricated from W-CuO powder mixture, Scr. Mater. 51 (2004) 677-681. https://doi.org/10.1016/j.scriptamat.2004.06.014

[2] F.T.N. Vüllers, R. Spolenak, From solid solutions to fully phase separated interpenetrating networks in sputter deposited "immiscible" W-Cu thin films, Acta Mater. 99 (2015) 213-227. https://doi.org/10.1016/j.actamat.2015.07.050

[3] A. Elaayed, W. Li, O.A. El Kady, et al., Experimental investigations on the synthesis of W-Cu nanocomposite through spark plasma sintering, J. Alloys Compd. 639 (2015) 373-380. https://doi.org/10.1016/j.jallcom.2015.03.183

Powder Metallurgy and Advanced Materials – RoPM&AM 2017   Materials Research Forum LLC
Materials Research Proceedings 8 (2018) 173-181   doi: http://dx.doi.org/10.21741/9781945291999-20

[4] Q. Zhou, P.W. Chen, Fabrication of W-Cu composite by shock consolidation of Cu-coated W powders, J. Alloys Compd. 657 (2016) 215-223. https://doi.org/10.1016/j.jallcom.2015.10.057

[5] W.T. Qiu, Y. Pang, Z. Xiao, et al., Preparation of W-Cu alloy with high density and ultrafine grains by mechanical alloying and high pressure sintering, Int. J. Refract. Met. Hard Mater. 61 (2016) 91-97. https://doi.org/10.1016/j.ijrmhm.2016.07.013

[6] S.H. Liang, L. Chen, Z.X. Yuan, et al., Infiltrated W-Cu composites with combined architecture of hierarchical particulate tungsten and tungsten fibers, Mater. Charact. 110 (2015) 33-38. https://doi.org/10.1016/j.matchar.2015.10.010

[7] Y. Yu, W. Zhang, H. Yu, Effect of Cu content and heat treatment on the properties and microstructure of W-Cu composites produced by hot extrusion with steel cup, Adv. Powder Technol. 26 (2015) 1047-1052. https://doi.org/10.1016/j.apt.2015.04.012

[8] C. Ding, X.C. Yuan, Z.B. Li, et al., Contact erosion characteristics in making process of SF6 circuit breakers, High Volt. Eng. 40 (2014) 3228-3232.

[9] Q. Zhang, X.H. Yang, B.Y. Liu, et al., Failure analysis of capacitor bank switch arcing contact for ultra-high voltage system, High Volt. Appar. 52 (2016) 27-32.

[10] Q. Zhang, S.H. Liang, B.Q. Hou, et al., The effect of submicron-sized initial tungsten powders on microstructure and properties of infiltrated W-25 wt% Cu alloys, Int. J. Refract. Met. Hard Mater. 59 (2016) 87-92. https://doi.org/10.1016/j.ijrmhm.2016.05.014

[11] Q. Zhang, S. Liang, L. Zhuo, Fabrication and properties of the W-30wt%Cu gradient composite with W@WC core-shell structure, Journal of Alloys and Compounds, 708 (2017) 796-803. https://doi.org/10.1016/j.jallcom.2017.03.064

[12] J. Johnson, R. German, Phase equilibria effects on the enhanced liquid phase sintering of tungsten-copper, Metall. Trans. A. 24 (1993) 2369–2377. https://doi.org/10.1007/BF02646516

[13] S.H. Hong, B.K. Kim, Fabrication of W–20 wt.% Cu composite nanopowder and sintered alloy with high thermal conductivity, Mater. Lett. 57 (2003) 2761–2767. https://doi.org/10.1016/S0167-577X(03)00071-5

[14] P. Chen, Q. Shen, G. Luo, M. Li, L. Zhang, Themechanical properties of W–Cu composite by activated sintering, Int. J. Refract. Met. Hard Mater. 36 (2013) 220–224. https://doi.org/10.1016/j.ijrmhm.2012.09.001

[15] Ijaz Ul Mohsin, Christian Gierl, Herbert Danninger, Sintering study of injection molded W–8%Ni–2%Cu compacts from mixed powders by thermoanalytical techniques, Int. Journal of Refractory Metals and Hard Materials, 29 (2011) 532–537. https://doi.org/10.1016/j.ijrmhm.2011.03.006

[16] J.S. Lee, T.H. Kim, Densification and microstructure of the nanocomposite W-Cu powders, Nanostruct. Mater. 6 (1995) 691–694. https://doi.org/10.1016/0965-9773(95)00152-2

[17] L.J. Kecskes, M.D. Trexler, B.R. Klotz, K.C. Cho, R.J. Dowding, Densification and structural change of mechanically alloyed W-Cu composites, Metall. Mater. Trans. A. 32A (2001) 2885–2893. https://doi.org/10.1007/s11661-001-1039-0

[18] F.A. Costa, A.G. Silva, U.U. Gomes, The influence of the dispersion technique on the characteristics of the W–Cu powders and on the sintering behavior, Powder Technol. 134 (2003) 123–132. https://doi.org/10.1016/S0032-5910(03)00123-2

Powder Metallurgy and Advanced Materials – RoPM&AM 2017   Materials Research Forum LLC
Materials Research Proceedings 8 (2018) 173-181   doi: http://dx.doi.org/10.21741/9781945291999-20

[19] M.H. Maneshian, A. Simchi, Z.R. Hesabi, Structural changes during synthesizing of nanostructured W–20 wt% Cu composite powder by mechanical alloying, Mater. Sci. Eng. A. 445–446 (2007) 86–93. https://doi.org/10.1016/j.msea.2006.09.005

[20] S.N. Alam, Synthesis and characterization of W–Cu nanocomposites developed by mechanical alloying, Materials Science and Engineering A 433 (2006) 161–168. https://doi.org/10.1016/j.msea.2006.06.049

[21] B.S. Murty, S. Ranganathan, Novel materials synthesis by mechanical alloying/milling, Int. Mater. Rev. 43 (1998) 101–141. https://doi.org/10.1179/imr.1998.43.3.101

[22] T.R. Malow, C.C. Koch, Mechanical properties in tension of mechanically attrited nanocrystalline iron by the use of the miniaturized disk bend test, Acta Mater. 46(18) (1998) 6459–6473. https://doi.org/10.1016/S1359-6454(98)00294-8

[23] M. Lungu, V. Tsakiris, E. Enescu, et al., Development of W-Cu-Ni Electrical Contact Materials with Enhanced Mechanical Properties by Spark Plasma Sintering Process Acta Physica Polonica A, 125 (2014) 327-330.

[24] M. Kaszuba, D. McKnight, M.T. Connah, et al., Measuring sub nanometre sizes using dynamic light scattering, J. Nanopart. Res. 10 (2008) 823. https://doi.org/10.1007/s11051-007-9317-4

Powder Metallurgy and Advanced Materials – RoPM&AM 2017   Materials Research Forum LLC
Materials Research Proceedings 8 (2018) 182-191   doi: http://dx.doi.org/10.21741/9781945291999-21

Wear behavior and microhardness of some W/Cu functionally graded materials obtained by spark plasma sintering

Claudiu NICOLICESCU[1, a *], Victor Horia NICOARĂ[2,b], Florin POPA[3,c], Traian Florin MARINCA[4,d]

[1,2]University of Craiova, Faculty of Mechanics, Department of Engineering and Management of the Technological Systems, Drobeta-Turnu Severin, Romania

[3,4]Technical University of Cluj-Napoca, Faculty of Materials and Environmental Engineering, Department of Materials Science and Engineering, Technical University of Cluj-Napoca, Cluj-Napoca, Romania

[a]nicolicescu_claudiu@yahoo.com, [b]victorczh@gmail.com, [c]florin.popa@stm.utcluj.ro, [d]traian.marinca@stm.utcluj.ro

Keywords: Functionally graded materials, Spark plasma sintering, Wear rate, Friction coefficient, Microhardness, Scanning electron microscopy, Optical microscopy

Abstract. This paper is focused on the elaboration of some W/Cu functionally graded materials (FGM) by spark plasma sintering (SPS) process, as well as on their characterization, from the wear behavior and microhardness point of view function of composition and sintering temperature. The raw materials used for the research were W/Cu mechanically alloyed powders for 20 hours, which were subjected to consolidation in three layers of compositions $W_{100-x}Cu_x$, where x is 25, 30 and 40 % wt. by SPS. The evolution of tribological parameters and microhardness function of the chemical composition and SPS temperature were investigated. Microhardness is influenced by the SPS temperature and composition of the layers namely, the highest value was attained for the sample sintered at 950 °C and layer 1 which consists in $W_{75}Cu_{25}$. The wear behavior is influenced by the composition of the layers and by ball testing material (100Cr6 and alumina).

Introduction

Copper alloys are frequently used in applications that require high electrical and thermal conductivity. In some applications that require strength and wear resistance it is necessary to alloy copper with others metals. Alloys based on copper and tungsten attract the attention due to the combination of properties such as low thermal expansion coefficient, high melting point, high strength and wear resistance conferred by tungsten with a high electrical and thermal characteristics conferred by copper [1-6].

The researches in the field of W/Cu alloys are focused on the controlling the microstructure by optimizing the composition or processing techniques [7-11]. Due to their insolubility and high differences between densities and melting points it's very difficult to produce W/Cu composite. There are different methods to produce W/Cu composite/nanocomposite namely: copper infiltration and liquid phase sintering [10, 12] which are considered classical methods, respectively new methods as mechanical alloying (MA), mechano-chemical processes (MCP) [13], mechano-thermochemical processing (MTP) [14], the thermo-mechanical method [15], wet-chemical methods [16] and spray drying [17].

Functional graded materials (FGM) based on W/Cu represent a new category of materials consisting in two or more layers, in which the microstructure and the composition vary from the

Powder Metallurgy and Advanced Materials – RoPM&AM 2017   Materials Research Forum LLC
Materials Research Proceedings 8 (2018) 182-191   doi: http://dx.doi.org/10.21741/9781945291999-21

top layer to the bottom layer and vice versa. This class of materials presents some advantages comparative to the single layer materials, namely: the properties are different in each layer, residual and thermal stresses are reduced and the fracture strength is optimized [18-20]. The main fields of applications of W/Cu FGM are: electrical contacts, plasma facing materials, heat sink materials, etc. [21]

In recent decades, Spark Plasma Sintering (SPS) became a popular sintering method which is widely used in fabrication process of different materials as: ceramics, cermets, metals, hard materials, composite materials and FGMs [22–31]. The SPS process is different by conventional sintering in that, the cold pre-pressed powder is sintered by discharge impact, spark plasma, Joule heat and intrinsic field effects produced by additional strong pulse currents through the powders. The SPS is assisted by a vertical uniaxial load in order to accelerate the process. The main advantages of the SPS process are: rapid heating (over 200 °C/min), short dwell times, lower temperatures and fine microstructures [32, 33].

The present work is focused on the elaboration of three layers W/Cu FGM by SPS process. The composition of the layers consisting in $W_{100-x}Cu_x$, where x is 25, 30 and 40 %weight. The influence of the SPS temperature and layers composition on the wear behavior and microhardness were studied.

Experimental work

Raw materials

Tungsten nanopowders (Fig. 1a) obtained by mechanical milling for 35 hours with particle size about 50 nm (according to Dynamic Laser Scattering analysis) and copper micrometric powders (type SE Pometon) with particle size around 10μm (Fig. 1b) were used as raw materials in order to prepare the mixtures.

Fig. 1. SEM images of the starting powders: a) tungsten nanopowders; b) copper powders

Tungsten nanopowders are agglomerated and have irregular shape (Fig. 1a) comparative with copper powders (Fig. 1b) which have dendritic shape, corresponding to electrolytic process.

Three types of mixtures with the following concentrations (% weight): 75W/25Cu, 70W/30Cu and 60W/40Cu were subjected to mechanical milling process for 20 hours in order to obtain composite powder. For mechanical milling (MM) process planetary ball mill Pulverisette 4 made by Fritsch was used. The milling parameters were: material of vials: stainless steel, vials volume: 250 ml, materials of balls: stainless steels, balls diameter: Φ=10 mm, balls to powder ratio: 2/1, rotational speed of the main disk: 400 rpm, rotational speeds of the vials: -800 rpm, milling medium: argon, milling type: dry.

The SEM images and element distribution maps obtained by EDX analysis show that inside of particles of all mixtures, the component elements are homogenous dispersed (Fig. 2).

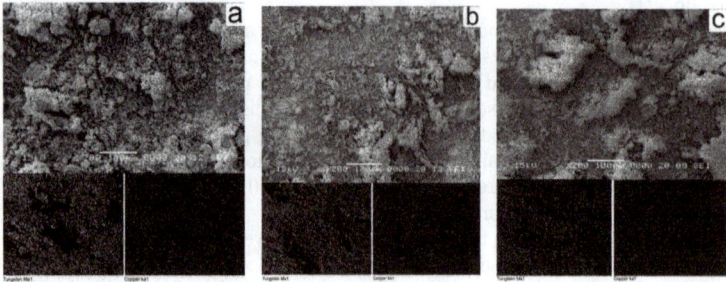

Fig. 2. SEM images and EDX analysis of the mixtures used for the research: a) 75W/Cu; b) 70W/Cu; c) 60W/Cu

Elaboration of functional graded materials
The mixtures obtained after MM process was used in order to obtain three layered functional graded materials (FGM) by spark plasma sintering (SPS) according to the flow chart presented in (Fig. 3).

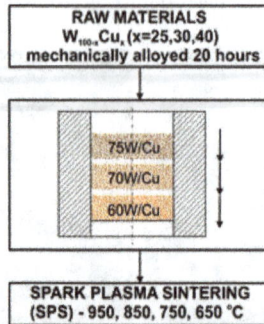

Fig. 3. Experimental flow chart

A thin graphite paper was put into the inner part of the die for lubrication and other two papers were put on the bottom and upper of the part in order to prevent the powder to stick on the graphite punches. The samples were sintered at four temperatures (950, 850, 750 and 650 °C) without dwell time and a pressure equal to 20 MPa using a SPS homemade system from Technical University of Cluj-Napoca. The heating rate was about 300 °C/min. After SPS the density of the samples was measured by the Archimedes' method. The samples were cut and polished in order to study the microstructural aspects and microhardness. The optical microscopy was made using an NIKON microscope with NIS ELEMENTS image software. SEM characterisation was performed using a JEOL JSM5600LV microscope equipped with EDX spectrometer (Oxford Instruments, INCA 200 software). The micro hardness was measured using a Namicon tester with a load of 9.8

Powder Metallurgy and Advanced Materials – RoPM&AM 2017   Materials Research Forum LLC
Materials Research Proceedings **8** (2018) 182-191   doi: http://dx.doi.org/10.21741/9781945291999-21

N and a dwell time of 15 seconds. The investigation of wear behavior was performed using a CSM Instruments tribometer TRB 01-2541 and a Taylor Hobson Surtronic 25+ profilometer. The parameters for wear testing were: type: ball on disk; load – 2N; testing method - linear; amplitude - 6mm; speed - 10cm/s; distance - 60 m; ball material – 100Cr6 and alumina; temperature - 25 °C.

Results and discussion
The relative density function of the SPS temperature is plotted in (Fig. 4).

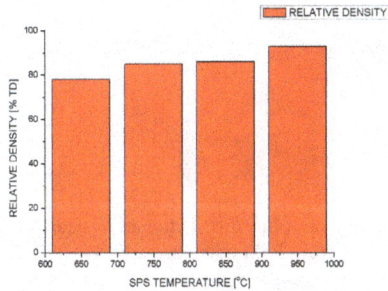

Fig. 4. Relative density function of the SPS temperature

The highest value of the density is attained at 950 °C and it is observed that it decrease with the decreasing of the sintering temperature.

Fig. 5 presents optical microscopy images of the samples (not etched) at the four sintering temperatures. The microstructures of the samples are homogenous and present structural gradient. The dark grey colour corresponds to the layer 1 with 75%W and so on. The thickness of the middle layer increases with the decreasing of the SPS temperature. The lowest value of the thickness (1216.85 μm) was attained for the sample sintered at highest temperature 950 °C. Higher density lead to lower thickness as expected.

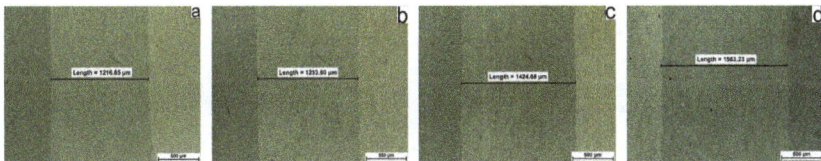

Fig. 5. Optical microscopy aspects of the three layered samples (75X): a) 950 °C; b) 850 °C; c) 750 °C; d) 650 °C

The EDX line distribution on the three layers, (Fig. 6) shows the distribution of W and copper function the scanning distance. It is observed that the content of W decrease from left (first layer – 75%W) to the right (third layer - 60%W). In the same time a slight increase in Cu is recorded, according the layer composition change. In (Fig. 7) are presented the interfaces of the three layers. Analysing (Fig. 6, 7) it is obvious that the graded structure was attained. The interfaces seems to assure continuity from one layer to another as it can be seen in (Fig. 7). The composition

Powder Metallurgy and Advanced Materials – RoPM&AM 2017   Materials Research Forum LLC
Materials Research Proceedings 8 (2018) 182-191   doi: http://dx.doi.org/10.21741/9781945291999-21

differences between the layers are small, probably due to the powder superficial mix at the interface. In (Fig. 8) is plotted the evolution of the microhardness.

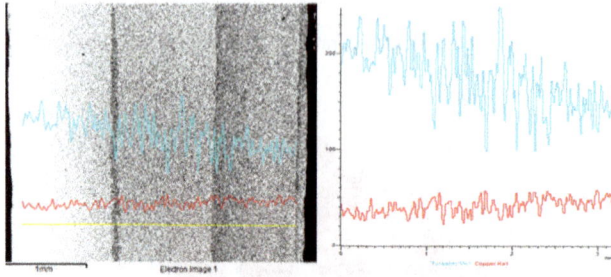

Fig. 6. SEM image (left) and EDX line distribution of W and Cu elements of the three layered W/Cu FGM obtained by SPS at 950 °C

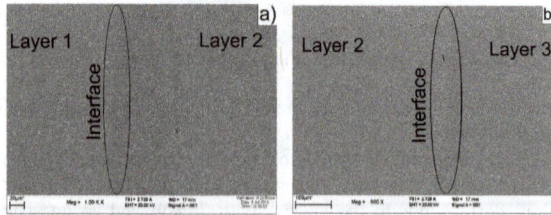

Fig. 7. SEM images of the interfaces of W/Cu FGM obtained by SPS at 950 °C: a) between layer 1 and layer 2; b) between layer 2 and layer 3.

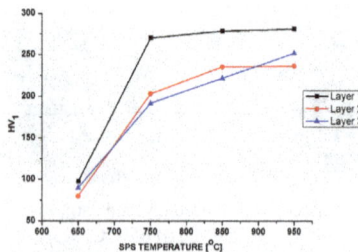

Fig. 8. Evolution of microhardness as a function of sintering temperature

The measurements of the microhardness were performed in all the three layers, even if only the layer 1 and 3 are important from this point of view, because only these are in contact with other materials. The bonding between W and copper particles influences the microhardness of the layers. The highest value of the microhardness was attained in the layer 1 (75W/Cu) sintered at 950 °C and it decreases with the decreasing of the W content.

The results on the wear tests are presented in (Table 1) and the optical light microscopy of the worns resulted after the tribological tests (with 100Cr6 ball) are presented in (Fig. 9 and 10). In (Fig. 11) are presented optical images of the two balls after tribological tests.

Table 1. Wear test results

T [°C]	Layer	Mean friction coefficient μ		Worn track section [μm²]		Worn cap diameter [μm]		Sample wear rate [mm³/n/m]		Partner wear rate [mm³/n/m]	
		Ball material		Ball material		Ball material		Ball material		Ball material	
		100Cr6	Alumina	100Cr6	Alumina	100Cr6	Alumina	100Cr6	Alumina	100Cr6	Alumina
950	Layer 1	0.463	0.311	2735	88	459.7	-	$136.7*10^{-6}$	$4.4*10^{-6}$	$6.101*10^{-6}$	-
	Layer 3	0.430	0.275	936.5	31.4	392.3	-	$46.82*10^{-6}$	$1.57*10^{-6}$	$3.233*10^{-6}$	-
850	Layer 1	0.370	0.268	2153.5	74.5	446.3	-	$107.7*10^{-6}$	$3.727*10^{-6}$	$5.421*10^{-6}$	-
	Layer 3	0.498	0.285	2426	50.6	473.3	-	$121.3*10^{-6}$	$2.528*10^{-6}$	$6.858*10^{-6}$	-
750	Layer 1	0.423	0.295	2090	29.6	473.8	-	$104.5*10^{-6}$	$1.48*10^{-6}$	$6.889*10^{-6}$	-
	Layer 3	0.461	0.287	457	59.1	151.4	-	$22.85*10^{-6}$	$2.955*10^{-6}$	$7.158*10^{-8}$	-
650	Layer 1	0.472	0.342	91.5	641	271.3	-	$4.572*10^{-6}$	$32.05*10^{-6}$	$7.393*10^{-7}$	-
	Layer 3	0.485	0.313	240.5	108.8	221.2	-	$12.03*10^{-6}$	$5.438*10^{-6}$	$3.268*10^{-7}$	-

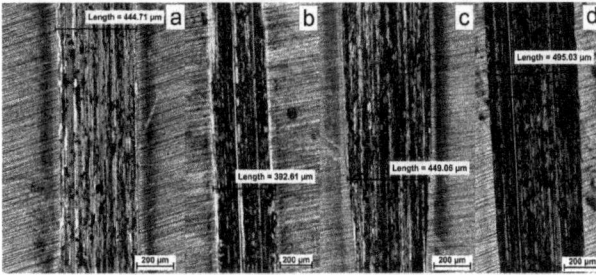

Fig. 9. Worn of the samples (75x): a) 950 °C layer 1; b) 950 °C layer 3; c) 850 °C layer 1; d) 850 °C layer 3

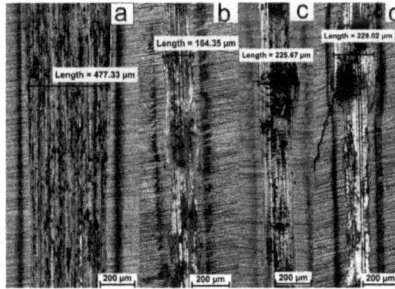

Fig. 10. Worn of the samples (75x): a) 750 °C layer 1; b) 750 °C layer 3; c) 650 °C layer 1; d) 650 °C layer 3

The tribological measurements were made only in the layer 1 and layer 3 because the material obtained is used for the fabrication of electrical contacts so, the middle layer is important only from the electrical and thermal point of view. The friction coefficient presents values in the range

Powder Metallurgy and Advanced Materials – RoPM&AM 2017   Materials Research Forum LLC
Materials Research Proceedings 8 (2018) 182-191   doi: http://dx.doi.org/10.21741/9781945291999-21

of 0.370-0.498 for the samples which were tested with 100Cr6 ball which are higher than the friction coefficients obtain in the case of alumina ball (0.268-0.342). These values are lower compared with the case of W-30wt%Cu which reported friction coefficient about $\mu=0.64$ [34] respectively $\mu=0.78$ and $\mu=0.56$ in the case of W-25wt%Cu [35]. As it can be seen from (Fig. 11), the worn cap of the 100Cr6 ball is larger compared with that of the alumina ball in all the cases.

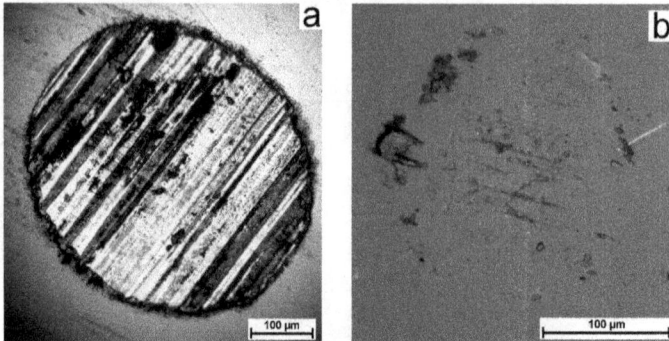

Fig. 11. Worn cap images of the balls after the test for the sample sintered at 950 °C, layer 1: a) 100Cr6 ball (150X); b) alumina ball (300X)

From (Fig. 9, 10) it can be observed that the wear is abrasive due to the presence of the ball material on the sample. The worns of the samples are in accordance with the partner worns (balls). The samples sintered at 650 °C present micro cracks after the wear testing (Fig. 10 c, d)

Summary
The experimental results obtained in this research lead to the following conclusions:
- it can be obtained materials with functional and structural gradient by stacking layers and using spark plasma sintering;
- the microhardness is influenced by the sintering temperature and composition of the layers namely, it increasing with the increasing of W content and SPS temperature;
- regarding the tribological tests, in the case of sintered materials because the presence of the pores which tend to fill with material of the ball, the wear rates aren`t in accordance with the microhardness. Because of that, the sample sintered at 650 °C presents the lowest values for the worn track sections and wear rates.

Acknowledgement
This work was supported by the strategic grant POSDRU/159/1.5/S/133255, Project ID 133255 (2014), co-financed by the European Social Fund within the Sectorial Operational Program Human Resources Development 2007 – 2013

References
[1]   K.   Lu,   The   future   of   metals,   Science,   328   (2010)   319–320. https://doi.org/10.1126/science.1185866

Powder Metallurgy and Advanced Materials – RoPM&AM 2017   Materials Research Forum LLC
Materials Research Proceedings 8 (2018) 182-191   doi: http://dx.doi.org/10.21741/9781945291999-21

[2] C.F. Zhu, F.P. Du, Q.Y. Jiao, et al., Microstructure and strength of pure Cu with large grains processed by equal channel angular pressing, Mater. Des. 52 (2013) 23–29. https://doi.org/10.1016/j.matdes.2013.05.029

[3] K. Maki, Y. Ito, H. Matsunaga, H. Mori, Solid-solution copper alloys with high strength and high electrical conductivity, Scripta Mater. 68 (2013) 777–780. https://doi.org/10.1016/j.scriptamat.2012.12.027

[4] D.H. Dai, D.D. Gu, Thermal behavior and densification mechanism during selective laser melting of copper matrix composites: simulation and experiments, Mater. Des. 55 (2014) 482–491. https://doi.org/10.1016/j.matdes.2013.10.006

[5] B.B. Liu, J.X. Xie, X.H. Qu, Fabrication of W–Cu functionally graded materials with high density by particle size adjustment and solid state hot press, Compos. Sci. Technol. 68 (2008) 1539–1547. https://doi.org/10.1016/j.compscitech.2007.10.023

[6] N. Selvakumar, S.C. Vettivel, Thermal, electrical and wear behavior of sintered Cu–W nanocomposite, Mater. Des. 46 (2013) 16–25. https://doi.org/10.1016/j.matdes.2012.09.055

[7] F.A. Costa, A.G.P. Silva, U.U. Gomes. The influence of the dispersion technique on the characteristics of the W–Cu powders and on the sintering behavior, Powder Technol. 134 (2003) 123-132. https://doi.org/10.1016/S0032-5910(03)00123-2

[8] M.H. Maneshian, A. Simchi, Solid state and liquid phase sintering of mechanically activated W–20wt.% Cu powder mixture, J. Alloys Compd. 463 (2008) 153-159. https://doi.org/10.1016/j.jallcom.2007.08.080

[9] J.L. Johnson, J.J. Brezovsky, R.M. German, Effects of tungsten particle size and copper content on densification of liquid-phase-sintered W-Cu, Metall Mater Trans A. 36A (2005) 2807-2814. https://doi.org/10.1007/s11661-005-0277-y

[10] J.L. Johnson, J.J. Brezovsky, R.M. German, Effect of liquid content on distortion and rearrangement densification of liquid-phase-sintered W-Cu, Metall Mater Trans A. 36A (2005) 1557-1565. https://doi.org/10.1007/s11661-005-0247-4

[11] X. Wei, D. Yu, Z. Sun, et al., Arc characteristics and microstructure evolution of W/Cu contacts during the vacuum breakdown, Vacuum 107 (2014) 83-89. https://doi.org/10.1016/j.vacuum.2014.04.005

[12] A.K. Bhalla, J.D. Williams, A Comparative Assessment of Explosive and Other Methods of Compaction in the Production of Tungsten-Copper Composites, Powder Metal. 1 (1976) 31–37. https://doi.org/10.1179/pom.1976.19.1.31

[13] J. Cheng, P. Song, Fabrication and characterization of W–15Cu composite powders by a novel mechano-chemical process, Mater. Sci. Eng. A. 488 (2008) 453–457. https://doi.org/10.1016/j.msea.2007.11.022

[14] S.H. Hong, B.K. Kim, Fabrication of W–20 wt % Cu composite nanopowder and sintered alloy with high thermal conductivity, Mater. Lett. 57 (2003) 2761–2767. https://doi.org/10.1016/S0167-577X(03)00071-5

Powder Metallurgy and Advanced Materials – RoPM&AM 2017   Materials Research Forum LLC
Materials Research Proceedings 8 (2018) 182-191   doi: http://dx.doi.org/10.21741/9781945291999-21

[15] Y. Li, X. Qu, Z.S. Zheng, et al., Properties of W–Cu composite powder produced by a thermo-mechanical method, Int. J. Refract. Metal. Hard Mater. 21 (2003) 259–264. https://doi.org/10.1016/j.ijrmhm.2003.08.001

[16] L. Wan, J. Cheng, P. Song, et al., Synthesis and characterization of W–Cu nanopowders by a wet-chemical method, Int. J. Refract. Metal. Hard Mater. 29 (2011) 429–434. https://doi.org/10.1016/j.ijrmhm.2011.01.006

[17] X. Shi, H. Yang, S. Wang, et al., Characterization of W–20Cu ultrafine composite powder prepared by spray drying and calcining-continuous reduction technology, Mater. Chem. Phys. 104 (2007) 235–239. https://doi.org/10.1016/j.matchemphys.2007.03.032

[18] E.L. Courtright, A review of fundamental coating issues for high temperature composites, Surf. Coat. Technol. 68–69 (1994) 116–125. https://doi.org/10.1016/0257-8972(94)90148-1

[19] X.C. Zhang, B.S. Xu, H.D. Wang, et al, Effects of compositional gradient and thickness of coating on the residual stresses within the gradedcoating, Mater. Des. 28(4) (2007) 1192–1197. https://doi.org/10.1016/j.matdes.2006.01.012

[20] A. Polat, O. Sarikaya, E. Celik, Effects of porosity on thermal loadings of functionally graded Y_2O_3–ZrO_2/NiCoCrAlY coatings, Mater. Des. 23(7) (2003) 641–644.

[21] J.J. Sobczak, L. Drenchev, Metallic Functionally Graded Materials: A Specific Class of Advanced Composites, Journal of Materials Science & Technology, Volume 29, Issue 4, April (2013) 297-316.

[22] H.P.Yuan, L.G. Li, Q. Shen, et al, In situ synthesis and sintering of ZrB_2 porous ceramics by the spark plasma sintering–reactive synthesis (SPS–RS) method. Int J Refract Met Hard Mater. 34 (2012) 413–417. https://doi.org/10.1016/j.ijrmhm.2012.01.007

[23] A. Teber, F. Schoenstein, F. Têtard, et al, The effect of Ti substitution by Zr on the microstructure and mechanical properties of the cermet $Ti_{1-x}Zr_xC$ sintered by SPS. Int J Refract Met Hard Mater. 31 (2012) 64–70. https://doi.org/10.1016/j.ijrmhm.2011.06.013

[24] J.M. Lee, J.H. Kim, S.H. Kang. Advanced W–HfC cermet using in-situ powder and spark plasma sintering. J Alloys Compd. 552 (2013) 14–19. https://doi.org/10.1016/j.jallcom.2012.09.116

[25] P. Feng, Y.H. He, Y.F. Xiao, et al., Effect of VC addition on sinterability and microstructure of ultrafine Ti(C, N)-based cermets in spark plasma sintering. J Alloys Compd. 460 (2008) 453–459. https://doi.org/10.1016/j.jallcom.2007.05.091

[26] L. Ding, D.P. Xiang, Y.Y. Li, et al., Effects of sintering temperature on fine-grained tungsten heavy alloy produced by high-energy ball milling assisted spark plasma sintering. Int J Refract Met Hard Mater. 33 (2012) 65–69. https://doi.org/10.1016/j.ijrmhm.2012.02.017

[27] A. Teber, F. Schoenstein, F. Têtard, et al., Effect of SPS process sintering on the microstructure and mechanical properties of nanocrystalline TiC for tools application. Int J Refract Met Hard Mater. 31 (2012) 64–70. https://doi.org/10.1016/j.ijrmhm.2011.06.013

[28] Z.H. Qiao, J. Räthel, L.M. Berger, et al., Investigation of binderless WC–TiC–Cr_3C_2 hard materials prepared by spark plasma sintering (SPS). Int J Refract Met Hard Mater. 38 (2013) 7–14. https://doi.org/10.1016/j.ijrmhm.2012.12.002

Powder Metallurgy and Advanced Materials – RoPM&AM 2017   Materials Research Forum LLC
Materials Research Proceedings 8 (2018) 182-191   doi: http://dx.doi.org/10.21741/9781945291999-21

[29] C.F. Hu, Y. Sakka, H. Tanaka, et al., Synthesis, microstructure and mechanical properties of (Zr, Ti)B2–(Zr, Ti)N composites prepared by spark plasma sintering. J Alloys Compd. 494 (2010) 266–270. https://doi.org/10.1016/j.jallcom.2010.01.006

[30] M. Eriksson, M. Radwan, Z.J. Shen, Spark plasma sintering of WC, cemented carbide and functional graded materials. Int J Refract Met Hard Mater. 36 (2013) 31–37. https://doi.org/10.1016/j.ijrmhm.2012.03.007

[31] H.B. Feng, Q.C. Meng, Y. Zhou, et al., Spark plasma sintering of functionally graded material in the Ti–TiB$_2$–B system. Mater Sci Eng. A. 397 (2005) 92–97. https://doi.org/10.1016/j.msea.2005.02.003

[32] X.L. Shi, H. Yang, S. Wang, Spark plasma sintering of W–15Cu alloy from ultrafine composite powder prepared by spray drying and calcining–continuous reduction technology, Mater Charact. 60 (2009) 133–137. https://doi.org/10.1016/j.matchar.2008.07.012

[33] X. Tang, H. Zhang, D. Du, et al., Fabrication of W– Cu functionally graded material by spark plasma sintering method, Int. Journal of Refractory Metals and Hard Materials, 42 (2014) 193–199. https://doi.org/10.1016/j.ijrmhm.2013.09.005

[34] Q. Zhang, S. Liang, L. Zhuo, Fabrication and properties of the W-30wt%Cu gradient composite with W@WC core-shell structure, Journal of Alloys and Compounds, 708 (2017) 796-803. https://doi.org/10.1016/j.jallcom.2017.03.064

[35] Q. Zhang, S.H. Liang, B.Q. Hou, et al., The effect of submicron-sized initial tungsten powders on microstructure and properties of infiltrated W-25 wt%Cu alloys, Int. J. Refract. Met. Hard Mater. 59 (2016) 87-92. https://doi.org/10.1016/j.ijrmhm.2016.05.014

Powder Metallurgy and Advanced Materials – RoPM&AM 2017   Materials Research Forum LLC
Materials Research Proceedings 8 (2018) 192-199   doi: http://dx.doi.org/10.21741/9781945291999-22

Morphology and mechanical characteristics of some TBCs used for internal combustion valves

Marius PANȚURU [1a,] Daniela CHICET [2b*], Ovidiu MOCĂNIȚA [1c], Marcelin
BENCHEA [1d] and Corneliu MUNTEANU [1e]

[1]Technical University "Gheorghe Asachi" of Iasi, Mechanical Engineering Department, 6 D.
Mangeron Blvd, Iasi, Romania

[2]Technical University "Gheorghe Asachi" of Iasi, Materials Science and Engineering Department,
41 D.Mangeron Blvd, Iasi, Romania

[a]mpanturu1967@yahoo.com, [b*]daniela.chicet@gmail.com, [c]ovidiumocanita@yahoo.com,
[d]marcelin_ben@yahoo.com, [e]cornelmun@gmail.com

Keywords: TBC, Internal combustion valves, Morphology, Mechanical characteristics

Abstract. Three types of commercial powders have been deposited on the inlet and outlet valve plates in order to increase their lifetime, but especially the temperature in the combustion chamber. The layers were coated by atmospheric plasma spray method. The coatings morphology was analysed using two complementary methods: scanning electron microscopy and X-ray diffraction. The mechanical characteristics analysed were: microhardness, modulus of elasticity and adhesion / cohesion of coatings using scratch tests. Following those tests it was observed that the coatings are physically suited for further tests as thermal barrier coatings (TBC) on the valve discs of internal combustion engines.

Introduction

The distribution system (especially the intake/evacuation areas) of the internal combustion engine is subjected, during its operation, to a series of very complex loads involving: mechanical impact and high frequency micro-slipping, high temperatures with a very large variation, presence of microparticles, etc. Another very important stress factor is the working pressure, which often in combination with other stresses causes damage to the valve disc and implicitly change the contact geometry of the seat of the valve. Taking into account all this, but also that the new regulations related to the emission of combustion gases will become more and more strict, we come up with the proposal to cover the valves discs with layers as thermal barrier.

Thermal barrier coatings have initially been used for gas turbine elements protection applications, in the specialized literature being available multiple studies on this type of use. [1-5] Starting from these studies, the range of applications has been expanded so that over the past 20 years, TBCs have found many other applications, one of which is covering the components of diesel engines in order to improve their thermal efficiency, to reduce weight by removing the cooling system, to increase the efficiency by lowering the amount of energy lost through thermal effect and to improve the durability of components [6,7].

Depending on the working conditions, different mechanisms of wear and destruction of TBCs become dominant. These coatings are in fact complex systems formed of the top layer of TBC, the intermediate layer with bonding function that supports the upper layer and the substrate, so that the properties of the whole system influence its lifetime in operation. By analysing the components, it is observed that in the case of the TBC top layer these properties are the microstructure, density, thickness, distribution of the micro-cracks and cohesion in the layer

Powder Metallurgy and Advanced Materials – RoPM&AM 2017   Materials Research Forum LLC
Materials Research Proceedings 8 (2018) 192-199   doi: http://dx.doi.org/10.21741/9781945291999-22

(between splats that form it). In the case of the bonding layer, it is the oxidation resistance, the density of the layer, its thickness and the surface roughness [8]. An equally important role for lifetime in operation is the difference in thermal expansion, the residual stresses of the system, but also its geometry. [9,10]

The most commonly element used and studied for this type of application is zirconia, because it exhibits a high coefficient of thermal expansion and low thermal conductivity, the last one due to the presence of micropores according to the studies of Zhou et al. [11] The most successful material used is currently yttrium-stabilized zirconia (YSZ) [12].

Thus, in the present study, three types of TBCs were deposited on the discs of the intake and exhaust valve are analysed by the atmospheric plasma jet deposition method (APS) and studied in order to observe their properties: the morphology - analysed by two complementary methods: scanning electron microscopy and X-ray diffraction, and the mechanical characteristics - microhardness, modulus of elasticity and adhesion / cohesion of coatings.

Materials and methods

The three types of coating systems with TBC role proposed for study were deposited on the discs of the intake and exhaust valves by the atmospheric plasma jet deposition method (APS) using the following materials (all of them are commercial powders, manufactured by Metco Oerlikon):
- the bonding layer, common for all samples was produced from Al_2O_3-NiAl powder;
- the top coat for sample 1 (S1), was produced from Cr_2C_3 – NiCr powder;
- the top coat for sample 2 (S2), was produced from MgZrO - NiCr powder;
- the top coat for sample 3 (S3), was produced from ZrO - CaO powder.

There were used as substrate discs of intake or exhaust valves, organized as 3 sets of four intake valves and four exhaust valves. The coatings were produced using the facility SPRAYWIZARD 9MCE for atmospheric plasma spraying.

The coating morphology was analysed using two complementary methods: scanning electron microscopy with the Quanta 200 3D microscope (FEI, The Netherlands, 2009) using the Low Field Detector at 1000x/5000x magnification or Z contrast and X-ray diffraction with the XPERT PRO MD (Panalitycal, Netherlands, 2009) diffractometer.

The mechanical characteristics analysed were: microhardness, modulus of elasticity (by indentation) and adhesion / cohesion of coatings using scratch tests, all tests being carried out with the UMTR 2M-CTR Microtribometer, using an indenter with diamond tip Rockwell type, and a force of 20N (for indentation), respectively 10N for scratch (Progressive Load Scratch Test mode).

Results

As mentioned before, the microstructure of the top coat is one of the elements that influence the lifetime and the functionality of the coating system. The secondary electron images of the samples surfaces realised at different magnification are presented in Fig. 1, 2 and 3.

In the case of all three samples, the specific structure of the coatings made by thermal deposition formed by splashes (resulted from the partial or total melting of the powders used), micro-cracks and pores of different sizes is observed.

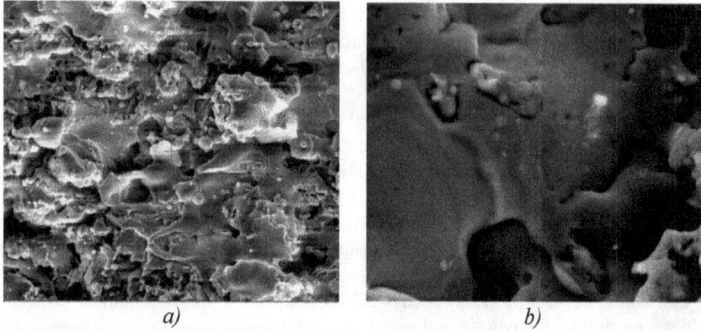

Fig. 1. Typical SEM images of S1 (Cr_2C_3 – NiCr) coatings surface morphology:a)1000x; b)5000x.

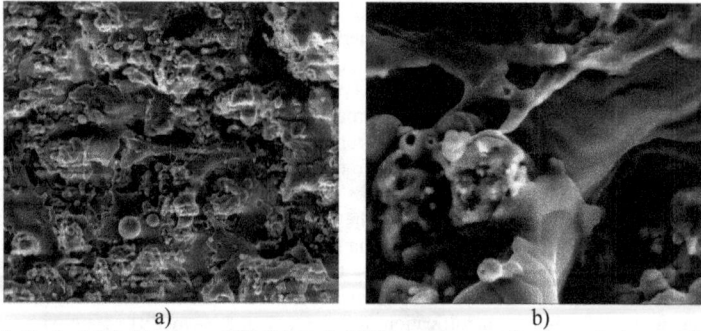

Fig. 2. Typical SEM images of S2 (MgZrO–NiCr) coatings surface morphology: a)1000x; b)5000x.

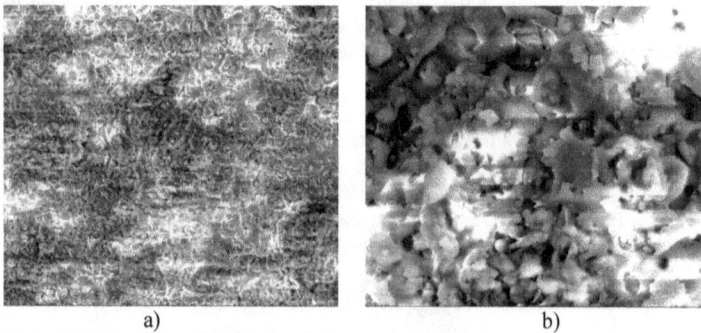

Fig. 3. Typical SEM images of S3 (ZrO - CaO) coatings surface morphology: a)1000x; b)5000x.

Fig. 4 presents the XRD patterns resulted for the three types of thermal barrier coatings studied in the present article.

Powder Metallurgy and Advanced Materials – RoPM&AM 2017   Materials Research Forum LLC
Materials Research Proceedings **8** (2018) 192-199   doi: http://dx.doi.org/10.21741/9781945291999-22

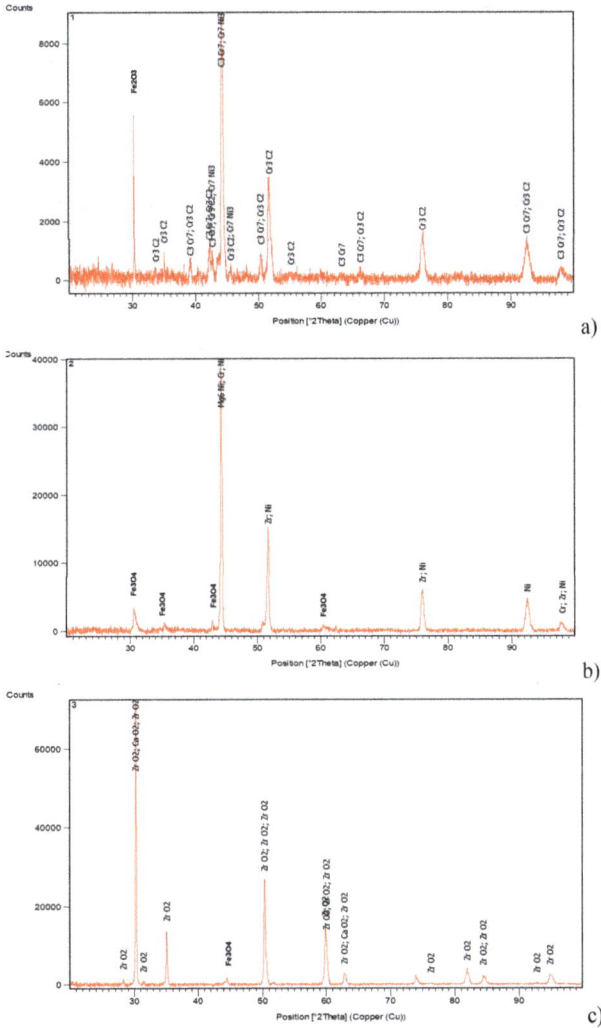

Fig. 4. XRD patterns: a) S1 (Cr_2C_3 – NiCr); b) S2 (MgZrO – NiCr); c) S3 (ZrO – CaO).

The following phases are observed on the XRD patterns:

- for the Cr_2C_3 – NiCr coating: chromium carbides (orthorhombic Cr_3C_2 and Cr_7C_3) and intermetallic compound (tetragonal δ-Cr_7Ni_3). The chromium carbides in Ni matrix are used with success at temperatures higher than 540 °C [13] and are sustained for the thermal barrier role from the intermediate layer based on Al_2O_3;

- for the MgZrO – NiCr coating: metallic and intermetallic compound of cubic structure (Cr, Mg_6Ni, Ni, β-Zr);
- for the ZrO – CaO coating: oxides (tetragonal CaO_2 and ZrO_2 and monoclinic ZrO_2).

There were observed also some iron-oxide contaminants (Fe_2O_3, Fe_3O_4) on the samples, caused by handling and storing the samples after dismantling them from the coating devices.

Table 1. Coatings microhardness and modulus of elasticity

Coating type	E value for 3 points (GPa)			Average E value (GPa)	Microhardness value in 3 points (GPa)			Average microhardness value (GPa)
1	69,06	53,57	67,88	63,50	1,297	0,938	1,306	1,180
2	61,03	53,74	51,13	55,3	0,817	0,711	0,721	0,749
3	72,94	82,29	55,12	70,11	1,369	1,515	1,038	1,307

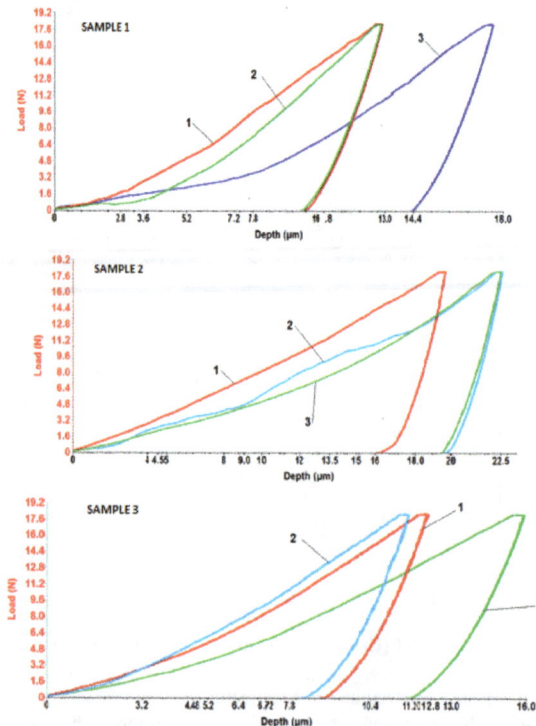

Fig. 5. ”Load – depth” indentation curves for three points (1, 2 and 3): a) S1; b) S2; c) S3.

Powder Metallurgy and Advanced Materials – RoPM&AM 2017   Materials Research Forum LLC
Materials Research Proceedings **8** (2018) 192-199   doi: http://dx.doi.org/10.21741/9781945291999-22

The microindentation results are presented in table 1 and in Fig. 5. In Table 1 are synthesized the measurements of coatings microhardness and modulus of elasticity resulted from the micro-indentation tests. Comparatively, the smallest value of the microhardness is observed in case of the sample 2 – the MgZrO – NiCr coating, which can be explained in terms of the lamellar and porous structure specific to the thermal spray coatings.

It is obvious that the "load – depth" indentation curves are not following a similar path, thus we can conclude that the microstructure is not perfectly homogenous and uniform, presenting surface variations.

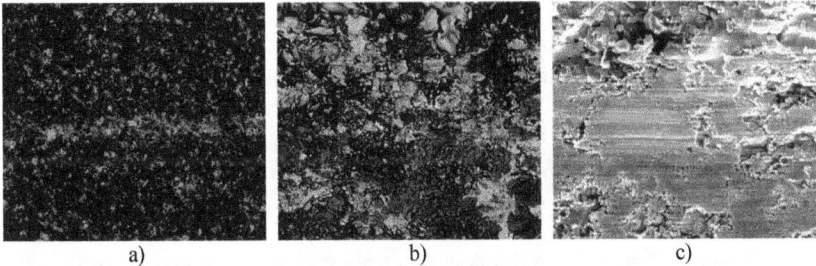

| a) | b) | c) |

Fig. 6. SEM images of the Cr_2C_3 - NiCr coating at the final (a, 100x) and start point (b, c, 1000x)

| a) | b) | c) |

Fig. 7. SEM images of the MgZrO–NiCr coating at the final (a, 100x) and start point (b,c, 1000x)

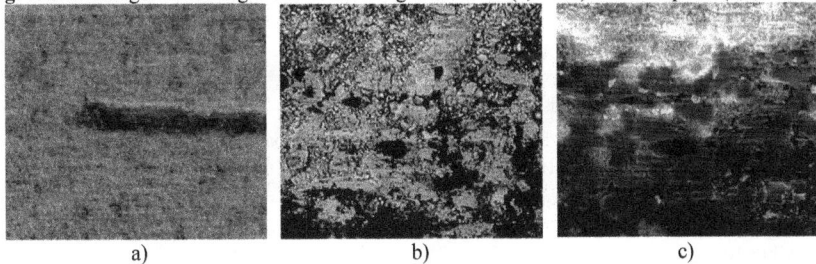

| a) | b) | c) |

Fig. 8. SEM images of the: ZrO – CaO coating at the final (a, 100x) and start point (b, c, 1000x)

Fig. 6, 7, 8 presents the scratch mark aspects resulted after the scratch tests, observed at the beginning and at the end of the scratch marks, for each of the three layers. It is observed that in the

Powder Metallurgy and Advanced Materials – RoPM&AM 2017   Materials Research Forum LLC
Materials Research Proceedings 8 (2018) 192-199   doi: http://dx.doi.org/10.21741/9781945291999-22

start points, where the applied force is maximum, the coatings exhibited some plastic deformations in case of samples 1 and 2 and some delaminations in case of sample 3.

Summary

The structure of these coatings is lamellar, specific to layers made by thermal deposition in the plasma jet, and the component phases in the three cases can provide both the resistance to thermal and mechanical stresses,

On the basis of the SEM images, we can state that in the case of samples 1 and 2 only slight deformation of the layer deposited as thermal barrier occurred; the bonding layer was not exposed. The same cannot be said in case of sample 3, where we observe on both the secondary electron images (a and c) and those made in phase contrast (b and d) that the bonding layer was exposed after the type test scratch. Following those tests it was observed that the coatings are physically suited for further tests as thermal barrier coatings on the valve discs of internal combustion engines.

References

[1] S.M. Meier, D.K. Gupta, The Evolution of Thermal Barrier Coatings in Gas Turbine Engine Applications, J. Eng. Gas Turbine Power Trans. 116 (1994) 250–257. https://doi.org/10.1115/1.2906801

[2] J.R. Nicholls, Advances in coating design for high-performance gas turbines, MRS Bull., 28 (2003) 659–670. https://doi.org/10.1557/mrs2003.194

[3] I. Gurrappa, A. Sambasiva Rao, Thermal barrier coatings for enhanced efficiency of gas turbine engines, Surf. Coat. Technol. 201 (2006) 3016–3029. https://doi.org/10.1016/j.surfcoat.2006.06.026

[4] M.J. Pomeroy, Coatings for gas turbine materials and long term stability issues, Mater. Des. 26 (2005) 223–231. https://doi.org/10.1016/j.matdes.2004.02.005

[5] G.W. Goward, Progress in coatings for gas turbine airfoils, Surf. Coat. Technol. 108–109 (1998) 73–79. https://doi.org/10.1016/S0257-8972(98)00667-7

[6] I. Kvernes, E. Lugscheider, Thick Thermal Barrier Coatings for Diesel Engines, Surf. Eng. 11 (1995) 296–300. https://doi.org/10.1179/sur.1995.11.4.296

[7] D.W. Parker, Mater. Des. 13 (1992) 345–351. https://doi.org/10.1016/0261-3069(92)90005-3

[8] M. Ekström, A. Thibblin, A. Tjernberg, C. Blomqvist, S. Jonsson, Evaluation of internal thermal barrier coatings for exhaust manifolds, Surf. Coat. Technol. 272 (2015) 198–212. https://doi.org/10.1016/j.surfcoat.2015.04.005

[9] R.A. Miller, Oxidation-Based Model for Thermal Barrier Coating Life, J. Am. Ceram. Soc. 67 (1984) 517–521. https://doi.org/10.1111/j.1151-2916.1984.tb19162.x

[10] K.A. Khor, Y.W. Gu, Thermal properties of plasma-sprayed functionally graded thermal barrier coatings, Thin Solid Films 372 (2000) 104–113. https://doi.org/10.1016/S0040-6090(00)01024-5

[11] C. Zhou, N. Wang, Z. Wang, S. Gong, H. Xu, Thermal cycling life and thermal diffusivity of a plasma-sprayed nanostructured thermal barrier coating, Scripta Mater. 51 (2004) 945–948. https://doi.org/10.1016/j.scriptamat.2004.07.024

Powder Metallurgy and Advanced Materials – RoPM&AM 2017   Materials Research Forum LLC
Materials Research Proceedings 8 (2018) 192-199   doi: http://dx.doi.org/10.21741/9781945291999-22

[12] U. Schulz, C. Leyens, K. Fritscher, M. Peters, B. Saruhan-Brings, O. Lavigne, J.-M. Dorvaux, M. Poulain, R. Mévrel, M. Caliez, Some recent trends in research and technology of advanced thermal barrier coatings, Aerosp. Sci. Technol. 7 (2003) 73–80. https://doi.org/10.1016/S1270-9638(02)00003-2

[13] Davis J. R. (ed.) (2004) Handbook of Thermal Spray Technology (pub.) ASM Int. Materials Park OH, USA.

Powder Metallurgy and Advanced Materials – RoPM&AM 2017   Materials Research Forum LLC
Materials Research Proceedings 8 (2018) 200-211   doi: http://dx.doi.org/10.21741/9781945291999-23

Composite material obtained by powder metallurgy with applications in the automotive industry

Cristina Ileana PASCU [1,a]*, Stefan GHEORGHE [1,b], Claudiu NICOLICESCU [1,c] and Daniela TARATA [1,d]

[1]Faculty of Mechanics, University of Craiova, Craiova, Romania

[a]i_pascu@yahoo.com, [b]stgheo@yahoo.com, [c]nicolicescu_claudiu@yahoo.com, [d]danielatarata@yahoo.com.

*i_pascu@yahoo.com

Keywords: Titanium hydride, Powder Metallurgy, Two-Steps Sintering, Multiple-Steps Sintering, Microstructure.

Abstract. Because of their great properties titanium and titanium alloys have been used in automotive industry, biomedical applications, aerospace industry, computer components, emerging applications, architecture of buildings, etc. In the last decade there has been revived interest in the utilization of the Powder Metallurgy (PM) route as a low-cost way for obtaining components from this alloys. This research presents the experimental results concerning the processing of Ti based alloy by Two-Steps Sintering and Multiple-Steps Sintering, techniques belonging to PM technology. The initial powder mixture consists in TiH_2 powder particles that have been combined with some metallic powders (Al, Mn, Sn, Zr) for improving the final mechanic-chemicals and functional properties for using in the automotive industry. As a result it was studied the physical-mechanical properties after sintering, the influence of the sintering temperature and time on the microstructural changes of the composite material based on titanium.

Introduction

Titanium alloys have multiple applications in diverse fields such as industrial and medical fields [1-4]. This is due to their excellent performances such as: low density, good corrosion resistance, non-magnetic properties, high specific strength, high chemical stability, resistant to high temperatures, etc. [5, 6]. The basic advantages of titanium alloys in terms of the automotive industry are the high strength to density, their low density, the outstanding corrosion resistance [7, 8]. In the automotive field, one of the greatest applications of titanium-based materials is for components of the internal combustion engine area that equip the vehicle (pistons, valves, connecting rod, crank caps, bolts, etc.) [9, 10]. Also, for modern jet turbine engines titanium alloys usually represent approximately 30% of the used materials, especially in the forward zone of the engine [11]. However, compared to other traditional materials, the major impediment represents the high cost of titanium [12]. Another disadvantage of titanium for applications in the automotive industry is its low tribological properties because of poor plastic shearing resistance and work hardening ability [13, 14].

Using inexpensive alloying elements (such as Sn, Mn, Fe, Cr, etc.) instead of expensive alloying metals (V, Nb, Mo, Zr, etc.) to improve the strengthen alloys is one of the methods to reduce the cost of manufacturing titanium alloys [15].

Due to the properties they possess, by alloying Ti with small amounts of aluminium percentage a Ti-Al (γ Ti-Al) alloy is obtained with excellent mechanical properties and corrosion resistance at

Powder Metallurgy and Advanced Materials – RoPM&AM 2017   Materials Research Forum LLC
Materials Research Proceedings 8 (2018) 200-211   doi: http://dx.doi.org/10.21741/9781945291999-23

temperatures above 700°C, which allows the replacement for traditional Ni based superalloy for turbine engine components [16].

Due to the very good wear behaviour of tin, by adding small amounts of Sn percentage based Ti-alloys are obtained, such as Ti-5Al-2.5Sn which is a medium-strength all alpha alloy, used for gas turbine engine and other applications that request oxidation resistance, good weld fabric ability and intermediate strength at temperatures up to 4800 °C [17].

Manganese is a metal that has Fe-like properties, being a good substitute for it, and improves wear resistance giving plasticity and good alloy elasticity, so adding small percentages of Mn to a Ti-based alloy gives remarkable properties [18, 19]. According to Wang [20], a novel near-α titanium alloy Ti-6.0Al-4.5Cr-1.5Mn was designed and prepared by the water-cooled copper crucible and present better mechanical strength compared other Ti-alloys which has been presented, but worse elongation because of the presence of Cr2Ti phase which may cause alloy's poor plasticity.

Carbon has Van der Waals bonds and has good wear behaviour, so, according to [21] the incorporation of a small amount of Al and C into the TiAlVSiCN will improve the hardness, wear and erosion resistance, and, also, lower production cost. Also, due to similar properties of Zr with those of Ti, in the last years, Ti-Zr alloys with zirconium contents ranging from 10 to 40 wt% have been investigated by melting process and were reported characteristics as excellent corrosion resistance with obvious applications in the automotive industry [22].

Another way of reducing the cost price in making alloys based on Ti is the use of Powder Metallurgy (PM) techniques [23, 24]. Accordingly to [25] a Ti-6Al-2Sn-4Zr-2Mo alloy produced by arc melting and powder metallurgy processes presents excellent mechanical properties and high resistance, in sections exposed to high temperatures. Toyota uses sintered titanium alloys Ti-6Al-4V/TiB and Ti-Al-Zr-Sn-Nb-Mo-Si/TiB in the intake and exhaust engine valves, respectively, are used from Toyota for own cars [26].

Among the PM method that appears to offer the greatest opportunity for real cost reductions it is considered that Two-Steps (TSS) and Multiple-steps (MSS) Sintering Techniques allow to obtain different compositions and microstructures of Ti products [27] for automotive components.

Conventional methods used for obtaining titanium alloys require special conditions of controlled atmosphere which result in a high cost of obtaining these alloys [28]. For this reason, in the last decade, many studies have focused for the producing of Ti-alloys by Powder Metallurgy using TiH_2 powder [29-34]. According to [35] a Ti-6Al-2Sn-4Zr-2Mo alloy was obtained by Powder Metallurgy using titanium hydride powders. This alloy based on TiH_2 powders proved an improvement of the mechanical properties and high corrosion resistance on high temperatures.

For this study, was proposed to obtain a titanium hydride alloy by adding small percentages of Al, Sn, Mn, Zr and C in order to improve the mechanical and tribological properties of this alloy, and further processing to be achieved by compacting and sintering, using the TSS and MSS techniques.

Experimental Procedures
Materials
For this research, titanium hydride (TiH_2) produced by Chemetall GmbH, has been used as initial material. The following chemical elements are present in TiH_2 composition: Titanium (min 95%), Hydrogen (min. 3.8%), Al (max.0.15%), Si (max.015%), Nitrogen (max.0.3%), Fe, Ni, Cl, Mg, C, Ni (all the last component are under 0.1%). In order to improve the mechanical and tribological characteristics in the initial powder were introduced by mixing for 60 min the following

Powder Metallurgy and Advanced Materials – RoPM&AM 2017   Materials Research Forum LLC
Materials Research Proceedings **8** (2018) 200-211   doi: http://dx.doi.org/10.21741/9781945291999-23

components: Mn powder, Sn powder, Alumix 321, 8% powder, Zr powder, and graphite powder. The mixing and homogenization of the mixture was carried out in a ball mill Planetary Mono Mill PULVERISETTE 6 classic line from Fritsch with ball: powders ratio 1:1, 250 ml grinding bowl, argon atmosphere, balls material stainless steel – 1.3541 ISO/EN/DIN codeX47Cr14, B50, rotational speed of main disk of 200 rpm. The weight ratio between the components is: 80%wt. TiH_2, 8%wt., Mn, 3%wt. Sn, 2%wt. Zr, 6%wt. Alumix321, and 1%wt. C. Particle size distributions, Fig. 1, was studied by dynamic laser scattering (DLS) using a 90Plus particle size analyser, Brookhaven Instruments Corporation, USA, equipped with 35 mW solid state laser, having 660 nm wavelength. The temperature was 25°C and the scattering angle was 90°. The dilution of the powders was made in water and the solution was subjected to ultrasonic treatment for 5 minutes to avoid flocculation of the particles.

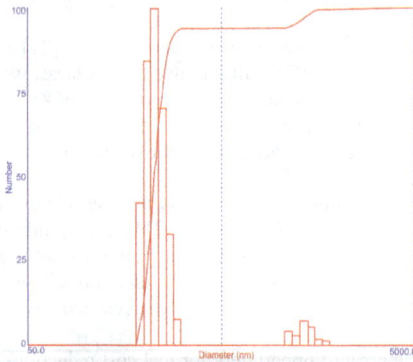

Fig. 1. Particle size distribution of the mixture.

Fig. 2. SEM image of the mixture

In Fig. 2 a SEM image of the mixture is shown. The mixture presents a bimodal particle size distribution, having particles with dimensions between (188-294 nm) respectively (1110-1730 nm). The highest number of particles (28% from the total number) have the dimension equal to 225 nm. The percent of the particles with higher size is lower than 1.5%. The mean hydrodynamic diameter is 291 nm. In Table 1 sub-micron dimensions and percentages of the powder particles accordingly with the grain-size distribution presented in Fig. 1 are shown.

Methods and Techniques
The homogenized mixture has been pressed by unilateral cold compaction at 600MPa compaction pressure, as cylindrical specimens with 12,05 mm diameter. An A009 electromechanical-computerized testing machine 100 kN equipped with TCSoft2004Plus software was used for unilateral cold compaction. The green density of each part has been determined as has been presented in a previous study [27]. The next step was the sintering of the green compacts. They have used four different sintering regimes using Powder Metallurgy techniques as following:
- V1 regime, using classical sintering at 1050 °C for 90 min dwell time;
- Two-steps sintering (TSS) regime, V2, at 1050 °C with dwell time of 15 min and, then, at 1000 °C for 75 min;
- TSS regime, V3, where the sintering was done at 1050 °C for 15 min and at 950 °C, for 75 min, Fig. 3;

- Multiple-steps sintering (MSS), V4, in three steps at 1050 °C for 15 min dwell time, 1000 °C for 20 min dwell time and 900 °C for 55 min dwell time, Fig. 4.

Table 1. Grain-size distribution and percentages of the powder particles

Dimension [nm]	Particle size distribution [%]
189	11.8%
206	23.6%
225	28%
246	19.6%
269	9.2%
294	1.9%
1110	1.1%
1213	0.8%
1326	2%
1448	1.4%
1583	0.3%
1729	0.3%

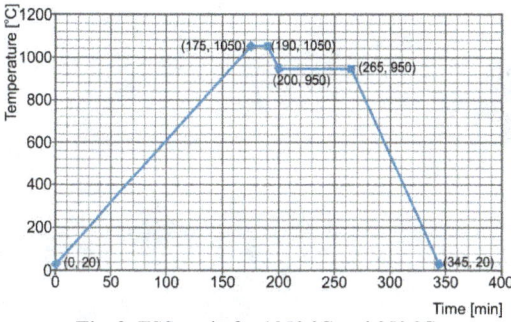

Fig. 3. TSS cycle for 1050 °C and 950 °C.

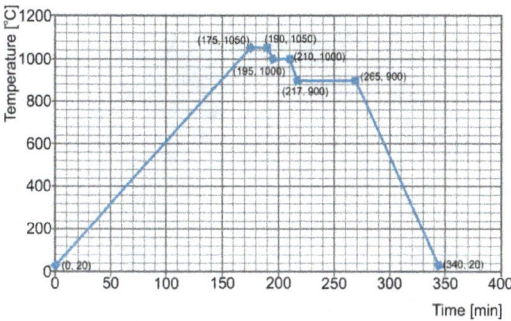

Fig. 4. MSS cycle at 1050 /1000/ 900 °C.

Powder Metallurgy and Advanced Materials – RoPM&AM 2017   Materials Research Forum LLC
Materials Research Proceedings **8** (2018) 200-211   doi: http://dx.doi.org/10.21741/9781945291999-23

The sintering treatments was done in Ar 99,9% sintering atmosphere. The samples were heated into a Nabertherm L3/11/C6 furnace with 6 °C/min heating rate for all four sintering cycles. For all 4 sintering cycles, the maximum temperature that was achieved was 1050 °C and the total cumulative dwell time was the same for all the specimens, 90 minutes, but, with different steps and sintering dwell times [36].

Results and discussion

The cylindrical sintered specimens were analysed by scanning electron microscopy (SEM) using a 5th generation Phenom desktop SEM platform. Superficial surface analysis has highlighted the specific porosity of particle parts and the roughness has values between 3.77-5.24 μm for Ra roughness parameter, as shown in Fig. 5.

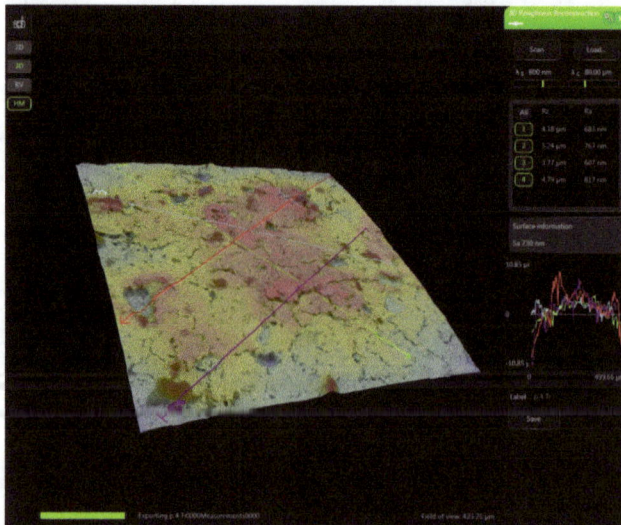

Fig. 5. 3D Surface roughness of sample for version V1 (sintering temperature 1050°C/90min).

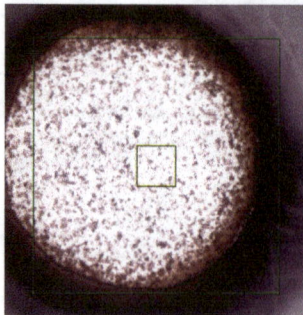

Fig. 6. Type of sintering sample for all the studied cases.

Powder Metallurgy and Advanced Materials – RoPM&AM 2017   Materials Research Forum LLC
Materials Research Proceedings **8** (2018) 200-211   doi: http://dx.doi.org/10.21741/9781945291999-23

The sample structures was physically examined with an integrated spectrometer for EDS analysis. This device has an optical magnification from 20x to 135x, SEM magnification from 80 up to 150.000x, resolution: < 10 nm and digital zoom: maximum 12x. Fig. 6 presents the type of sample which was analysed by SEM in this study.

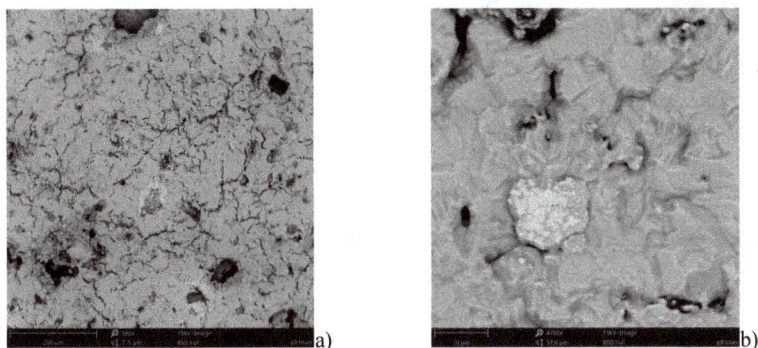

Fig. 7. Scanning electron microscopy (SEM) of the sample obtained after version V1 (sintering temperature 1050 °C/90 min): a – x200 μm; b – x10 μm.

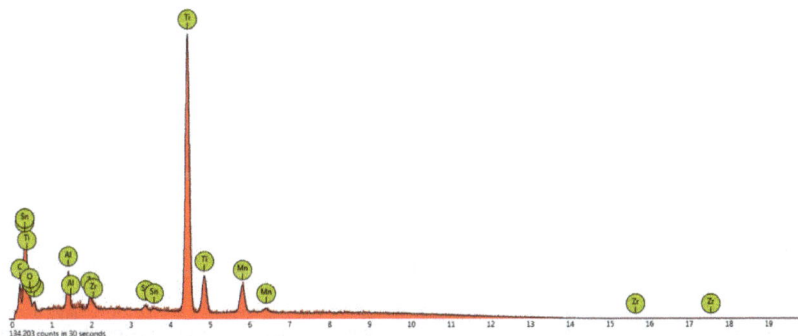

Fig. 8. Spectrometry elemental analysis of the sample obtained after version V1 (sintering temperature 1050 °C/90 min).

Fig. 7, 9, 11 and 13 present SEM images for all four studied cases, with two orders of magnitude for each sample. Also, the spectrometry elemental analysis for all versions are presented in Figures 8, 10, 12, 14. The shown structures are in nonchemical etching metallographic state. At the sintering temperature used titanium forms solid solutions type β-Ti with all the alloying elements used, as shown in the equilibrium diagrams. The finishing structure is influenced by de sub-micron dimensions of the particles, like is presented in Fig. 1. The

Powder Metallurgy and Advanced Materials – RoPM&AM 2017   Materials Research Forum LLC
Materials Research Proceedings **8** (2018) 200-211   doi: http://dx.doi.org/10.21741/9781945291999-23

resulting structures are due of this phenomenon and highlight compact structures with some mass agglomerations of carbon, Fig. 15 spot 3 and 4.

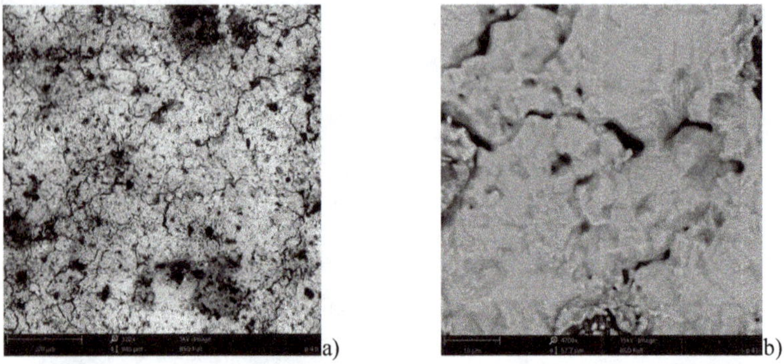

Fig. 9. Scanning electron microscopy (SEM) of the sample obtained after version V2 (sintering temperature 1050 °C/15 min – 1000 °C/75 min): a – x200 μm; b – x10 μm.

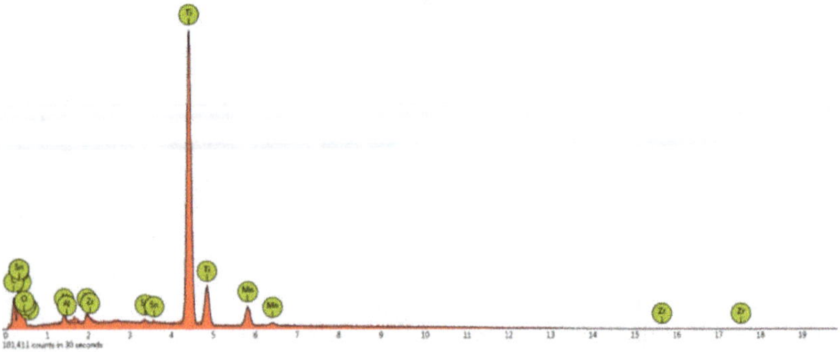

Fig. 10. Spectrometry elemental analysis of the sample obtained after version V2 (sintering temperature 1050 °C/15 min – 1000 °C/75 min).

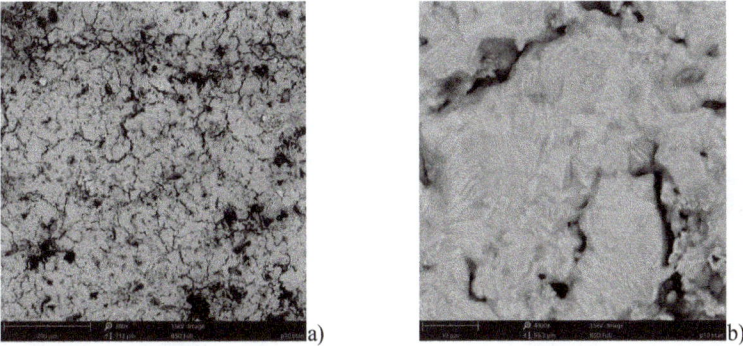

Fig. 11. Scanning electron microscopy (SEM) of the sample obtained after version V3 (sintering temperature 1050 °C/15-950 °C/75 min): a – x200 μm; b – x10 μm.

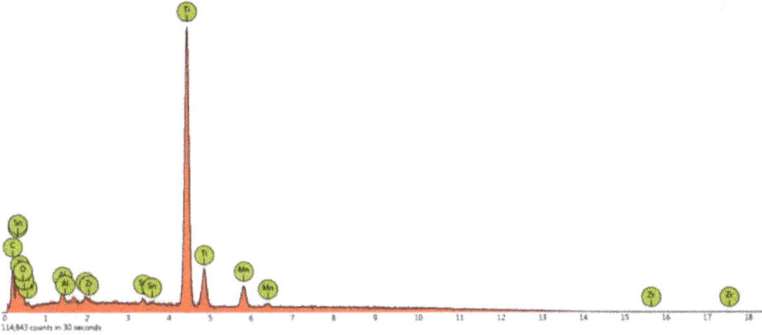

Fig. 12. Spectrometry elemental analysis of the sample obtained after V3 (Sintering temperature 1050 °C/15-950 °C/75 min).

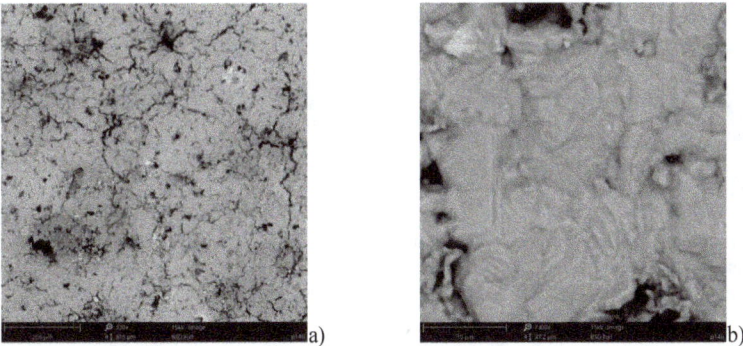

Fig. 13. Scanning electron microscopy (SEM) of the sample obtained after version V4 (sintering temperature 1050 °C/15-1000 °C/20 min-900 °C/55 min): a – x200 μm; b – x10 μm.

Powder Metallurgy and Advanced Materials – RoPM&AM 2017   Materials Research Forum LLC
Materials Research Proceedings **8** (2018) 200-211   doi: http://dx.doi.org/10.21741/9781945291999-23

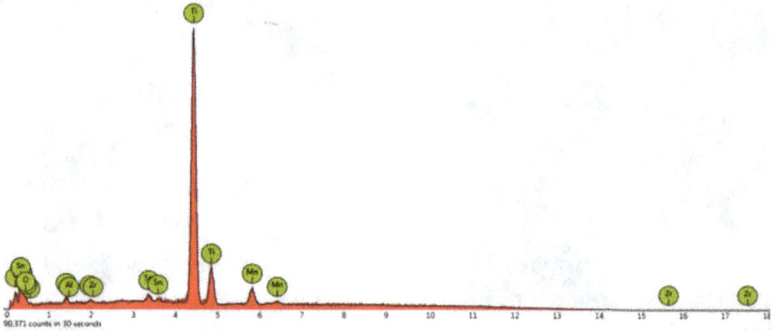

Fig. 14. Spectrometry elemental analysis for version V4 (sintering temperature 1050 °C/15-1000 °C/20 min-900 °C/55 min): a – x200 μm; b – x10 μm.

SEM analysis reveal that the volume of surface pores is reduced during the sintering process and the space between the particles shrinks during the sintering process. Due to this phenomena, the contacted particle surfaces fuse together during sintering. Also, the spectral analysis used has identified the presence of all alloying elements, as well some traces of oxygen are visible. A further reduction process probably is needed [38].

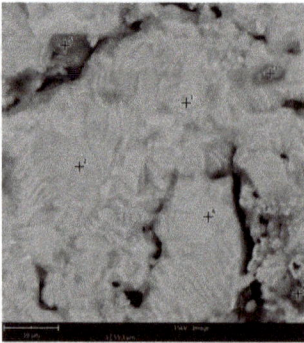

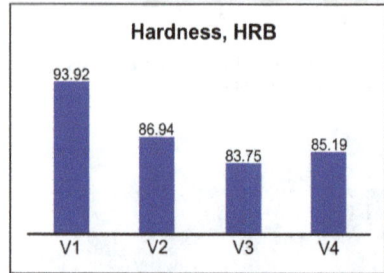

Fig. 15. SEM analysis with indication of spots investigated by spectral analysis.

Fig. 16. HRB Hardness of the sintered samples.

The density and HV micro-hardness were also determined. It has been observed that the density of the sintered parts ranged between 3.59-3.65 g/cm^3 [36], reached over 90% of the theoretical density of the titanium hydride-based alloy, which means that it has been produced the densification process. This fact, according to [24, 39] allows the use of Ti-based alloy in the manufacture of components in the automotive field. The hardness of the specimens was measured after sintering by the Vickers method and the results were equated in HRB units. The samples were metallographic prepared and were analysed with a Nikon Eclipse MA 100 microscope which is equipped with NIS ELEMENTS acquisition data software. The Vickers microhardness HV1 was determined using an NAMICON 400-DTS tester. The dwell time was 15 seconds. As shown in Fig. 16, the average values of microhardness exceed 80 HRB, which is comparable to the known hardness values of pure titanium and 6Al-4V titan alloy very used in automotive and aerospace applications [37]. The maximum value is 93.92 HRB in regime 1 (sintering 1050 °C with 90-minute hold), followed by 86.94, 85.19 and 83.75HRB for regime 2 (1050°C/15 min-1000°C/75 min), V4 (1050°C/15 min-1000°C/20 min-900°C/55 min), V3 (1050°C/15 min-950°C/75 min). As with the behaviour of this sintering alloy, where the density increases were higher for the regime 4 than for the regime 3, the same type of behaviour is found in terms of material hardness. The explanation lies in the fact that the intermediate dwell at 1000 °C in the regime 4 helps to increase the hardness even if the final 900 °C dwell is lower, in this case the 950 °C range used in the regime 3.

Summary
The time for mixing powder in mechanical mill (1 hour) allows for the mixture to have sub-micron mean hydrodynamic diameter (291 nm). The 600MPa mechanical pressure is good enough to have a compact samples with a good green density. Sintering temperature of 1050 °C combined with different smaller dwells allows the samples to improve their density. The structures of sintered titanium alloy are influenced by the sub-micron dimensions of powder and the time of mechanical alloying time too. The HRB hardness has a maximum value equal with 94 HRB then the sintering temperature is 1050 °C with 90 minutes hold. Alloying elements such as Sn and graphite improve the wear resistance of titanium based alloys that are destined for the automotive industry [26]. Also, by using TiH_2 as initial material the densification process seems to be achieved at 1050 °C, the density and hardness of the sintered alloy are comparable to those presented in [37, 39] for materials used in the field of automotive components.

References
[1]   M. Qian, Cold compaction and sintering of titanium and its alloys for near-net shape or preform fabrication, Int. J. Powder Metall. 46 (2010) 29-44.

[2]   M.J. Donachie, Titanium: a technical guide. 2nd ed. Materials Park (OH): ASM International; 2000.

[3]   G. Lütjering, J.C. Williams. Titanium. Heidelberg, Germany: Springer; 2007.

[4]   M.N. Gungor, M.N. Imam M.A. Froes F.H, Innovations in Titanium Technology, Warendale: Wiley's Publishing; 2007.

[5]   R. Boyer, G. Welsch, E.W. Collings, Materials Properties Handbook: Titanium Alloys, ASM International, 1994.

[6]   I.M. Robertson, G.B. Schaffer, Review of densification of titanium based powder systems in press and sinter processing, Powder Metall., 53 (2010) 146-162. https://doi.org/10.1179/174329009X434293

[7]   H. Fujii, K. Takahashi, Y. Yamashita, Application of Titanium and Its Alloys for Automobile Parts, Nippon Steel Tech. Report., 38 (2003) 70-75.

[8]   A. Mayyas, A. Qattawi, M. Omar, D. Shan, Design for sustainability in automotive industry: A comprehensive review, Renew. Sust. Energ. Rev., 16 (2012) 1845-1862. https://doi.org/10.1016/j.rser.2012.01.012

[9]   O. Schauerte, Titanium in Automotive Production, Adv. Eng. Mat., 5 (2003) 411-418. https://doi.org/10.1002/adem.200310094

Powder Metallurgy and Advanced Materials – RoPM&AM 2017   Materials Research Forum LLC
Materials Research Proceedings 8 (2018) 200-211   doi: http://dx.doi.org/10.21741/9781945291999-23

[10]   F.H. Froes, H. Friedrich, J. Kiese, D. Bergoint, Titanium in the Family Automobile: the Cost Challenge, Trans Tech Publications: Zurich; Switzerland; 2010.

[11]   G.R. Watkins, Development of a High Temperature Titanium Alloy for Gas Turbine Applications, PhD. Thesis, The University of Sheffield, March 2015.

[12]   R.B. Boyer, Attributes, Characteristics, and Applications of Titanium and Its Alloys, JOM, 62 (2010) 35-43. https://doi.org/10.1007/s11837-010-0071-1

[13]   A. Ohidul, M. Haseeb, ASMA. Response of Ti–6Al–4V and Ti–24Al–11Nb alloys to dry sliding wear against hardened steel, Tribol. Int. 35 (2002) 357–362. https://doi.org/10.1016/S0301-679X(02)00015-4

[14]   K.M. Chen, Y. Zhou, X. X. Li, Q. Y. Zhang, L. Wang, S.Q. Wang, Investigation on wear characteristics of a titanium alloy/steel tribo-pair, Adv Mater Res-Switz. 65 (2015) 65-73. https://doi.org/10.1016/j.matdes.2014.09.016

[15]   Z. Kailiang, N. Gui, T. Jiang, M. Zhu, X. Lu, J. Zhang, C. Li, The Development of the Low-Cost Titanium Alloy Containing Cr and Mn Alloying Elements, Metall. Mater. Trans. A 45A (2014) 1761-1766.

[16]   V. Henriques, J. de Oliveira, E. Diniz, A. Dutra. et al., Development of Techniques for Gamma Ti-Al Production, 22nd SAE Brasil International Congress and Display, 2013, https://doi.org/10.4271/2013-36-0392.

[17]   R. Boyer, G. Welsch, E.W. Collings, Materials Properties Handbook: Titanium Alloys, ASM International; USA; 2007.

[18]   D. Lehmhus, M. Busse, A. Herrman, K. Kayvantash, Structural Materials and Processes in Transportation, Wiley-VCh Verlag GmbH & Co, Weinheim, Germany, 2013.

[19]   M.J. Gázquez, J.P. Bolívar, R. Garcia-Tenorio, F. Vaca, A Review of the Production Cycle of Titanium Dioxide Pigment, Mater. Sci. Appl., 5 (2014) 441-458. https://doi.org/10.4236/msa.2014.57048

[20]   H. Wang, S. Wang, P. Gao, C.H. Li, Microstructure and mechanical properties of a novel near-α titanium alloy Ti6.0Al4.5Cr1.5Mn, Adv Mater Res-Switz, 67 (2016) 170-174. https://doi.org/10.1016/j.msea.2016.06.083

[21]   J. Lin, R. Wei, F. Ge, M. Lei, TiSiCN and TiAlVSiCN nanocomposite coatings deposited from Ti and Ti-6Al-4V targets, Surf Coat Techn. 2017.

[22]   T. Luz, V. Henriques, J. de Oliveira, E. Diniz, Production of Ti-Zr Alloy by Powder Metallurgy, SAE Technical Paper 2013.

[23]   F.H.S. Froes, Powder metallurgy of titanium alloys. In: Chang I, Zhao Y, editors. Advances in powder metallurgy. Cambridge: Woodhead Publishing; 2013. p. 202–240. https://doi.org/10.1533/9780857098900.2.202

[24]   C.A. Lavender, V.S. Moxson, V.A. Duz, Cost-Effective Production of Powder Metallurgy Titanium Auto Components for High-Volume Commercial Applications, 2010, http://www.pnl.gov/main/publications/external/technical_reports/PNNL-19932.pdf.

[25]   R. Pereira, V. Henriques, J. de Oliveira, E. Diniz, Development of Production Techniques for Aerospace Titanium Alloys, SAE Technical Paper 2013, https://doi.org/10.4271/2013-36-0370.

[26]   C. Veiga, J.P. Davim, A.J.R. Loureiro, Properties and applications of titanium alloys: A brief review, Rev.Adv.Mater.Sci. 32 (2012) 14-34.

[27]   C.I. Pascu, S. Gheorghe, C. Nicolicescu, Aspects about Sintering Behaviour of a Titanium Hydride Powder based Alloy used for Automotive Components, Applied Mech. Mater. 823 (2016) 467-472.

[28]   E. do Nascimento Filho, V. Henriques, J. de Oliveira, E. Diniz, Microstructural Study of Ti-6Al-4V Produced with TiH_2 Powder, SAE Technical Paper 2012, https://doi.org/10.4271/2012-36-0197.

[29]   O. Ivasishin, V. Moxson, Low-cost titanium hydride powder metallurgy, in: Ma Qian and F. H. Froes (Est), Titanium hydride powder metallurgy, Science, Technology and Applications, Elsevier Inc., Library of the Congress, New York, 2015; 117–148. https://doi.org/10.1016/B978-0-12-800054-0.00008-3

[30]   I.M. Robertson, G.B. Schaffer, Comparison of sintering of titanium and titanium hydride powders, Powder Metall. 53 (2010) 12-19. https://doi.org/10.1179/003258909X12450768327063

[31]   V. Henriques, J. de Oliveira, E. Diniz, T. Lemos, Densification of titanium alloys obtained by powder metallurgy, SAE Technical Paper 2010, https://doi.org/10.4271/2010-36-0235.

[32]   H.T. Wang, M. Lefler, Z.Z. Fang, T. Lei, S. M. Fang, J.M. Zhang, Q. Zhao, Titanium and Titanium Alloy via Sintering of TiH_2, Key Eng. Mat. 436 (2010) 157-163. https://doi.org/10.4028/www.scientific.net/KEM.436.157

[33]   Z.Z. Fang, P. Sun, H. Wang, Hydrogen Sintering of Titanium to Produce High Density Fine Grain Titanium Alloys, Adv. Eng. Mat. 14 (2012) 383-387. https://doi.org/10.1002/adem.201100269

[34]   D.W. Lee, H.S. Lee, J.H. Park, S.M. Shin, J.P. Wang, Sintering of Titanium Hydride Powder Compaction, 2nd International Materials, Industrial, and Manufacturing Engineering Conference, MIMEC2015, Procedia Manufacturing 2 (2015) 550 – 557.

[35]   H. Sargentini, T. Lemos, A. Henriques, J. Faria, Development of titanium alloys production for high temperatures applications, SAE Technical Paper 2010, https://doi.org/10.4271/2010-36-0170.

[36]   C.I. Pascu, S. Gheorghe, D. Tarata, C. Nicolicescu, C. Miritoiu, Study about the Influence of Two-Steps Sintering (TTS) Route for an Alloy based on Titanium, Applied Mech. Mater. 880 (2018) 256 - 261.

[37]   G. Ozkan, V. Gülsoy, V. Günay, T. Baykara, R.M. German, Injection Molding of Mechanical Alloyed Ti¬Fe¬Zr Powder, Mater. Transactions, 53 (2012) 1100-1105. https://doi.org/10.2320/matertrans.M2012031

[38] Y. Xia, Z.Z. Fang, T. Zhang, Y. Zhang, P. Sun, Z. Huang, Deoxigenation of titanium hydride with calcium hydride, Proceedings of the 13th World Conference on Titanium, San Diego, 2015; 867-875.

[39] S.J. Park, A. Arockiasamy, H. El Kadiri, W. Joost, Production of Heavy Vehicle Components from Low-Cost Titanium Powder, Contract No.: DE-FC-26-06NT42755, Mississippi State University, 2014, http://energy.gov/sites/prod/files/2014/04/f14/4_automotive_metals-titanium.pdf.

Powder Metallurgy and Advanced Materials – RoPM&AM 2017   Materials Research Forum LLC
Materials Research Proceedings 8 (2018) 212-218   doi: http://dx.doi.org/10.21741/9781945291999-24

Adhesion analysis for niobium nitride thin films deposited by reactive magnetron sputtering

Florina Maria ȘERDEAN [1,a] *, Violeta Valentina MERIE [2,b], Gavril NEGREA [2,c], Horea George CRIȘAN [1,d]

[1] Technical University of Cluj-Napoca, Department of Mechanical Systems Engineering, 103-105 Muncii Avenue, 400641, Cluj-Napoca, Romania

[2] Technical University of Cluj-Napoca, Department of Materials Science and Engineering, 103-105 Muncii Avenue, 400641, Cluj-Napoca, Romania

[a]Florina.Rusu@omt.utcluj.ro, [b]Violeta.Merie@stm.utcluj.ro, [c]Gavril.Negrea@ispm.utcluj.ro, [d]Horea.Crisan@auto.utcluj.ro

* corresponding author

Keywords: Niobium nitride; Thin films; Atomic force microscope; Surface energy; Adhesion.

Abstract. This paper focuses on determining the adhesion force for four samples of niobium nitride thin films using the atomic force microscope, samples which were deposited by reactive magnetron sputtering on silicon substrates. The main objective is to investigate the influence of the deposition parameters on the adhesion force and other important parameters. Hence, the nitrogen flow rate was varied in order to identify its influence for this type of thin films. Several tests were performed in multiple points of each sample using the spectroscopy in point mode of the atomic force microscope. Using the obtained experimental values and two existing mathematical models for the adhesion force, the surface energy of the niobium nitride thin films was estimated.

Introduction

Adhesion forces at micro- and nano- scale have been the focus of many researchers' work due to their influence on failure mechanisms in Micro-Electro-Mechanical Systems (MEMS). Due to the intensive growth of MEMS, issues such as reliability and long-term durability have gained importance. At micro-scale such aspects are influenced by the fact that the surface forces become more significant than the body forces. Therefore, it is vital to study characteristics such as roughness or adhesion force, that lead to permanent surface attachment, also known as stiction. Along the years several thin films have been investigated in order to determine the adhesion force and other properties (mechanical, tribological, etc.) [1-2]. Due to their applicability in MEMS industry, thin films still present interest for researchers [3-4] and several methods have been used to deposit niobium nitride (NbN) thin films [5-7].

One of the high-resolution methods that allows to measure the adhesion force is the atomic force microscopy (AFM). The procedure consists in moving the tip of the AFM cantilever close to the sample surface. Due to the existing attraction forces they come in contact (the snap-in phenomenon occurs). The loading phase continues and the cantilever is deflected. During the unloading when the cantilever goes back to its initial position the contact between the tip and the sample is broken (the pull-off phenomenon occurs). The AFM software provides a curve that allows to determine the adhesion force (a representative example is given in Fig. 1).

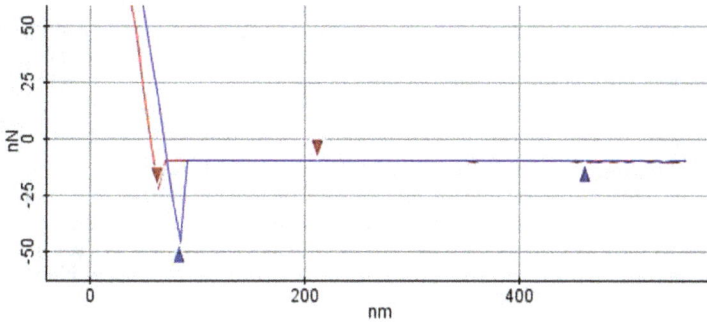

Fig. 1. An example of the force vs. z scan curve for the contact between the AFM tip and the sample surface.

In the open literature there are investigations regarding the influence of fabrication parameters on crystallization, microstructure, and surface composition of NbN thin films [8] and methods such as scratch tests have been used to determine the adhesion [9]. In contrast with the existing research, the present paper encompasses the research regarding the influence of the nitrogen flow rate on the adhesion and other characteristics of NbN thin films. The experimental investigations were conducted using the spectroscopy in point mode of the AFM.

Materials and experimental procedure
Materials
This paper is focused on studying four samples consisting in solid thin films of niobium nitride. They were obtained by depositing one layer of niobium nitride on a silicon Si (100) substrate by reactive magnetron sputtering using a Nb target with a purity of 99.95%.

The deposition was conducted in a high vacuum chamber containing a mixture of argon and nitrogen. Using a constant discharge current I_d, a constant argon flow rate Q_{Ar} and a constant pressure P, all samples were deposited for the same period of time at ambient temperature. The obtained thickness of the deposited thin films was around 0.32 μm and it was determined using a JEOL 5600LV electron microscope. As shown in Table 1, the nitrogen flow rate was varied between 0.25 cm^3/min and 1.75 cm^3/min.

Table 1. Deposition conditions for investigated samples.

Thin film	T (°C)	Time (min)	P (mtorr)	I_d (mA)	Q_{Ar} (cm^3/min)	Q_{N2} (cm^3/min)
Sample 1	23	20	2.2	300	40	0.25
Sample 2	23	20	2.2	300	40	0.75
Sample 3	23	20	2.2	300	40	1.25
Sample 4	23	20	2.2	300	40	1.75

Powder Metallurgy and Advanced Materials – RoPM&AM 2017   Materials Research Forum LLC
Materials Research Proceedings 8 (2018) 212-218   doi: http://dx.doi.org/10.21741/9781945291999-24

Motivated by the applicability of niobium nitride thin films experimental tests were conducted in order to determine the adhesion force. The software of the AFM used to find the adhesion forces provides several important characteristics that present interest such as roughness together with its characteristic statistics (see for example Fig. 2), the snap-in force or the adhesion energy (see for example Fig. 3 where the results are presented for the third conducted measurement).

Experimental procedure

The experimental procedure regarding the four thin films at nano-scale was conducted in the Micro & Nano Systems Laboratory from the Technical University of Cluj-Napoca using a XE 70 AFM. The investigations were performed at a temperature of 23 °C, a relative humidity of 35 % and a scanning frequency of 0.75 Hz. The cantilever used for the tests is a NSC35C cantilever which, according to the specifications provided by the manufacturer has a width of 35 μm, a length of 130 μm, a thickness of 2 μm, a force constant of 5.4 N/m and a resonance frequency of 150 kHz. The tip radius of the cantilever is smaller than 20 nm. The XEI Image Processing Tool for SPM (Scanning Probe Microscopy) Data was used for data interpretation.

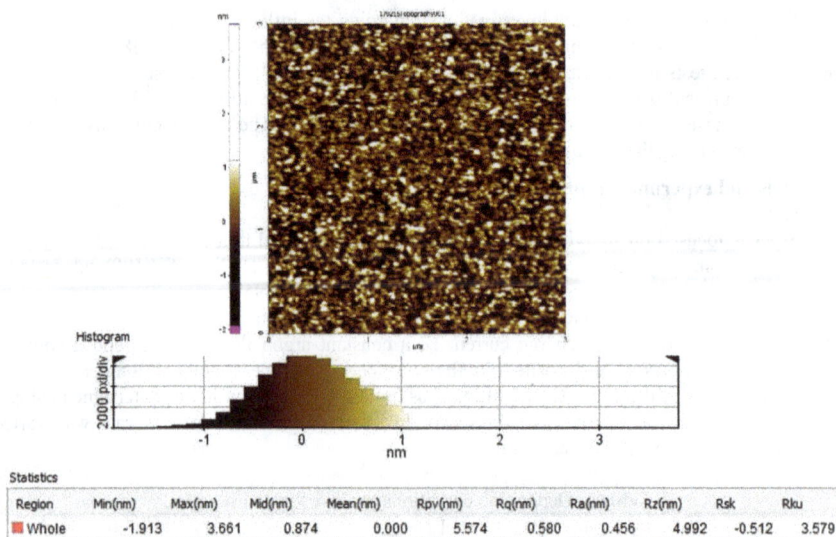

Region	Min(nm)	Max(nm)	Mid(nm)	Mean(nm)	Rpv(nm)	Rq(nm)	Ra(nm)	Rz(nm)	Rsk	Rku
Whole	-1.913	3.661	0.874	0.000	5.574	0.580	0.456	4.992	-0.512	3.579

Fig. 2. The surface of Sample 3 and the statistical parameters provided for its roughness by the AFM software.

The technique used to determine the adhesion force (the pull-off force) between the AFM tip and the NbN thin films, as well as other parameters of interest such as the snap-in force and adhesion energy is the spectroscopy in point mode of the AFM. This technique returns as output together with values for different parameters, some AFM experimental curves similar to the one presented in Fig. 3 which present both the loading and the unloading phases of the procedure. The values of the adhesion forces between the AFM tip and each of the four NbN thin films were

Powder Metallurgy and Advanced Materials – RoPM&AM 2017   Materials Research Forum LLC
Materials Research Proceedings **8** (2018) 212-218   doi: http://dx.doi.org/10.21741/9781945291999-24

obtained based on these curves and determined automatically by the software (see in Fig. 3 the value for the pull-off force). The measurements of the adhesion force were conducted in 5 random points for each sample and the average value was computed. The points were chosen by studying the 3D image of each sample (see for example Fig. 4) so that any defects are avoided. This procedure was repeated for roughness, snap-in force and adhesion energy.

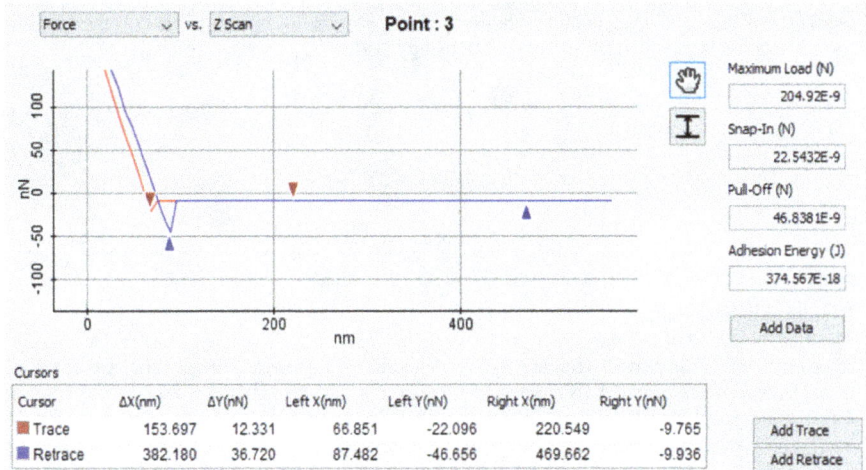

| Force | vs. | Z Scan | | **Point : 3** |

	Maximum Load (N)
	204.92E-9
	Snap-In (N)
	22.5432E-9
	Pull-Off (N)
	46.8381E-9
	Adhesion Energy (J)
	374.567E-18
	Add Data

Cursors

Cursor	ΔX(nm)	ΔY(nN)	Left X(nm)	Left Y(nN)	Right X(nm)	Right Y(nN)	
Trace	153.697	12.331	66.851	-22.096	220.549	-9.765	Add Trace
Retrace	382.180	36.720	87.482	-46.656	469.662	-9.936	Add Retrace

Fig. 3. The force vs. z scan curve together with the charactristics provided by the AFM software for an experimental test conducted on Sample 3.

Fig. 4. 3D image of the third sample.

Powder Metallurgy and Advanced Materials – RoPM&AM 2017   Materials Research Forum LLC
Materials Research Proceedings 8 (2018) 212-218   doi: http://dx.doi.org/10.21741/9781945291999-24

Theoretical models
One of the main factors that influence the adhesion is the type of contact between the surfaces. This contact has been modelled in different ways along the years of research in order to accurately estimate the adhesion force. Two of the most known and used models are the JKR (Johnson-Kendall-Roberts) model and the DMT (Derjaguin-Müller-Toporov) model. These models were established due to the need to remove the differences between the values when applying experimental tests and when applying the Hertz theory for low loads. In both models the adhesion force is expressed as a function of the surface energy [10]:

$$F_{ad}^{JKR} = \frac{3}{2}\pi\gamma R \qquad (1)$$

$$F_{ad}^{DMT} = 2\pi\gamma R \qquad (2)$$

where γ is the surface energy and R is the radius of the AFM tip. These models are used for the contact between the AFM tip and a surface with low roughness.

Results and discussions
The main objective of this paper is to study the influence of the nitrogen flow rate on the adhesion force and other important characteristics. Therefore, for each sample the experimental measurements were repeated five times and the average of the obtained values is presented in Table 2 for roughness, attraction forces and adhesion energy, respectively. For the adhesion energy all five measurements together with the average and standard deviation are included in Table 3.

Table 2. Average experimental values for all investigated samples.

Thin film	Roughness (nm)	Attraction forces (nN)	Adhesion energy $(J \cdot 10^{-18})$
Sample 1	0.531	33.25	630
Sample 2	0.679	12.80	838
Sample 3	0.456	21.62	362
Sample 4	1.252	14.42	223

Table 3. Experimental values for the adehsion force.

Thin film \ Measurement	1 (nm)	2 (nm)	3 (nm)	4 (nm)	5 (nm)	Average (nm)	Standard Deviation (nm)
Sample 1	75.61	86.21	89.53	69.94	77.92	79.84	7.97
Sample 2	57.34	62.38	68.41	76.95	70.14	67.04	7.51
Sample 3	40.13	46.84	43.17	49.26	48.97	45.67	3.94
Sample 4	40.13	37.66	35.45	29.05	34.15	35.29	4.16

Powder Metallurgy and Advanced Materials – RoPM&AM 2017   Materials Research Forum LLC
Materials Research Proceedings **8** (2018) 212-218   doi: http://dx.doi.org/10.21741/9781945291999-24

As it can be seen the first sample has the highest value for the adhesion force. The adhesion force is strongly influenced by the surface free energy which is specific for each sample, therefore influenced by the nitrogen flow rate used at deposition. We believe that a decrease of the surface free energy of the thin films occurs with the increase of the nitrogen flow rate. Hence, it can be noted that using the same deposition parameters the adhesion force decreases up to almost 56% with the increase of the nitrogen flow rate.

By analyzing the data presented in the same table it can be seen that the other studied parameters do not follow a strictly increasing or strictly decreasing trend. The roughness values first increase, then a drop can be noted for the third sample and in the end for the sample obtained at the highest nitrogen flow rate, the highest average value for the roughness was obtained.

The attraction forces (snap-in forces) for the second sample are lower than for the first one, then the values increase to a higher value for the third sample and then decrease again for the fourth. The adhesion energy is characterized by an increase in values when comparing the first two samples, followed by a dramatic decrease for the last two samples.

Table 4. Surface energy for all investigated samples.

γ (J/m^2)	Sample 1	Sample 2	Sample 3	Sample 4
Minimum value	0.706	0.593	0.404	0.312
Maximum value	0.941	0.790	0.538	0.416

Using the average values obtained experimentally for the adhesion force and the two mathematical models presented above, the surface energy was estimated and the computed values are presented in Table 4.

Summary

The elaboration using in general different deposition parameters such as in this case the nitrogen flow rate determines different material properties possibly due to the preferential orientation of the atoms planes. The results of this study concerning the influence of the nitrogen flow rate on the properties of NbN thin films deposited by reactive magnetron sputtering on silicon substrate have shown it influences all studied parameters. The obtained results allowed even to quantify the degree of influence on the adhesion force.

The conducted experimental work has not indicated a certain direct correlation such as direct proportionality or inverse proportionality or even the same or inverse trend between the nitrogen flow rate and the other investigated parameters besides the adhesion force. However, the results show that the studied parameters are influenced by the increase of nitrogen flow rate when the other deposition parameters were kept constant. The roughness and the attractive forces have an inverse trend variation. Moreover, the tests proved a direct correlation for the case of the adhesion force values which decreased with the nitrogen flow rate increase.

The XE 70 AFM used for the experimental investigations and the XEI Image Processing Tool for SPM data used for the interpretation of the data allowed to determine the roughness of each of the four samples. Due to the low roughness values mathematical models suitable for the contact between a sphere and a plane were used to model the contact between the AFM tip and the sample surfaces. Based on these models the surface energy of the NbN thin films was computed. The obtained values indicate that it can vary up to three times depending on the nitrogen flow rate.

Powder Metallurgy and Advanced Materials – RoPM&AM 2017   Materials Research Forum LLC
Materials Research Proceedings **8** (2018) 212-218   doi: http://dx.doi.org/10.21741/9781945291999-24

Acknowledgement
The results presented in this paper were obtained with the support of the Technical University of Cluj-Napoca through the research Contract no. 2003/12.07.2017, Internal Competition CICDI-2017. The authors would like to thank Professor Marius Pustan and Professor Corina Bîrleanu for easing the access to the atomic force microscope.

References

[1] Z. Stanimirović and I. Stanimirovic , Mechanical properties of MEMS materials, Micro Electronic and Mechanical Systems, book edited by: Kenichi Takahata, p. 572, December 2009, INTECH, Croatia.

[2] A. D. Pogrebnjak, O. V. Bondar, G. Abadias, V. Ivashchenko, O. V. Sobol, S. Jurga and E. Coy, Structural and mechanical properties of NbN and Nb-Si-N films: Experiment and molecular dynamics   simulations,   Ceram.   Int.,   42   (2016)   11743-11756. https://doi.org/10.1016/j.ceramint.2016.04.095

[3] Y. Lei et al., Fabrication of niobium titanium nitride thin films with high superconducting transition temperatures and short penetration lengths, IEEE Trans. Appl. Supercond., 15 (2005) 44-48. https://doi.org/10.1109/TASC.2005.844126

[4] C.-L. Huang et al., Characteristics of reactively sputtered niobium nitride thin films as diffusion barriers for Cu metallization, Electron. Mater. Lett., 9 (2013) 593-597. https://doi.org/10.1007/s13391-012-2173-0

[5] M. Ziegler et al., Superconducting niobium nitride thin films deposited by metal organic plasma-enhanced atomic layer deposition, Supercond. Sci. Technol., 26 (2013) 025008-5. https://doi.org/10.1088/0953-2048/26/2/025008

[6] Y. Ufuktepe et al., Superconducting niobium nitride thin films by reactive pulsed laser deposition, Thin Solid Films, 545 (2013) 601-607. https://doi.org/10.1016/j.tsf.2013.08.051

[7] F. Mercier et al., Niobium nitride thin films deposited by high temperature chemical vapor deposition,   Surf.   Coat.   Technol.,   260   (2014)   126-132. https://doi.org/10.1016/j.surfcoat.2014.08.084

[8] J.E. Alfonso, J. Buitrago, J. Torres et al., Influence of fabrication parameters on crystallization, microstructure, and surface composition of NbN thin films deposited by rf magnetron sputtering, J. Mater. Sci., 45 (2010) 5528-5533. https://doi.org/10.1007/s10853-010-4612-3

[9] Y. J. Totik, Investigation of the adhesion of NbN coatings deposited by pulsed dc reactive magnetron sputtering using scratch tests, Coat. Technol. Res., 7 (2010) 485-492. https://doi.org/10.1007/s11998-009-9200-6

[10] D.S. Grierson, E.E. Flater, R.W. Carpick, Accounting for the JKR–DMT transition in adhesion and friction measurements with atomic force microscopy, J. Adhes. Sci. Technol., 19 (2005) 291–311. https://doi.org/10.1163/1568561054352685

Powder Metallurgy and Advanced Materials – RoPM&AM 2017   Materials Research Forum LLC
Materials Research Proceedings 8 (2018) 219-227   doi: http://dx.doi.org/10.21741/9781945291999-25

Reduction of the substitutional disorder by heat treatments in Mn$_{2-x}$Co$_x$VAl Heusler alloys

Radu GAVREA [1,a], Marin COLDEA [1,b], Lucian BARBU-TUDORAN [2,3,c],
Olivier ISNARD [4,5,d], Viorel POP [1,e,*], and Diana BENEA [1,f]

[1]Babes-Bolyai University, Faculty of Physics, 400084 Cluj-Napoca, Romania

[2]Electron Microscopy Center, Faculty of Biology & Geology, Babes-Bolyai University, RO-400006, Cluj-Napoca, Romania

[3]National Institute for Research and Development of Isotopic and Molecular Technologies, 400293 Cluj Napoca, Romania

[4]Université Grenoble Alpes, Institut Néel, Grenoble, F 38042, France

[5]CNRS, Institut Néel, 25 rue des Martyrs, F-38042 Grenoble, France

[a]gavrea_radu@yahoo.com, [b]marin.coldea@phys.ubbcluj.ro, [c]lucian.barbu@itim-cj.ro, [d]lolivier.isnard@grenoble.cnrs.fr, [e,*]viorel.pop@phys.ubbcluj.ro, [f]diana.benea@phys.ubbcluj.ro

Keywords: Heusler compounds, Substitutional disorder, X-ray diffraction, Neutron diffraction

Abstract. We report on the preparation and the atomic disorder reduction by annealing in Mn$_{2-x}$Co$_x$VAl Heusler alloys. The degrees of the B2 and L2$_1$ atomic ordering for the as-cast samples, obtained from intensity ratios of (200) and (111) peaks respectively related to (220) peak of the X-ray patterns, are significantly improved after annealing at 700 - 800 °C for 72 h. The diminution of the substitutional disorder is essential in these types of compounds, as the half-metallic character and the magnetic properties are primarily influenced by this factor.

Introduction

Heusler alloys are ternary intermetallic compounds of the L2$_1$ structure with stoichiometric composition X$_2$YZ, where X and Y are usually two different transition metals and Z is a nonmagnetic element [1]. Earlier studies have shown that Mn$_2$VAl Heusler alloy is a half-metallic ferrimagnet [2-5]. This compound is characterized by an antiparallel coupling between the Mn and V magnetic moments, the total spin moment being 2 μB per formula unit [2, 3]. The high Curie temperature of 760 K [3] makes it interesting for spintronic applications. The spin compensation in Mn$_{2-x}$Co$_x$VAl alloy was induced by progressive substitution of Co for Mn and a fully compensated ferrimagnetic behavior has been experimentally obtained for the MnCoVAl alloy [4]. The presence of a considerable atomic disorder in the Mn2VAl compound due to the intermixing of the V and Al atoms has been reported [5]. Previous studies have shown that the magnetic properties and the half-metallic character of these Heusler alloys are strongly influenced by the crystallographic disorder [1, 3, 6, 7]. The aim of the present work is to reduce the substitutional disorder by heat treatments in Mn$_{2-x}$Co$_x$VAl Heusler alloys. For the evaluation of the atomic ordering in the full Heusler alloys, the Takamura's model has been used [8]. In order to determine and to adjust the ordering parameters defined in this model, X-ray diffraction (XRD), differential scanning calorimetry (DSC) and neutron diffraction studies have been performed.

Powder Metallurgy and Advanced Materials – RoPM&AM 2017   Materials Research Forum LLC
Materials Research Proceedings 8 (2018) 219-227   doi: http://dx.doi.org/10.21741/9781945291999-25

Experimental details

The $Mn_{2-x}Co_xVAl$ (x= 0, 0.2, 0.6, 1) ingots were prepared by induction melting under a purified Ar atmosphere of the starting components Mn (99.95 wt %), Al (99.999 wt %), V (99.99 wt %) and Co (99.99 wt %). An excess of 3 wt % of manganese was added to the stoichiometric mixture in order to compensate for preferential Mn evaporation during the melting processes. The samples were turned and remelted repeatedly in order to ensure homogeneity. The water-cooled copper crucible ensured a rapid cooling of the alloys after melting. The samples were wrapped in tantalum foil, sealed in quartz tubes and subsequently annealed in an Ar atmosphere for 72 hours. The stoichiometry of our as-cast samples was investigated using the energy dispersive X-ray analysis (EDX). The crystal structure of the alloys was investigated at room temperature by using a Brüker D8 Advance diffractometer using Cu Kα radiation. The structural transformations in the 50 – 1000 °C temperature range were identified from differential scanning calorimetry under Ar atmosphere with a temperature ramp rate of 20 °C/min. The cooling was performed at 20 °C/min controlled by forced air cooling (Q600 TA Instruments). The neutron diffraction investigations have been performed at the Institute Laue-Langevin, Grenoble, France, using the high intensity powder diffractometer D1B [9] exploiting the wavelength of 0.128 nm and 0.252 nm respectively, which were selected by Ge and pyrolytic graphite monochromator respectively. The diffraction patterns were indexed by using the FULLPROF program [10].

Results and discussions

The EDX measurements on the as-cast MnCoVAl sample are shown in Fig. 1. The quantity for each element is given in atomic percent which, for ideal MnCoVAl alloy stoichiometry, should be 25 at % for each element. The previous studies showed that 5 wt % excess of Al should be added in order to compensate the weight loss due to the evaporation of Al [11].

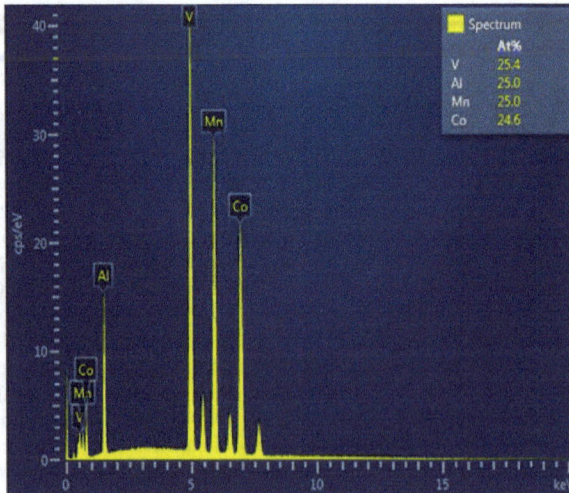

Fig. 1. EDX spectrum of the as-cast MnCoVAl alloy.

Powder Metallurgy and Advanced Materials – RoPM&AM 2017   Materials Research Forum LLC
Materials Research Proceedings **8** (2018) 219-227   doi: http://dx.doi.org/10.21741/9781945291999-25

By using no additional Al and by adding 3 wt % Mn, in our samples the weight loss of the final materials is less than 1 wt %. As can be observed in Fig.1, the stoichiometry of our sample is in good agreement with the desired Heusler-type structure, taking into account the measurement errors.

The $Mn_{2-x}Co_xVAl$ alloys crystallize in an ideal full Heusler ($L2_1$) structure (Fig. 2), were the Mn/Co atoms occupy the 8c positions at (1/4 1/4 1/4) and (3/4 3/4 3/4), V occupy the 4a positions at (0 0 0) and Al occupy the 4b positions at (1/2 1/2 1/2) [3].

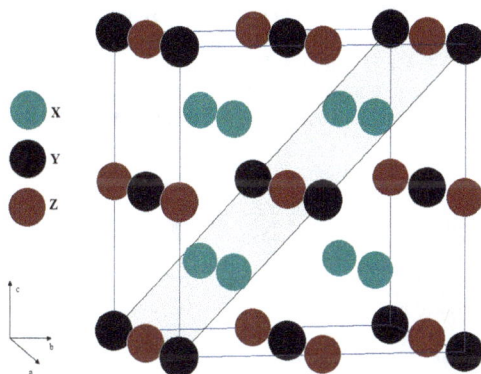

Fig. 2. X_2YZ $L2_1$-type crystal structure of Heusler alloys.

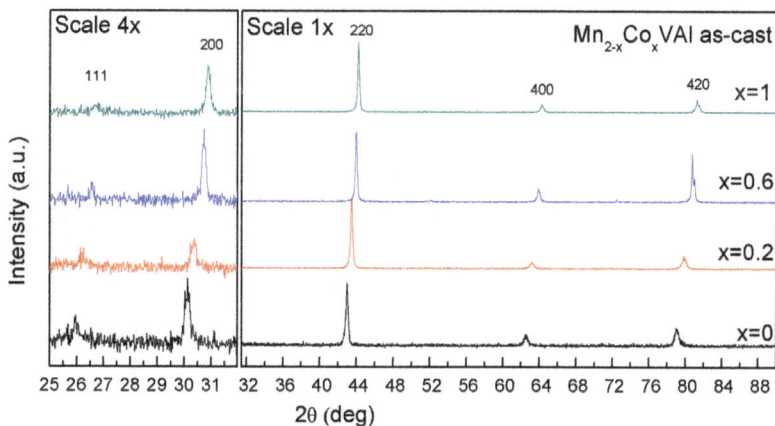

Fig. 3. Room temperature X-ray diffraction pattern of the as-cast $Mn_{2-x}Co_xVAl$ samples. The data are normalized to the intensity of the (220) reflection.

Powder Metallurgy and Advanced Materials – RoPM&AM 2017   Materials Research Forum LLC
Materials Research Proceedings 8 (2018) 219-227   doi: http://dx.doi.org/10.21741/9781945291999-25

The X-ray diffraction patterns at room temperature of the as-cast $Mn_{2-x}Co_xVAl$ (x = 0, 0.2, 0.6 and 1) alloys are shown in Fig.3. The XRD patterns prove that the as-cast alloys crystallize in a single phase, corresponding to X_2YZ Heusler type structure, cubic space group $Fm\bar{3}m$ (spatial group no. 225), where the Mn and Co atoms occupy the 8c Wyckoff sites (X), while V and Al atoms are placed on the 4a (Y) and 4b (Z) crystal site, respectively (see Fig.2.). The (111) and (200) superlattice diffraction lines from the XRD patterns prove that all the $Mn_{2-x}Co_xVAl$ alloys exhibit a stable $L2_1$ structure of full Heusler alloys. The extinction of the reflection from the (111) plane indicates an intermixing between the V and Al atoms. Also, if all Mn, V and Al atoms get intermixed, both super-latttice reflections (111) and (200) would disappear [1, 4, 8].

We employed the Takamura's model to investigate the substitutional disorder in our $Mn_{2-x}Co_xVAl$ (x = 0, 0.2, 0.6 and 1) alloys. In this model, two types of ordering parameters have been defined to describe the intermixing between the atomic positions. The S_{B2} order parameter describes the probability of Mn atoms to occupy the X sites (8c) in the X_2YZ full Heusler alloys, being defined:

$$S_{B2} = \frac{n_{Mn\,on\,X} - n_{Mn\,on\,X}^{random}}{n_{Mn\,on\,X}^{full\,order} - n_{Mn\,on\,X}^{random}} \quad (1)$$

The second order parameter S_{L21} describes the probability of V to occupy the Y position in the X_2YZ full Heusler alloys:

$$S_{L21} = \frac{n_{V\,on\,Y} - n_{V\,on\,Y}^{random}}{n_{V\,on\,Y}^{full\,order} - n_{V\,on\,Y}^{random}} \quad (2)$$

Table 1. Structural parameters including S_{B2} and S_{L21} ordering parameters unit cell constant and the site occupation for the as-cast $Mn_{2-x}Co_xVAl$ samples.

Co content (x)	Atoms	X site	Y site	Z site	S_{B2}	S_{L21}	a_{lat} (nm)
0	Mn	1.84	0.08	0.08	0.847	0.60	0.59087
	V	0.08	0.14	0.78			
	Al	0.08	0.78	0.14			
0.2	Mn/Co	1.96	0.02	0.02	0.953	0.68	0.58880
	V	0.02	0.84	0.14			
	Al	0.02	0.14	0.84			
0.6	Mn/Co	1.80	0.10	0.10	0.807	0.52	0.58533
	V	0.10	0.74	0.16			
	Al	0.10	0.16	0.74			
1.0	Mn/Co	1.86	0.07	0.07	0.855	0.52	0.58115
	V	0.07	0.75	0.18			
	Al	0.07	0.18	0.75			

Powder Metallurgy and Advanced Materials – RoPM&AM 2017   Materials Research Forum LLC
Materials Research Proceedings 8 (2018) 219-227   doi: http://dx.doi.org/10.21741/9781945291999-25

The S_{B2} and S_{L21} ordering parameters are related to the site occupation in the $L2_1$ structure. Using the Takamura's extended order model for Heusler compounds [8] the S_{B2} and S_{L2} ordering parameters are calculated from the peaks ratios of the XRD patterns as follows:

$$\frac{I_{200}}{I_{220}} = (S_{B2})^2 \frac{I_{200}^{full-order}}{I_{220}^{full-order}} \qquad (3),$$

$$\frac{I_{111}}{I_{220}} = \left[S_{L2_1}\left(\frac{3-S_{B2}}{2}\right)\right]^2 \frac{I_{111}^{full-order}}{I_{220}^{full-order}} \quad (4)$$

where $\frac{I_{200}}{I_{220}}$ and $\frac{I_{111}}{I_{220}}$ are the experimental intensity ratios obtained between (200), (111) and (220) diffraction peaks respectively and $\frac{I_{200}^{full-order}}{I_{220}^{full-order}}$, $\frac{I_{111}^{full-order}}{I_{220}^{full-order}}$ are the diffraction intensity ratios for the ideal structure. The calculated values of S_{B2} and S_{L21} ordering parameters together with the lattice parameters and the site occupation for the as-cast $Mn_{2-x}Co_xVAl$ samples are given in Table 1. Previously we have shown that any changes of these parameters will significantly influence the magnetic and electronic properties of $Mn_{2-x}M_xVAl$ (M = Co or Cu) Heusler systems [6, 7]. As can be seen in Table 1, a substantial degree of mixing between the V and Al is present in all our samples. Also, a smaller but non-negligible intermixing between Mn, V and Al is present on the 8c sites.

In order to decrease the substitutional disorder in our samples, we performed heat treatments. The temperature for the annealing has been settled using the differential scanning calorimetry (DSC) investigations shown in Fig. 4. The DSC curve of Mn_2VAl sample indicates the presence of two new reversible phase transitions (endothermic peaks shown by arrows on heating curve and exothermic peaks on the cooling curve respectively) around 800 ° C. The ordering degree was improved by heat treatment at 700 °C avoiding the above two phase transitions. The DSC curve of MnCoVAl sample doesn't show any phase transitions up to 900 °C, so the heat treatment at 800 °C results in an improving of the ordering parameters for this sample.

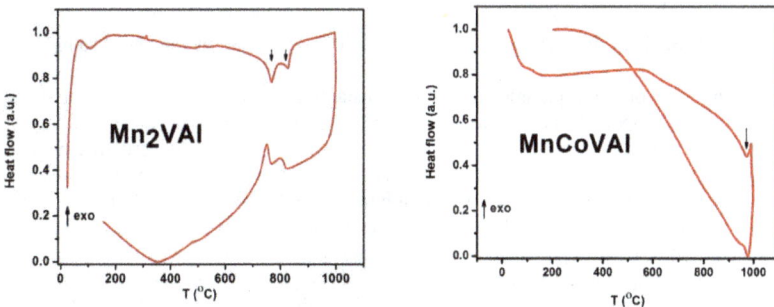

Fig. 4. DSC curves for the Mn_2VAl (left) and MnCoVAl (right) samples.

Powder Metallurgy and Advanced Materials – RoPM&AM 2017   Materials Research Forum LLC
Materials Research Proceedings **8** (2018) 219-227   doi: http://dx.doi.org/10.21741/9781945291999-25

The (111) and (200) superlattice diffraction lines from the XRD patterns at room temperature of the annealed $Mn_{2-x}Co_xVAl$ (x = 0, 0.2, 0.6 and 1) alloys are shown in Fig.5. For comparison are given the XRD patterns of as-cast samples. All the annealed samples crystallize in a Heusler type structure ($Fm\overline{3}m$ space group no. 225). The increase of 2θ angle for the diffraction peaks shows the decrease of the lattice parameters after annealing.

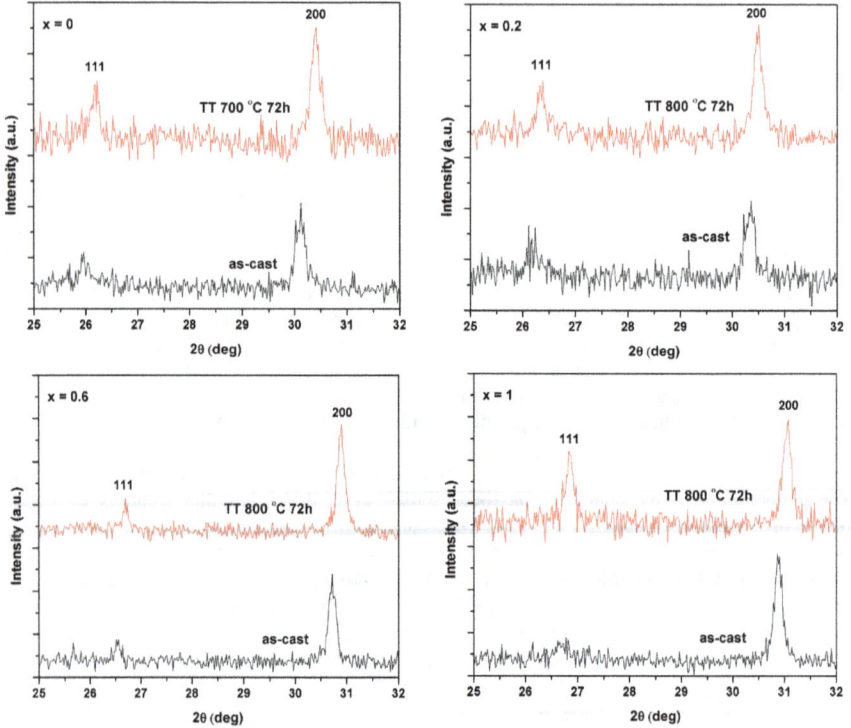

Fig. 5. X-ray diffraction patterns of $Mn_{2-x}Co_xVAl$ as-cast and annealed samples.

The S_{B2} and S_{L21} ordering parameters, lattice constants and the site occupation for the annealed $Mn_{2-x}Co_xVAl$ compounds are shown in Table 2. A substantial increase of the S_{B2} ordering parameter compared with the as-cast samples is observed. Also, the S_{L21} ordering parameters are improved for all the $Mn_{2-x}Co_xVAl$ samples. Accordingly, the probability of Mn and Co atoms to occupy the 8c sites is close to 100%, whilst an intermixing between Al and V atoms is still present for all investigated samples. Therefore, the annealing of the as-cast samples leads to an improvement of the order parameters. The evolution of the lattice constant vs. Co content in the $Mn_{2-x}Co_xVAl$ compounds, both as-cast and annealed samples, is shown in Fig. 6. The variation of the lattice constant vs. Co content in the $Mn_{2-x}Co_xVAl$ alloys is in agreement with the relationship between the Co and Mn metallic radii (r_{Co}=0.125 nm, r_{Mn}=0.127 nm). The monotonous decrease of the lattice constant proves that the Co for Mn substitution occurs in the 8c site. The lattice constant follows the same trend for both cases (as-cast and annealed samples) by decreasing with increasing of the Co content. The contraction of the lattice parameter for the annealed samples suggests that the heat treatment increases the order in our samples and reduces the internal stress. Also, the values of the lattice parameter are in agreement with others results reported in literature (5.92 Å) [3,12,13].

Table 2. The lattice parameters, S_{B2} and S_{L21} ordering parameters and the site occupation for the annealed $Mn_{2-x}Co_xVAl$ samples.

Co content (x)	Atoms	X site	Y site	Z site	S_{B2}	S_{L21}	a_{lat} (nm)
0	Mn	1.99	0.005	0.005	0.995	0.66	0.58967
	V	0.005	0.83	0.165			
	Al	0.005	0.165	0.83			
0.2	Mn/Co	1.99	0.005	0.005	0.997	0.77	0.58765
	V	0.005	0.90	0.095			
	Al	0.005	0.095	0.90			
0.6	Mn/Co	1.99	0.005	0.005	0.981	0.59	0.58304
	V	0.005	0.79	0.205			
	Al	0.005	0.205	0.79			
1.0	Mn/Co	1.98902	0.00549	0.00549	0.989	0.959	0.58003
	V	0.00549	0.97918	0.01533			
	Al	0.00549	0.01533	0.97918			

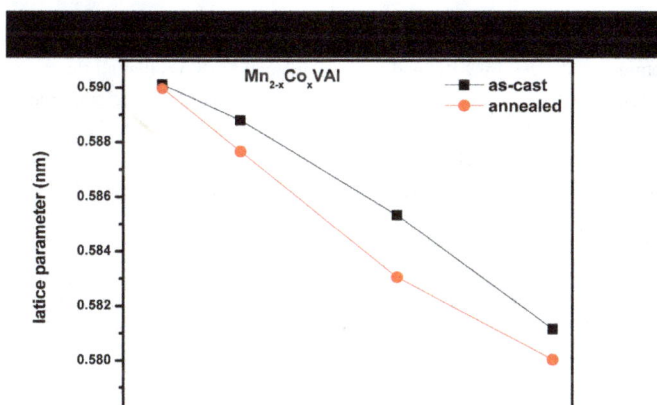

Fig. 6. The lattice constant vs. the Co content in the $Mn_{2-x}Co_xVAl$ Heusler compounds.

The neutron diffraction experiments have been employed in order to investigate the crystallographic structural properties of our samples. The powder neutron diffraction patterns measured at 300 K for Mn_2VAl (x = 0) for the as-cast sample is given in Fig. 7 as example. The experimental data and the calculated fitting curves are indicated by red dots and solid black lines, respectively, while the blue lines at the bottom show the difference between them. The fitting curves of the Mn_2VAl and MnCoVAl alloys have been calculated by considering the full Heusler structure in cubic space group $Fm\overline{3}m$ (spatial group no. 225). From these preliminary results we may conclude that the neutron diffraction measurements confirm that our samples have a single-phase structure without any impure phases as can be observed also from XRD measurements.

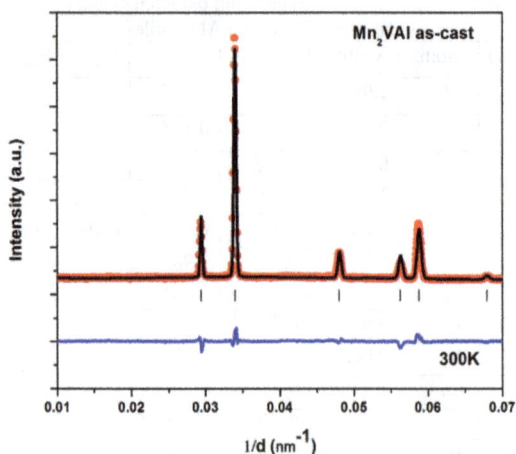

Fig. 7. Room temperature neutron diffraction pattern for the as-cast $Mn_{2-x}Co_xVAl$ sample (x = 0)

Summary
Bulk $Mn_{2-x}Co_xVAl$ (x = 0, 0.2, 0.6 and 1) alloys have been prepared by induction melting. The crystal structure investigated by XRD and neutron diffraction shows that all the analyzed samples are single phases belonging to the stable $L2_1$ (spatial group no. 225) structural order. The degrees of $L2_1$ and B2 order have been calculated from the peaks ratios of the XRD patterns using the Takamura's extended order model for Heusler compounds. Annealing of as-cast samples leads to an increase of the order parameters, which could also be inferred from the reduction of the lattice parameters.

References

[1] M. Meinert, J. M. Schmalhorst, G. Reiss, E. Arenholz, Ferrimagnetism and disorder of epitaxial $Mn_{2-x}Co_xVAl$ Heusler compound thin films, J. Phys. D: Appl. Phys. 44 (2011) 215003. https://doi.org/10.1088/0022-3727/44/21/215003

[2] I. Galanakis, K. Ozdogan, E. Sasioglu, B. Aktas, Doping of Mn_2VAl and Mn_2VSi Heusler alloys as a route to half-metallic antiferromagnetism, Phys. Rev. B 75 (2007) 092407. https://doi.org/10.1103/PhysRevB.75.092407

[3] C. Felser, A. Hirohata, Heusler alloys - Properties Growth and Applications, Springer Series in Materials Science, Springer International Publishing, Switzerland, 2016.

Powder Metallurgy and Advanced Materials – RoPM&AM 2017   Materials Research Forum LLC
Materials Research Proceedings 8 (2018) 219-227   doi: http://dx.doi.org/10.21741/9781945291999-25

[4] B. Deka, A. Srinivasan, R. K. Singh, B. S. D. C. S. Varaprasad, Y. K. Takahashi, K. Hono, Effect of Co substitution for Mn on spin polarization and magnetic properties of ferrimagnetic Mn_2VAl, J. Alloys Compd. 662 (2015) 510-515. https://doi.org/10.1016/j.jallcom.2015.12.089

[5] Y. Yoshida, M. Kawakami, T. Nakamichi, Magnetic Properties of a Ternary Alloy $Mn_{0.5}V_{0.5-y}Al_y$, J. Phys. Soc. Japan 50 (1981) 2203-2208. https://doi.org/10.1143/JPSJ.50.2203

[6] R. Gavrea, A. Bolinger, V.Pop, O. Isnard, M.Coldea, D. Benea, Phys. Status Solidi B (2017).

[7] R. Gavrea, C. Leostean, M. Coldea, O Isnard, V. Pop and D. Benea, Effects of Co for Mn substitution on the electronic properties of $Mn_{2-x}Co_xVAl$ as probed by XPS, Intermetallics, 93 (2018) 155-161. https://doi.org/10.1016/j.intermet.2017.12.003

[8] Y. Takamura, R. Nakane, S. Sugahara, Analysis of $L2_1$-ordering in full-Heusler Co_2FeSi alloy thin films formed by rapid thermal annealing, J. Appl. Physics 105 (2009) 07B109.

[9] ILL yellow book, available at www.ill.fr.

[10] J. Rodriguez-Carvajal, Recent advances in magnetic structure determination by neutron powder diffraction, Physica B, 192 (1993) 55-69. https://doi.org/10.1016/0921-4526(93)90108-I

[11] C. Jiang, M. Venkatesan, J. M. D. Coey, Transport and magnetic properties of Mn_2VAl: Search for half-metallicity, Solid State Comm. 118 (2001) 513-516. https://doi.org/10.1016/S0038-1098(01)00151-X

[12] T. Nakamichi, C.V. Stager, Phenomenological formula of NMR satellite of Heusler alloys and magnetic structure of Mn_2VAl, J. Magn. Magn. Materials 31(1983) 85. https://doi.org/10.1016/0304-8853(83)90165-8

[13] K. R. A. Ziebeck and P. J. Webster, A neutron diffraction and magnetization study of Heusler alloys containing Co and Zr, Hf, V or Nb, J. Phys. Chem. Solids 35 (1974) 1-7. https://doi.org/10.1016/0022-3697(74)90002-X

Keywords index

Materials Research Forum LLC
doi: http://dx.doi.org/10.21741/9781945291999

Powder Metallurgy and Advanced Materials – RoPM&AM 2017   Materials Research Forum LLC
Materials Research Proceedings 8 (2018)   doi: http://dx.doi.org/10.21741/9781945291999

Author index

Powder Metallurgy and Advanced Materials – RoPM&AM 2017   Materials Research Forum LLC
Materials Research Proceedings 8 (2018)   doi: http://dx.doi.org/10.21741/9781945291999

Powder Metallurgy and Advanced Materials – RoPM&AM 2017　　　Materials Research Forum LLC
Materials Research Proceedings **8** (2018)　　　doi: http://dx.doi.org/10.21741/9781945291999

Powder Metallurgy and Advanced Materials – RoPM&AM 2017   Materials Research Forum LLC
Materials Research Proceedings 8 (2018)   doi: http://dx.doi.org/10.21741/9781945291999